# Catalytic Materials: Relationship Between Structure and Reactivity

A C S   S Y M P O S I U M   S E R I E S **248**

# Catalytic Materials: Relationship Between Structure and Reactivity

**Thaddeus E. Whyte, Jr.,** EDITOR
*Catalytica Associates, Inc.*

**Ralph A. Dalla Betta,** EDITOR
*Catalytica Associates, Inc.*

**Eric G. Derouane,** EDITOR
*Mobil Technical Center*

**R. T. K. Baker,** EDITOR
*Exxon Research & Engineering Co.*

Based on the 1983 State-of-the-Art Symposium sponsored by
the Division of Industrial and Engineering Chemistry,
San Francisco, California,
June 13–16, 1983

American Chemical Society, Washington, D.C. 1984

**Library of Congress Cataloging in Publication Data**

Catalytic materials.
  (ACS symposium series, ISSN 0097-6156; 248)

  "Based on the 1983 state-of-the-art symposium
sponsored by the Division of Industrial and
Engineering Chemistry, San Francisco, California, June
13-16, 1983."

  Includes bibliographies and indexes.

  1. Catalysts—Congresses.

  1. Whyte, Thaddeus E., 1937-     . II. American
Chemical Society. Division of Industrial and
Engineering. III. Series.

QD505.C387     1984     541.3'95     84-2776
ISBN 0-8412-0831-X

# ACS Symposium Series

## M. Joan Comstock, *Series Editor*

### *Advisory Board*

# FOREWORD

The ACS SYMPOSIUM SERIES was founded in 1974 to provide a medium for publishing symposia quickly in book form. The format of the Series parallels that of the continuing ADVANCES IN CHEMISTRY SERIES except that in order to save time the papers are not typeset but are reproduced as they are submitted by the authors in camera-ready form. Papers are reviewed under the supervision of the Editors with the assistance of the Series Advisory Board and are selected to maintain the integrity of the symposia; however, verbatim reproductions of previously published papers are not accepted. Both reviews and reports of research are acceptable since symposia may embrace both types of presentation.

# CONTENTS

# PREFACE

T<small>HE CHARACTERIZATION OF CATALYST STRUCTURES</small> has undergone revolutionary developments in recent years. Powerful novel techniques and instrumentation are now used to analyze catalyst structure before, during, and after use. Many of these advances are responsible for placing the field of catalysis on an improved scientific basis. These developments have resulted in a better understanding of catalytic phenomena, and therefore improvements in commercial catalysts and the discovery of new systems. The application of advanced electronics and computer analysis has optimized many of these analytical tools. These developments are especially evident in spectroscopy, zeolite structure elucidation, and microscopy; several other techniques have also been developed. Thus, the difficult goal of unraveling the relationships between the structure and reactivity of catalytic materials is finally within reach.

Spectroscopic developments have accelerated advances in the field of catalysis. This volume analyzes the impact on catalyst structure and reactivity of EXAFS, SIMS, Mössbauer, magic-angle spinning NMR (MASNMR), and electron-energy-loss vibrational spectroscopy. Many of these techniques are combined with other analytical tools such as thermal decomposition and temperature-programmed reactions.

The major effect of new advanced techniques on catalyst structure is found in zeolite catalysis. NMR techniques, especially MASNMR, have helped to explain aluminum distribution in zeolites and to increase our understanding of critical parameters in zeolite synthesis and crystallization. MASNMR, combined with TEM, STEM, XPS, and diagnostic catalytic reaction probes, has advanced our knowledge of the critical relationship between the structure and reactivity patterns of zeolites in the chemical fuels industry. Throughout the symposium upon which this book is based, many correlations were evident between theoretical quantum mechanical calculations and the structures elucidated by these techniques.

Improvements in the resolution and versatility of microscopic techniques have come about rapidly. TEM, STEM, and high-resolution electron microscopy have helped the catalytic chemist to analyze the effects of metal-support interactions and particle-size effects—developments that will probably lead to improvements in commercial technologies. Several novel analytical methods, arising from very clever experimentation, were discussed at the symposium. These included photoacoustic spectroscopy, inelastic tunneling spectroscopy, and Rutherford back-scattering spectrometry.

Thus the catalytic scientist is now very close to achieving the goal of establishing clear relationships between the structure and reactivity of catalytic materials. The next breakthrough will be the deliberate design of specific structures to improve the activity and selectivity of catalytic materials.

THADDEUS E. WHYTE, JR.
Catalytica Associates, Inc.
Mountain View, California

RALPH A. DALLA BETTA
Catalytica Associates, Inc.
Mountain View, California

ERIC G. DEROUANE
Mobil Technical Center
Princeton, New Jersey

R. T. K. BAKER
Exxon Research & Engineering Co.
Clinton, New Jersey

January 1984

# SPECTROSCOPY

# Thermal Decomposition of Iron Pentacarbonyl on Titania

## Genesis of Fe/TiO$_2$ Catalysts

J. PHILLIPS
Department of Chemical Engineering, The Pennsylvania State University, University Park, PA 16802

J. A. DUMESIC
Department of Chemical Engineering, University of Wisconsin-Madison, Madison, WI 53706

Mössbauer spectroscopy and volumetric gas phase analysis were used to study the nature of surface species formed during the decomposition of Fe(CO)$_5$ on cleaned titania powder. The titania was pretreated so as to produce samples with different hydroxyl group and Ti$^{3+}$ concentrations. It was found that the presence of Ti$^{3+}$ did not affect the nature of the iron species which formed. The extent of decomposition was found to be proportional to the hydroxyl group density. On all surfaces, low temperature (383 K) decomposition led to the formation of an Fe$^{2+}$ species and an Fe$^{\circ}$ species (possibly Fe(CO)$_2$), both of which were probably associated with surface hydroxyl groups. These species were very highly dispersed on the support. High temperature decomposition (673 K) led to nearly complete conversion of the metal to the Fe$^{2+}$ species. A very small fraction of the iron apparently sintered to form metallic iron particles during the high temperature treatment. Prolonged reduction of these sample in hydrogen at ca. 700 K led to the formation of small metallic iron particles (less than 9 nm in size).

One area of current interest in heterogeneous catalysis is understanding the effects of metal-support interactions on the catalytic properties of Group VIII metals. Recently it has been found that the interaction between TiO$_2$ supports and metal particles is particularly strong. This interaction has been denoted as a "strong metal-support interaction"(SMSI). Indeed, Tauster et al. (1) showed that following a high temperature reduction (ca. 770 K) in hydrogen, noble metals (e.g., Ru, Rh, Pd, Os, Ir, Pt) supported on TiO$_2$ do not adsorb hydrogen or carbon monoxide. On other refractory supports (e.g., SiO$_2$, Al$_2$O$_3$), and even on TiO$_2$ following

0097-6156/84/0248-0003$06.00/0

low temperature reduction in hydrogen, particles of these metals
adsorb hydrogen and carbon monoxide readily. Another area of
research in catalysis is the use of metal carbonyl clusters for
the preparation of supported metal catalysts (e.g., 2,3). These
clusters offer the potential opportunity for preparing catalysts
with novel properties (e.g., having particularly high dispersions).
Of importance in this respect is the interaction of the support
with the various metal species formed during decarbonylation of
the clusters. Such species include metal subcarbonyl species,
metal cations and metal crystallites.

The present paper focuses on the interactions between iron
and titania for samples prepared via the thermal decomposition of
iron pentacarbonyl. (The results of ammonia synthesis studies
over these samples have been reported elsewhere (4).) Since it
has been reported that standard impregnation techniques cannot be
used to prepare highly dispersed iron on titania (4), the use of
iron carbonyl decomposition provides a potentially important catalyst preparation route. Studies of the decomposition process as
a function of temperature are pertinent to the genesis of such
Fe/TiO$_2$ catalysts. For example, these studies are necessary to
determine the state and dispersion of iron after the various
activation or pretreatment steps. Moreover, such studies are
required to understand the catalytic and adsorptive properties of
these materials after partial decomposition, complete decarbonylation or hydrogen reduction. In short, Mössbauer spectroscopy
was used in this study to monitor the state of iron in catalysts
prepared by the decomposition of iron carbonyl. Complementary
information about the amount of carbon monoxide associated with
iron was provided by volumetric measurements.

Experimental

TiO$_2$ Preparation

TiO$_2$ powder obtained from the Cabot corporation (Cab-O-Ti) was
used in these investigations. This material has a surface area
of between 50 and 70 m$^2$/g and is reportedly 99.9% pure TiO$_2$. This
material was 'cleaned' using a method similar to that of Munuera
et al. (5) and Cornaz et al. (6). This method reportedly removes
the organic contaminants from titania. The possible presence of
organic contaminants led Gebhardt and Herrington (7) to raise
questions regarding the accuracy of early studies of titania
surface structure. In brief, the titania was cleaned using the
following sequence of steps: (i) heating the fresh powder in a
pyrex vessel to 670 K (or higher) in flowing oxygen, (ii) pumping
on the powder (10$^{-2}$Pa) for one hour while heating the sample to
570 K, (iii) boiling the partially dehydroxylated TiO$_2$ in deionized
water for ten minutes, and (iv) drying the TiO$_2$ (12 hours) at
380 K. This procedure produces organic free, completely hydroxy-

lated $TiO_2$ according to Munuera et al. (5). For this study, the
purity of the cleaned material was tested by heating to 600 K
under vacuum ($10^{-2}Pa$). It was found that the cleaned material
remained white. Powder which was not cleaned in this manner
turned blue-black when evacuated at 600 K. Any color change
following low temperature outgassing was attributed by Gebhardt
and Herrington (7) to the preferential movement of carbonaceous
impurities to the surface, or the reduction of the surface by
these impurities.

## Pellet Production


Thin pellets of $TiO_2$ were produced for these experiments because
of the high degree of non-resonant adsorption of γ-rays by $TiO_2$.
Specifically, it was found that no more than 0.4 g of the material
could be used in a one-inch diameter pellet if the Mössbauer
spectra were to be collected in a reasonable period of time
(e.g., 24 h). The pressures required to produce stable pellets
of this size were sufficiently high (> 8000 lbs/in$^2$) to effectively
decrease the apparent diffusivity of iron carbonyl throughout the
titania sample. That is, following the decomposition of $Fe(CO)_5$
on pure titania pellets it was found that the pellets were discolored on the outermost surfaces (black) and completely white
inside. An alternative technique was developed for the production of pellets used in this study. These pellets were produced
by pressing approximately 0.4 g of cleaned $TiO_2$ powder between two
one-inch diameter Grafoil discs (i.e., discs of oriented graphite
from Union Carbide). One Grafoil disc was removed from the pellet
before use. This left the sample with a $TiO_2$ to graphite surface
area ratio of about 12 to 1. The presence of Grafoil did not
contribute to the subsequent decomposition of iron carbonyl since
no trace of the well-characterized Mössbauer spectrum of $Fe(CO)_5$
decomposed on Grafoil (8-9) was ever observed in these studies.
It was found that stable pellets could be produced in this manner
with only 1000 lbs/in$^2$ of pressure. These pellets were found to
be uniformly discolored following the decomposition of $Fe(CO)_5$.

## Mössbauer Spectroscopy


The Mössbauer spectroscopy cell used in this investigation has
been described in detail elsewhere (9). The Mössbauer spectroscopy instruments and the fitting routine used are also described
elsewhere (9). All isomer shifts are reported relative to
metallic iron at room temperature.
    $TiO_2$ pellet samples were pretreated once placed inside the
cell. The pretreatments given to each sample are listed in Table
I. Following the last of these treatments, the sample was cooled
to room temperature and the cell evacuated to a pressure of

TABLE I.  Results of Computer
Decomposed on

| Sample # | Pretreatment | Surface type | Decomposition temperature |
|----------|--------------|--------------|---------------------------|
| 1 | 600 K, $10^{-2}$Pa.-4 hrs. | I | 373 K<br>673 K<br>oxidized[b]<br>re-reduced[c] |
| 2 | 723 K, $10^{-2}$Pa.-5 hrs. | II | 383 K<br>673 K |
| 3 | 723 K, $10^{-2}$Pa.-10 hrs.<br>723 K, $O_2$ flow-1 hr.<br>723 K, $H_2$ flow-5 hrs. | III | 373 K<br>673 K |
| 4 | 723 K, $10^{-2}$Pa.-4 hrs. | II | 383 K<br>723 K |
| 5 | 723 K, $10^{-2}$Pa.-6 hrs.<br>723 K, $O_2$ flow-1 hr.<br>723 K, $H_2$ flow-4 hrs. | III | 383 K |

[a]Normalized to sample #1 after high temperature (673 K)
decomposition.
[b]Exposed to the ambient laboratory atmosphere (see Figure 1D).

[c]Reduced at 650 K in flowing 90%CO/10%CO$_2$.

[d]Mossbauer parameters:  isomer shift, $\delta$, and quadrupole
splitting, $\Delta E_Q$, in units of mm/s.

Fitting Mössbauer Spectra of Fe(CO)$_5$
Pretrated Samples of TiO$_2$

| Relative[a] spectral area | Mössbauer Parameters[d] | | | | |
|---|---|---|---|---|---|
| | Fe$^{2+}$ | | | Zero-Valent Iron | |
| | $\Delta E_Q$ | $\delta$ | Area(%) | $\delta$ | Area(%) |
| 0.58 | 1.8 | +1.04 | 66 | -0.29 | 34 |
| 1.0 | 1.8 | +1.04 | 96 | -0.12 | 4 |
| --- | 1.0 | +1.04 | 90 | -0.08 | 10 |
| 0.23 | 1.6 | +1.02 | 59 | -0.27 | 41 |
| 0.30 | 1.9 | +1.04 | 96 | -0.15 | 4 |
| 0.33 | 1.9 | +1.09 | 70 | -0.28 | 30 |
| 0.49 | 1.6 | +1.0 | 99 | --- | --- |
| 0.33 | 1.7 | +1.06 | 58 | -0.33 | 42 |
| 0.55 | 1.7 | +1.05 | 96 | -0.12 | 4 |
| 0.22 | 2.0 | +1.0 | 89 | 0.24 | 11 |

$10^{-3}$Pa.  Fe(CO)$_5$ was then admitted from a 200 cm$^3$ bulb and the decomposition procedure (see Table I) started.

## Volumetric Gas Phase Analysis

A detailed description of the apparatus used for volumetric studies is given elsewhere (8).  This apparatus is equipped with a precision pressure gage (Texas Instruments) which allows quantitative determination of the number of gaseous molecules in calibrated volumes.  A sample of TiO$_2$ (ca. 1 g) was first pre-treated in a pyrex cell (see Table II), cooled to room temperature and then evacuated to a pressure of $10^{-4}$Pa.  A known quantity of Fe(CO)$_5$ (ca. 100 μmol) was next admitted to the sample cell con-taining the TiO$_2$.  The amount of Fe(CO)$_5$ adsorbed on the samples in this manner never exceeded the expected density of "residual" hydroxyl groups (see discussion section and Table II).  The sample cell was next heated between 370 and 380 K (see Table II). The temperature of the sample cell was kept at the decomposition temperature until the system pressure stabilized ($\Delta P/P < .0005$ in one hour).  Typically, this process took 20 hrs.  Using the final pressure and the ideal gas law the CO/Fe ratio on the surface was determined (see Table II).  Mass spectroscopic analysis of the gas following the completion of one study showed the gas to be primarily CO with traces (less than 1%) of other gases such as H$_2$, H$_2$O, and CO$_2$.  It was also determined, using control experi-ments, that CO does not adsorb appreciably on the surface of cleaned TiO$_2$ at 373 K.

## Results

## Mössbauer Spectroscopy

The results of Mössbauer spectroscopy investigations of Fe(CO)$_5$ decomposition on TiO$_2$ samples pretreated in three different fashions are given in Table I.  These three samples were pretreated in a manner intended to produce different populations of Ti$^{3+}$ ions and hydroxyl groups on the surface of the support.  This is explained in the discussion section.

Seven representative Mössbauer spectra are presented.  The five spectra of Figure 1 are a sequence recorded following various treatments of sample 1, the fully hydroxylated sample.  Spectrum 1A was recorded at liquid nitrogen temperature after Fe(CO)$_5$ had been admitted to the sample cell.  This is essentially a spectrum of frozen Fe(CO)$_5$.  Spectrum 1B was recorded at room temperature after the sample had been heated to 350 K for 10 hours.  This spectrum consists of two components:  a zero-valent iron species (34%) and an Fe$^{2+}$ species (66%).  The numbers in parentheses represent the fraction of the total spectral area produced by the

TABLE II.  Results of Volumetric Gas-Phase Analysis for Determination of the CO/Fe Stoichiometry on Titania Samples After Decomposition of $Fe(CO)_5$

| Sample # | Pretreatment | Surface[a] type | Decomposition temperature | Surface CO/Fe ratio | Final[b] iron concentration |
|---|---|---|---|---|---|
| 1 | 600 K, $10^{-2}$ Pa.-3 hrs. | I | 383 K | 1.30 | 0.40 $Fe/nm^2$ |
| 2 | 600 K, $10^{-2}$ Pa.-3 hrs. | II | 373 K | 0.97 | 0.30 $Fe/nm^2$ |
| 3 | 725 K, $10^{-3}$ Pa.-4 hrs.<br>725 K, $O_2$ flow-1 hr.<br>725 K, $H_2$ flow-5 hrs. | III | 383 K | 0.85 | 0.25 $Fe/nm^2$ |
| 4 | 725 K, $10^{-3}$ Pa.-3 hrs.<br>725 K, $O_2$ flow-1 hr.<br>725 K, $H_2$ flow-4 hrs. | III | 373 K | 1.35 | 0.25 $Fe/nm^2$ |

[a]See discussion section for explanation.

[b]Calculated from the amount of $Fe(CO)_5$ admitted into the sample cell.

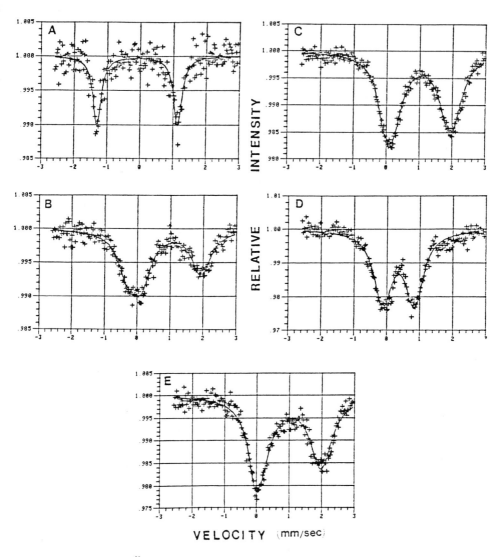

Figure 1.  Mössbauer Spectra of Fe(CO)$_5$ Decomposed on
Hydroxylated Titania (Sample 1 in Table I).  (A) Spectrum
recorded at 77 K immediately after admission of Fe(CO)$_5$,
(B) Spectrum recorded at room temperature (R.T.) following
decomposition at 373 K, (C) Spectrum recorded at R.T.
following decomposition at 673 K, (D) Spectrum recorded at
R.T. following exposure to ambient atmosphere, (E) Spectrum
recorded at R.T. following reduction in a flowing 90%CO/
10%CO$_2$ gas mixture.

given component. The recoil-free fraction of each component may
be different, and it is thus impossible to determine the atomic
ratios from the spectral ratios. Spectrum 1C was recorded at
room temperature after the sample had been heated to 673 K in
vacuum. This spectrum is almost entirely that of an $Fe^{2+}$ species
(96%). There is also a singlet species with a slightly negative
isomer shift which is due to zero-valent iron (4%). Spectrum 1D
was recorded at room temperature after the sample had been exposed
to the ambient laboratory atmosphere. This is the spectrum of an
$Fe^{3+}$ species which shows that exposure to air will fully oxidize
all of the surface iron species. Spectrum 1E was recorded after
the sample was reduced in a flowing $90\%CO/10\%CO_2$ gas mixture for
10 hours at 653 K. This reduction procedure restores most of the
iron to the $Fe^{2+}$ state (90%); however, some of the spectral area
(10%) is due to a zero-valent iron species. The two spectra
of Figure 2 were recorded following treatments of sample 3, the
most dehydroxylated sample, and are intended to demonstrate that
the nature of $Fe(CO)_5$ decomposition on partially dehydroxylated
$TiO_2$ is similar to the decomposition process on fully hydroxylated
$TiO_2$. That is, spectrum 2A recorded following decomposition at
383 K on sample 3 is similar to spectrum 1B which was recorded
following decomposition at 383 K on sample 1. Spectrum 2B,
recorded at room temperature following decomposition at 673 K on
sample 3, is similar to spectrum 1C which was recorded after
decomposition at 673 K on sample 1.

Volumetric Gas Phase Analysis

The results of this work are given in Table II. It can be seen
that the CO/Fe surface ratio varies from sample to sample. This
probably results from the presence of two surface species, in
different relative amounts, on each sample. This is explained in
the discussion section.

Discussion

It was found in this work that the concentration of hydroxyl
species on the surface determines the extent of $Fe(CO)_5$ decomposition on $TiO_2$. The concentration of reduced titanium cations
($Ti^{3+}$) on the surface apparently has no effect on the process.
Neither concentration affects the nature of the species formed.
To see that this is true, qualitative models of the titania surface produced following each of the three pretreatments used in
this study (i.e., evacuation at 600 K, evacuation at 720 K, and
hydrogen reduction at 720 K) are developed below on the basis of
earlier studies of $TiO_2$ surfaces (5-6,10-22). Surface Type I:
Following outgassing at about 600 K the $TiO_2$ surface should be
almost entirely free of molecular water (except on the rutile
fraction), but about one half of the surface should be covered

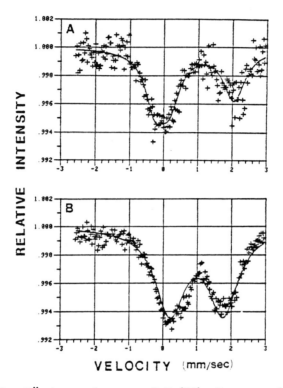

Figure 2. Mössbauer Spectra of Fe(CO)$_5$ Decomposed on
Dehydroxylated Titania (Sample 3 in Table I). (A) Spectrum
recorded at R.T. following decomposition of Fe(CO)$_5$ and
373 K, (B) Spectrum recorded at R.T. following decomposition
at Fe(CO)$_5$ at 673 K.

with hydroxyl groups. Furthermore, only a limited number of $Ti^{3+}$ ions should exist in the surface. These samples are white. Surface Type II: After outgassing at 720 K the $TiO_2$ surface should be entirely free of molecular water and only a small number of hydroxyl groups (ca. 0.5 $OH/nm^2$) should remain on the surface. An appreciable amount of $Ti^{3+}$ may also exist in such samples. Indeed, materials treated at this temperature are gray, suggesting the presence of reduced forms of titania. Surface Type III: Hydrogen reduction at 720 K probably produces a surface in which there is no molecular water and only a small number of hydroxyl groups. Furthermore, the surface following this treatment may have a high concentration of $Ti^{3+}$ species. Materials treated in this manner were found to be pastel blue in color. The surface "type" of each sample is given in Table I and II.

The spectra of Figure I suggest that $Ti^{3+}$ sites apparently do not affect the nature of the surface species formed by thermal decomposition of $Fe(CO)_5$. A visual comparison of these spectra (decomposition on a Type III surface) following identical treatments (1B and 2A, 1C and 2B) shows very little difference. The data in Table I also indicate that the various species present on the surface are the same after equivalent iron carbonyl decomposition procedures on Type I, II and III surfaces. Thus, the hope that $Ti^{3+}$ ions produced by pretreatments might stabilize a zero-valent iron species following low temperature reduction through a strong metal-support interaction proved to be groundless. It should be noted that one explanation for this behavior is that the $Ti^{3+}$ cations formed during Type II and III treatments are located primarily within the titania support and not at the surface (23-24).

From the results of Table I it can be suggested that the decomposition of $Fe(CO)_5$ is associated with hydroxyl groups on the support surface. That is, there is one major difference between the spectra collected on hydroxylated (Type I surface) samples and those collected on partially dehydroxylated (Type II and III surfaces) samples, namely the total spectral area. In Table I, it is shown that the total spectral area of sample 1 (hydroxylated) following high temperature decomposition (673 K) is more than twice that of samples 2 and 3 (partially dehydroxylated) following the same treatment. This is probably due to the fact that samples 2 and 3 have a lower surface hydroxyl density than sample 1. (There is also a greater total area for the spectrum of sample 1 than for the spectra of samples 2 and 3 following low temperature decomposition. However, two species are present following low temperature decomposition, and the relative recoil-free fractions of these species are not known. This makes a quantitative comparison difficult.) Indeed, Burwell, Brenner and Bowman (25-28), in extensive studies of $Mo(CO)_6$ decomposition on $\gamma-Al_2O_3$, found that there is a definite "capacity" of any $\gamma-Al_2O_3$ surface. That is, no more than about a 2% loading is possible on a hydroxylated $\gamma-Al_2O_3$ surface. On less hydroxylated surfaces the loading was found to be lower, Moreover, Burwell and Brenner postulate (25) that following low temperature (373 K) decomposition, one of the following species is formed:

$$(\sigma\text{-OH})_3 \cdot Mo(CO)_3 \text{ or } (\sigma\text{-O}^-)(\sigma\text{-OH})_2 \cdot Mo(CO)_3 \qquad (3)$$

where $\sigma$ represents a surface site. Upon further heating they suggest that the following reaction may occur (26-28):

$$Mo(CO)_3(ads) + 2\sigma\text{-OH} \xrightarrow{540 \text{ K}} (\sigma\text{-O}^-)_2Mo + 3CO + H_2 \qquad (4)$$

and finally at very high temperatures complete oxidation can occur:

$$(\sigma\text{-O}^-)_2Mo + 2OH^- \rightarrow (\sigma\text{-O}^-)MoO_2 + H_2 \text{ (gas)} \qquad (5)$$

A similar role of hydroxyl groups in the decomposition of $Fe(CO)_5$ could explain the lower total loading (smaller spectral area) on the more severely dehydroxylated $TiO_2$ samples. This idea will be developed more fully in the following sections.

## Low Temperature Decomposition (< 383 K)

On all samples it was found, using Mössbauer spectroscopy, that following low temperature decomposition two surface species exist, one which produces a quadrupole split spectrum and the other a spectral singlet. Based on the isomer shift and quadrupole splitting of the former one can identify this species as $Fe^{2+}$. Based on the slightly negative isomer shift of the latter, one can identify the other species as zero-valent iron. As suggested earlier, both species are probably associated with surface hydroxyl groups. The oxide ($Fe^{2+}$) species is probably a totally decarbonylated species which has chemically interacted with the surface hydroxyls. The metallic species is probably a subcarbonyl (see Volumetric Gas Phase Analysis section) which is bound to an hydroxyl site. It is not possible to identify this species on the basis of the Mössbauer spectra alone. The negative isomer shift singlet could belong to a number of different species. However, the identification of the zero-valent species as a subcarbonyl is consistent with other studies of metal carbonyl decomposition. Brenner et al. (29), for example, suggested a general formula for the decomposition of metal (M) carbonyls on hydroxylated supports:

$$M(CO)_j + n(\sigma\text{-OH} \xrightarrow{\text{Heat}} (\sigma\text{-O}^-)_n M^{n+} + (n/2)H_2 + jCO \qquad (6)$$

Based on the Mössbauer spectra, one can assign n the value of 2 for the $Fe(CO)_5/TiO_2$ system. The suggestion that both an $Fe^{2+}$ and a subcarbonyl species exist on the surface at the same time is consistent with the study of iron carbonyl decomposition on $\gamma\text{-Al}_2O_3$ conducted by Brenner and Hucul (30). These workers found, on the basis of the amount of hydrogen evolved, that following complete decomposition (873 K) the value of n (see Equation (6)) for iron decomposed on $\gamma$-Al$_2$O$_3$ was 2. They found that hydrogen evolved continuously from 373 K to 900 K, leading them to suggest that both subcarbonyl species (Fe(CO)$_2$) and Fe$^{2+}$ species exist on the surface of $\gamma$-Al$_2$O$_3$ over this range of temperatures. This is also consistent with the work of Basset et al. (31) who found that the decomposition of Fe$_3$(CO)$_{12}$ on MgO at 393 K led to the production of both Fe$^{2+}$ and an Fe$^\circ$ species.

## High Temperature Decomposition (673 K)

Following decomposition at 673 K there are apparently two species of iron on the TiO$_2$ surface: Fe$^{2+}$ (90% of the spectral area) with the same Mössbauer parameters as the Fe$^{2+}$ species formed during low temperature decomposition, and a zero-valent iron species with a slightly negative isomer shift. The preponderance of Fe$^{2+}$ on the surface probably results from the completion of the interaction between surface hydroxyls and subcarbonyl species (see Equation (6)). The zero-valent species is tentatively identified as belonging to a metallic iron particle. Another possibility is that it results from an iron-titanium species (Fe$_x$Ti). Tatarchuck and Dumesic (32) observed the formation of such a complex with an isomer shift of -0.14 mm/sec when iron metal on titania was reduced in flowing hydrogen at 875 K. However, this is a higher temperature than that used in the present study. Furthermore, an ammonia synthesis reaction probe study conducted by Santos et al. (4) showed that strong interaction between titania and iron particles formed by the decomposition of Fe(CO)$_5$ did not take place until reduction was carried out at temperature above 773 K. That is, when the reduction temperature was increased to 773 K the apparent activation energy and reaction order with respect to ammonia pressure increased significantly in magnitude. In contrast, when the reduction temperature in those studies was 713 K, (still above the temperature used in this study) the activation energy and reaction order were found to be similar to those of iron particles supported on MgO (32). It is thus suggested that the zero-valent species formed during high temperature (673 K) treatment in the present study is probably due to small metallic iron particles. These particles produce a singlet Mössbauer spectra at 300°K suggesting that they are less than 9 nm in diameter. Further support for this hypothesis is derived from a Mössbauer spectra modelling program, described in detail elsewhere (22), which accounts for both superparamagnetic relaxation and collective excitation. Using this program we were able to show that iron oxide (90% of spectral area) and 9 nm metallic iron particles (10% of spectral area) together could indeed produce a spectrum similar to that of Figure 1C. Larger particles would not produce the observed spectra. This is shown in Figure 3. Following oxidation (Figure 1D) and re-reduction (Figure 1E)

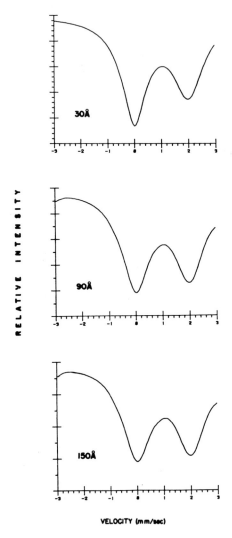

Figure 3. Assuming log normal particle distributions and
an anisotropy energy constant of $1.10^6$ ergs/cm$^3$ and a $\sigma$
value of 0.25 the above model spectra were generated using
a program described elsewhere (22). The average particle
size of each distribution used is given at the lower left
of each spectrum.

a zero valent iron species is again present.    This species is also
identified as small metallic iron particles.

## Volumetric Gas Phase Analysis

According to the above model developed for the interpretation of
the Mössbauer spectra, iron is "mono-dispersed" on the $TiO_2$ surface following low temperature thermal decomposition of $Fe(CO)_5$,
either as an oxidized species ($Fe^{2+}$) or as a subcarbonyl species.
A measure of the metallic iron dispersion following low temperature decomposition would provide a good test of this model.
Ideally, CO or $H_2$ adsorption would be used to determine the iron
dispersion.    However, for this system both $H_2$ and CO isotherms
were judged to be unsuitable.    The stoichiometry of hydrogen
adsorption on highly dispersed iron is not simple (34).    Furthermore, to adsorb hydrogen the CO must first be removed.    This also
makes CO isotherms impossible.    The removal of CO already present
would require a high temperature decomposition, resulting in the
oxidation of all the iron.    Thus, the most convenient technique
available for approximating the metallic  iron dispersion is to
determine the number of CO molecules which are retained by the
sample following decomposition of the $Fe(CO)_5$ (8).
In Table II, it can be seen that decomposition at low temperatures left a CO/Fe ratio on the surface of between 0.8 and 1.3.
A reasonable explanation for the high CO to iron ratios on the
surfaces, given the fact that Mössbauer spectroscopy studies show
that a large fraction of the iron is oxidized, is that iron
subcarbonyl species are stable on the surface at 383 K.    This is
true since CO does not adsorb strongly on iron oxides (35).
Indeed, assuming that about one-half of the iron is oxidized (as
would be suggested from the Mössbauer spectra), and using 1.0 as
the measured value of the CO to iron ratio, leads to the conclusion that $Fe(CO)_2$(ads) is present on the surface.    In fact, this
is the species that Brenner and Hucul postulated to be stable when
iron carbonyl is decomposed on γ-alumina (29).    The large variation in the measured CO to iron ratio can be explained by the
fact that the metallic iron to iron oxide ratio varies from sample
to sample.    The fact that no hydrogen gas was detected in this
study (as would be produced during the production of oxidized
species according to Equation 6) can be explained by the assumption
that the hydrogen produced during the oxidation reaction is
readsorbed by the titania at this relatively low temperature.    In
short, the results of volumetric gas phase analysis support the
model developed on the basis of Mössbauer spectroscopy.

## Summary

The results of this study suggest that the thermal decomposition

of iron pentacarbonyl on $TiO_2$ leads to the production of highly
dispersed iron species. These iron surface species appear to be
associated with surface hydroxyl groups. Reduced metal cations
($Ti^{3+}$) produced via high temperature outgassing or chemical reduction apparently play no role in the decomposition process. Low
temperature decomposition ($\leq$ 380 K) leads to the production of
both metallic iron species (possibly $(\sigma-HO)_3 \cdot Fe(CO)_2$) and $Fe^{2+}$
species. High temperature decomposition (ca. 670 K) leads to the
conversion of almost all the iron to the $Fe^{2+}$ species. At this
temperature a small fraction of the iron apparently forms small
metallic iron particles. Upon prolonged treatment in hydrogen
at temperatures near 700 K, the $Fe^{2+}$ on these samples can be
reduced to form small metallic iron particles (e.g., 9 nm in size),
as discussed elsewhere (4).

## Acknowledgment

The authors are grateful to Luis Aparicio for assistance with the
experimental work.

## References

1. Tauster, S.; Fung, S. C.; Garten, R. L. J. Am. Chem. Soc.
   1978, 100, 170.
2. Phillips, J.; Dumesic, J. A. Review - accepted for publication in Applied Catalysis.
3. Ugo, R. Catal. Rev.1975, 11, 225.
4. Santos, J.; Phillips, J.; Dumesic, J. A. J. Catal. 1983, 81,
   147.
5. Munuera, G.; Moreno, F.; Gonzalez, F. In "Seventh Int. Sym.
   on the Reactivity of Solids," (J. S. Anderson, M. W. Roberts,
   and F. S. Stone, eds.) p. 681, Chapman & Hall, London (1972).
6. Cornaz, P. F.; Van Hoof, J. H. C.; Pluijm, F. J.; Schuit,
   G. C. A. Disc. Faraday Soc. 1966, 41, 290.
7. Gebhardt, J.; Herrington, K. J. Phys. Chem. 1958, 62, 120.
8. Phillips, J.; Dumesic, J. A. Appl. of Surf. Sci. 1981, 7, 215.
9. Phillips, J.; Clausen, B.; Dumesic, J. A. J. Phys. Chem. 1980
   84, 1814.
10. Gravelle, P. L.; Juillet, F.; Meriaudeau, P.; Teichner, S. J.
    Disc. Faraday Soc. 1971, 52, 140.
11. Iyengar, R. D.; Codell, M.; Karra, J. S.; Turkevich, J. J. Am.
    Chem. Soc. 1966, 88, 5055.
12. Mashchenko, A. I.; Kasanskii, V. B.; Pariskii, G. B.; Sharapov,
    V. N. Kinetika i Kataliz, 1967, 8, 853.
13. Boehm, H. P. Adv. in Cat. 1969, 16, 179.
14. Primet, M.; Pichat, P.; Mathieu, M. V. C. R. Acad. Sci. Paris
    1968, B267, 799.
15. Iyengar, R. D.; Codell, M. Advan. Colloid Interface Sci. 1972,
    3, 365.
16. Jackson, P.; Parfitt, G. D. Trans. Far. Soc. 1971, 67, 2469.

17. Munuera, G.; Stone, F. S. Disc. Faraday Soc. 1971, 52, 205.
18. Kiselev, A. V.; Uvarov, A. V. Surf. Sci. 1967, 6, 399.
19. Primet, M.; Basset, J.; Mathier, M. V.; Prettre, M. J. Phys. Chem. 74, 2868.
20. Kaluza, U.; Boehm, H. P. J. Cat. 1971, 22, 347.
21. Lake, I. J. S.; Kemball, C. Trans. Faraday Soc. 1967, 63,
22. Phillips, J. Ph.D. Thesis, University of Wisconsin-Madison (1981).
23. Sexton, B. A.; Hughes, A. E.; Foger, K. J. Catal. 1982, 77, 85.
24. Chien, S. H.; Shelimov, B. N.; Resasco, D. E.; Lee, E. H.; Haller, G. L. J. Catal. 1982, 77, 301.
25. Burwell, R. L., Jr.; Brenner, A. In "Catalysis, Heterogeneous and Homogeneous," (B. Delmon and G. Jannes, eds.) p. 157, Elsevier, Amsterdam (1975).
26. Brenner, A.; Burwell, R. L., Jr. J. Catal. 1978, 52, 364.
27. Bowman, R. G.; Burwell, R. L., Jr. J. Catal. 1980, 52, 463.
28. Brenner, A.; Burwell, R. L., Jr. J. Catal. 1978, 52, 353.
29. Brenner, A.; Hucul, D. A.; Hardwick, S. J. Inorg. Chem. 1979, 18, 147B.
30. Brenner, A.; Hucul, D. A. Inorg. Chem. 1979, 18, 2836.
31. Hugues, F.; Bussiere, P.; Basset, J. M.; Courmereuc, D.; Chauvin, Y.; Bonnevoit, L.; Oliver, D. In "Proc. of the 7th Int. Cong. on Cat., Tokyo, 1980," (T. Seiyama and K. Tanabe, eds.) p. 418, Elserier, Amsterdam (1981).
32. B. Tatarchuk, and J. A. Dumesic, J. Catal. 1980, 70, 323.
33. Dumesic, J. A.; Topsøe, H.; Khammouma, S.; Boudart, M. J. Cat. 1975, 37, 503.
34. Tøpsoe, H.; Dumesic, J. A.; Topsøe, N.; Bohlbro, H. In "Proc. of the 7th Int. Cong. on Cat., Tokyo, 1980," (T. Seiyama and K. Tanabe, eds.) p. 247, Elsevier, Amsterdam (1981).
35. Udovic, T. J. Ph.D. Thesis, University of Wisconsin-Madison (1982).

RECEIVED October 31, 1983

# Secondary Ion Mass Spectrometry of the Ethylene/Ru(001) Interaction

L. L. LAUDERBACK[1] and W. N. DELGASS

School of Chemical Engineering, Purdue University, West Lafayette, IN 47407

Thermal desorption spectroscopy (TDS) and secondary
ion mass spectrometry (SIMS) studies show that the
interaction of ethylene with Ru(001) at 323 K is
accompanied by substantial dissociation and desorption of hydrogen producing an adlayer consisting
primarily of elemental carbon but also containing
small amounts of adsorbed hydrogen and hydrocarbon
species. It is shown that the hydrocarbon containing secondary ions seen in SIMS can be directly related to the identity of the small amounts of hydrocarbon species present. In particular, the SIMS
data provide direct evidence for the presence of
molecular ethylene and acetylenic complexes following
adsorption at 323 K and the hydrogenation of acetylenic complexes to ethylene upon heating to $\sim$650 K.

The relationships between secondary ions observed in secondary ion
mass spectrometry (SIMS) and the corresponding surface species responsible for these ions are of central importance in applying
this technique to the identification of adsorbates and reaction
intermediates on surfaces. Mounting evidence suggests that adsorbed molecular species can be emitted during SIMS by mechanisms
which preserve their molecular structure in the adsorbed state
(1-11). Thus, SIMS has high potential as a technique for the
direct, in situ observation of molecular adsorbates and reactive
intermediates. Benninghoven et al. (11), for example, have recently applied SIMS to study amino acids deposited on clean polycrystalline Cu and Ag foils by a molecular beam technique. In
virtually all cases, the main characteristics of the positive and
negative ion spectra consisted of parent-like peaks at $(M+1)^+$,
$(M-1)^+$ and $(M-1)^-$ where M is the molecular weight of the parent

[1]Current address: Department of Chemical Engineering, University of Colorado, Boulder,
CO 80309.

0097-6156/84/0248-0021$06.00/0
© 1984 American Chemical Society

amino acid.  These results were interpreted in terms of intact emission of precursors that were preformed on the surface prior to ion bombardment.

In order to investigate the possibility that parent-like molecular secondary ions might undergo atomic reorganization and atomic exchange during emission, we recently studied an Ru(001) surface containing equal amounts of molecularly adsorbed $C^{18}O$ and $^{13}CO$ ($\underline{10}$).  A complete lack of isotope mixing was observed in the cationized CO secondary ion (RuCO+), confirming that the CO groups present in RuCO+ do not undergo atomic exchange and are thus formed only by mechanisms which preserve their structure in the adsorbed state.

The concept of intact emission of adsorbed molecular species for identifying reaction intermediates is also well illustrated in several recent studies.  Benninghoven and coworkers ($\underline{2-4},\underline{12}$) used SIMS to study the reactions of $H_2$ with $O_2$, $C_2H_4$ and $C_2H_2$ on poly-polycrystalline Ni.  For the $C_2H_4$/Ni interaction, for example, direct relationships could be established between characteristic secondary ions and the presence of specific surface complexes ($\underline{12}$).  In another study, Drechsler et al. ($\underline{13}$) used SIMS to identify NH(ads) as the active intermediate during temperature-programmed decomposition of $NH_3$ on Fe(110).

In this paper, we present the results of an investigation of the interaction of ethylene with Ru(001) at 323 K and examine the relationships between surface species formed by this interaction and secondary ions emitted in SIMS.

Most of the ethylene that interacts with Ru(001) at 323 K produces a nondesorbable carbon layer.  This result is similar to that for the interaction of $C_2H_4$ with Ni, which produces a surface carbide at temperatures between about 300-600 K ($\underline{14}$).  SIMS results suggest, however, the presence of small amounts of molecularly adsorbed ethylene, acetylenic and other hydrocarbon complexes in addition to the nondesorbable carbon layer.

Correlations between surface species and emitted secondary ions are based on characterization of the surface adlayer by adsorption and thermal desorption measurements.  It is shown that the secondary ion ratios $RuC^+/Ru^+$ and $Ru_2C^+/Ru_2^+$ can be quantitatively related to the amount of nondesorbable surface carbon formed by the dissociative adsorption of ethylene.  In addition, emitted hydrocarbon-containing secondary ions can be directly related to hydrocarbon species on the surface, thus allowing a relatively detailed analysis of the hydrocarbon species present.  The latter results are consistent with ejection mechanisms involving intact emission and simple fragmentation of parent hydrocarbon species.

Experimental

All experiments were carried out in an ion pumped stainless steel ultrahigh vacuum chamber with a base pressure of about $1 \times 10^{-10}$

torr. Primary $Ar^+$ ions were generated by a Riber Cl 50 ion gun
and secondary ions were detected with a Riber Q156 quadrupole mass
spectrometer equipped with a 45° energy prefilter. The mass spectrometer is also equipped with an ionization filament for residual
gas analysis and thermal desorption measurements. All experiments
were performed with 5 KeV $Ar^+$ ions impinging on the sample surface
at a 45° polar angle measured from the surface normal. The primary ion current density was $5 \times 10^{-8}$ amps/cm$^2$.

The Ru single crystal was oriented by Laue x-ray back-scattering to within 1° of the Ru(001) plane, cut by a diamond saw and
mechanically polished. After being etched in hot aqua regia for
about 15 min, the crystal was spot welded to two tantalum heating
wires which were connected to two stainless steel electrodes on a
sample manipulator. The temperature was monitored by a Pt/Pt-10%
Rh thermocouple which was spot welded to the back of the crystal.
In this configuration, temperatures up to 1700 K could be routinely achieved. The surface cleaning procedure, which was similar to that used by Madey et al. (15) involved many heating and
cooling cycles up to 1600 K in $5 \times 10^{-7}$ torr of oxygen followed by
heating in vacuum 2-5 times to 1700 K to remove surface oxygen.
Surface cleanliness was verified by Auger electron spectroscopy
(AES) and SIMS.

### Results and Discussion

Adsorption and Thermal Desorption Measurements. In order to corre-
late emitted SIMS ions with the identity of various surface species formed by the ethylene/Ru(001) interaction, we first characterize the resulting adlayer by adsorption and thermal desorption
measurements.

Thermal desorption measurements were carried out by first
exposing the clean surface to a 15 L dose of ethylene at 323 K.
The mass spectrometer signal of the desired species was then monitored while increasing the sample temperature at a rate of approximately 65 K/sec by application of a constant heating voltage. This
relatively high heating rate was required in order to obtain measurable signals from the rather small amounts of desorbing species.
The resulting desorption spectra for $H_2$, $C_2H_4$ and $C_2H_6$ are shown
in Figure 1. For purposes of comparison, the hydrogen desorption
spectrum following a 5.0 L exposure of hydrogen to the clean surface at 310 K is also shown in Figure 1. Similar measurements of
$CH_4$, $C_3H_6$ and $C_3H_8$ revealed that these species were not desorbed
at levels above the detectable limits of the mass spectrometer.

The hydrogen desorption spectrum associated with the ethy-
lene/Ru(001) interaction shows a low temperature desorption peak
appearing in the same temperature region as the hydrogen peak due
to desorption of hydrogen from the clean surface. In addition, a
broad high temperature hydrogen desorption peak appears extending
from about 550 K to 900 K. The overlap of the low temperature
hydrogen peak with that for desorption of hydrogen from the clean

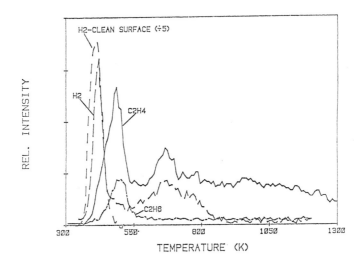

Figure 1.  $H_2$, $C_2H_4$ and $C_2H_6$ TDS spectra following a 15 L $C_2H_4$ exposure at 323 K and a $H_2$ TDS spectrum for the clean surface following a 5 L $H_2$ exposure at 310 K.

surface strongly suggests that the interaction of ethylene with
Ru(001) is accompanied by the dissociation of some hydrogen which
then bonds to the metal surface. The broad high temperature hydrogen desorption peak associated with the ethylene/Ru(001) interaction but not with hydrogen on the clean surface is indicative of
the dehydrogenation and subsequent desorption of hydrogen associated with hydrocarbon complexes on the surface. The breadth of
this peak suggests the presence of a relatively wide distribution
of C-H binding energies.

The ethylene desorption spectrum shows a desorption peak at
500 K and a higher temperature peak at 650 K. The later ethylene
peak begins to appear at about the same temperature as the high
temperature hydrogen desorption peak. In addition, a high temperature desorption tail is also observed, extending from the
trailing edge of the high temperature desorption peak. The ethane
desorption spectrum reveals a single desorption peak appearing in
the same temperature region as the low temperature ethylene desorption peak.

The simultaneous appearance of ethane with the low temperature ethylene desorption peak suggests that the low temperature
ethylene peak may correspond to the desorption of a molecularly
adsorbed ethylene species that can also undergo hydrogenation and
subsequently desorb as ethane. If this is the case, the effective
activation energies for desorption and hydrogenation of the molecular ethylene species are approximately equal.

In considering the nature of the high temperature ethylene
desorption peak, we note that this peak occurs simultaneously with
the high temperature hydrogen desorption peak. This suggests the
possibility that the desorbing ethylene might be formed by hydrogenation of surface complexes in which the reactive hydrogen is
supplied by the same dehydrogenation reactions that also give rise
to the desorption of hydrogen. In other words, the dehydrogenation of various hydrocarbon complexes may give rise to surface
hydrogen species that can desorb as hydrogen as well as react with
other hydrocarbon complexes, e.g. with acetylenic complexes, to
produce ethylene. Recent preliminary thermal desorption experiments carried out in this laboratory show self hydrogenation of
acetylene to ethylene in this temperature region.

These results are compared with the temperature dependence of
emitted ions in SIMS below, but first we consider quantitative
analysis of the surface layer.

To estimate the amounts of the various species desorbing in
Figure 1, we calibrated the mass spectrometer signal by measuring
the CO thermal desorption spectrum corresponding to a CO-saturated
surface at 300 K and set the area under the resulting desorption
curve equal to 0.58 ML as derived from the LEED results of
Williams et al. (16). Based on this calibration and after correcting for the ionization probabilities and the $1/(mass)^2$ transmission dependence of the mass spectrometer, the total amounts of
$H_2$, $C_2H_4$ and $C_2H_6$ that correspond to Figure 1 are 0.003, 0.002 and

0.0006 ML respectively. Although there is a considerable uncertainty in these numbers, particularly in the case of $H_2$ where it was necessary to retune the mass spectrometer, it is quite clear that only a very small fraction of the surface is covered with hydrogen and desorbable hydrocarbon species following a 15 L exposure to $C_2H_4$ at 323 K.

The amount of nondesorbable surface carbon produced by the ethylene/Ru(001) interaction was determined as a function of ethylene exposure by measuring, for each ethylene dose, the total amount of CO produced in a series of temperature-programmed reactions (TPR) designed to titrate the surface carbon by reaction with adsorbed oxygen. In these measurements, the sample was exposed to 6.0 L $O_2$ at 323 K following exposure of the clean surface to the desired dose of ethylene at 323 K. The CO mass spectrometer signal was then monitored while the sample temperature was increased at a linear rate of 6 K/sec until the CO formation ceased at about 800 K. The sample was then cooled to 323 K and dosed with another 6.0 L $O_2$ followed by another TPR. The procedure was repeated until no more CO was formed. SIMS measurements following this procedure confirmed that all of the surface carbon was removed. Contributions due to formation of $CO_2$ were neglected since measurements of the rate of $CO_2$ formation showed it to be approximately 100 times lower than the rate of CO formation. We note, as described elsewhere (17), that the loss of carbon by diffusion into the bulk can be neglected for the temperatures used in these measurements. The carbon surface coverage corresponding to a given ethylene dose was then estimated from the total area under all CO TPR curves using the previously described calibration. The measured carbon surface coverage as a function of ethylene exposure is shown in Figure 2. Carbon coverage increases rapidly for exposures up to about 5 L, after which the rate of carbon deposition begins to decline. Carbon deposition still occurs at a low rate, however, even after 15 L, which corresponds to a carbon surface coverage of 1.1 ML. From the initial slope of the coverage vs. exposure curve, the initial reaction probability for carbon deposition per incident ethylene molecule is estimated to be approximately 0.73. Comparing the 1.1 ML coverage of carbon following a 15 L dose, as shown in Figure 2, to the very small amounts of desorbable hydrogen, ethylene and ethane described previously clearly shows that adsorption of ethylene at 323 K is accompanied by substantial dissociation and subsequent desorption of hydrogen leading to an adlayer consisting primarily of nondesorbable carbon but also containing very small amounts of various hydrogen and hydrocarbon species.

## SIMS

The positive ion SIMS spectrum of the surface following a 15 L dose of ethylene at 323 K is shown in Figure 3. The spectrum in the $Ru^+$ and $Ru_2^+$ mass regions is characteristic of the Ru isotope

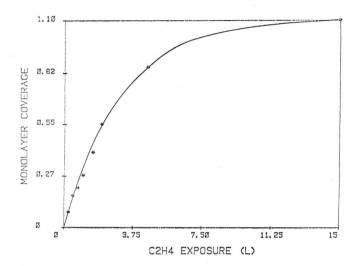

Figure 2.    Carbon coverage as a function of $C_2H_4$ exposure at 323 K.

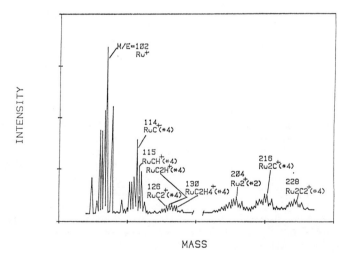

Figure 3.    Positive SIMS spectrum following a 15 L $C_2H_4$ dose at 323 K.

distribution, which allows easy identifications of the Ru- and
$Ru_2$-containing ion clusters. The observed Ru containing ions are
identified in this figure with the mass peak corresponding to the
most abundant Ru isotope (m/e = 102). Aside from the $Ru^+$ and $Ru_2^+$
ions, the spectrum is dominated by ions of the type: $Ru_xC_y(x,y =$
1,2), reflecting the high coverage of carbon resulting from the
dissociative adsorption of ethylene. Quite surprisingly, however,
measurable amounts of $RuCH^+$, $RuC_2H^+$ and $RuC_2H_4^+$ ions are also observed, indicating the presence of the very low coverage of hydrogen and hydrocarbon species indicated by the thermal desorption
spectra of Figure 1. The general relationship of the $RuCH^+$,
$RuC_2H^+$ and $RuC_2H_4^+$ ions to the desorbable species seen in Figure 1
is confirmed by the disappearance of these ions in spectra taken
after heating the sample to 763 K, as shown in Figure 4. The substantial reduction in intensity for the $Ru_2C_2^+$ ion also indicates a
possible relationship of this ion to desorbable hydrocarbon species or perhaps to the presence of C-C bonds on the surface. The
persistence of the $RuC^+$, $RuC_2^+$ and $Ru_2C^+$ ion species in Figure 4
confirms a general relationship of these ions to nondesorbable
surface carbon.

The corresponding negative ion spectrum following a 15 L $C_2H_4$
dose at 323 K is shown in Figure 5. In this spectrum the negative
ions $C^-$, $C_2^-$, $CH^-$, $C_2H^-$ and $C_2H_2^-$ appear. The negative hydrocarbon
ions again indicate high sensitivity to the small amounts of
hydrogen and hydrocarbon species on the surface. These ions also
disappear after the sample is heated to 763 K (Figure 6). The
persistence of the $C^-$ and $C_2^-$ ions after heating associates them
with the nondesorbable carbon. The $O^-$ ion observed in Figures 5
and 6 is due to an extremely high sensitivity of this ion to surface oxygen and represents only a trace oxygen impurity present at
levels below the detectable limits of AES and positive ion SIMS.

Exposure Dependence of SIMS. Additional insight into the relationships between emitted secondary ions and species present on
the surface is obtained by examining the exposure dependence of
the secondary ions and the relationship of this dependence to the
change in carbon coverage with exposure as shown in Figure 2.

The change in secondary ion yields for $Ru^+$ and $Ru_2^+$ along with
the ion yield ratio $Ru_2^+/Ru^+$ as a function of ethylene exposure is
shown in Figure 7. The $Ru^+$ and $Ru_2^+$ ion yields initially increase
to a maximum at about 0.8 L, corresponding to a carbon coverage of
0.2 ML, and then decrease with additional exposure until leveling
off at about 5.0 L. The initial increase in the ion yields for
$Ru^+$ and $Ru_2^+$ is indicative of an increase in the ionization probability and to an increasing work function. This is probably
caused by the formation of Ru-C dipoles. The subsequent decline
of the $Ru^+$ and $Ru_2^+$ ion yields for exposures greater than 0.8 L is
the expected result of a net decrease in the sputter yield due to
coverage of the surface with carbon. We note that these results
are analogous to the variation in the $Ni^+$ ion yield with acetylene

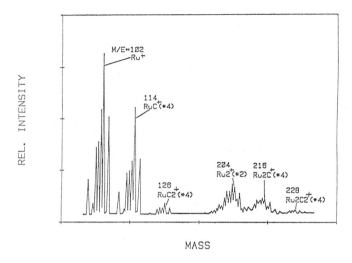

Figure 4. Positive ion SIMS spectrum of the surface follow-ing a 15 L $C_2H_4$ exposure, heating to 763 K and cooling to 323 K.

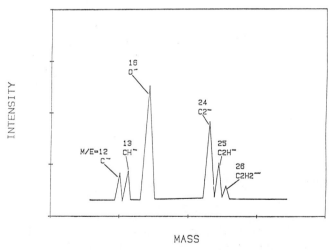

Figure 5. Negative ion SIMS spectrum following a 15 L $C_2H_4$ dose at 323 K.

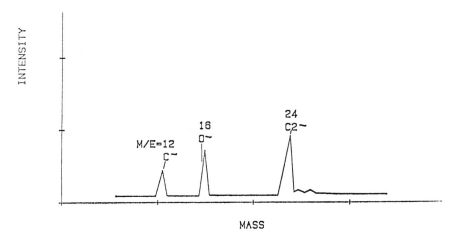

Figure 6.   Negative ion SIMS spectrum of the surface following a 15 L $C_2H_4$ exposure, heating to 763 K and cooling to
323 K.

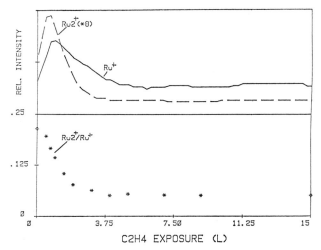

Figure 7.   Relative $Ru^+$ and $Ru_2^+$ ion yields and the $Ru_2^+/Ru^+$
in ratio as a function of $C_2H_4$ exposure at 323 K.

exposure recently reported by Benninghoven (12) for polycrystalline in nickel. Figure 7 also shows a rapid decline in the
$Ru_2^+/Ru^+$ ion ratio with exposure, an effect also observed during CO
and $O_2$ adsorption on Ru(001).

Figure 8 shows the relative change in ion ratios for $RuC^+/Ru^+$,
$Ru_2C^+/Ru_2$, $RuC_2H_4^+/Ru^+$ and $RuCH^+/Ru^+$ as a function of exposure.
For purposes of comparison, the relative change in carbon coverage
with exposure, based on Figure 2, is also shown. The secondary
ions are expressed as ratios in this figure in an attempt to cancel out effects due to changes in ionization probability as described above. The exposure dependences of the relative $RuC^+/Ru^+$
and $Ru_2C^+/Ru_2^+$ ion ratios, previously shown to be related to nondesorbable surface carbon, are essentially identical within experimental error to the relative change in the carbon surface
coverage, indicating a direct quantitative relationship between
these ion ratios and the amount of carbon on the surface. The
relative $RuC_2H_4^+/Ru^+$ and $RuCH^+/Ru^+$ ion ratios increase almost linearly with exposure to about 5.0 L and then level off to a nearly
constant value for higher exposures. The nearly identical behavior of the $RuC_2H_4^+/Ru^+$ and $RuCH^+/Ru^+$ ratios suggests that these
ions might be formed from either the same hydrocarbons species,
from different hydrocarbon species present in the same relative
proportions at all coverages, or both. Results to be described in
the next section suggest that $RuC_2H_4^+$ is formed by intact emission
of molecular ethylene while about 50% of the $RuCH^+$ intensity is
due to fragmentation of molecular ethylene and about 50% is derived from less hydrogenated hydrocarbon complexes.

Figure 9 shows the change in secondary ion yields for $C^-$, $C_2^-$,
$CH^-$, $C_2H^-$ and $C_2H_2^-$ as a function of ethylene exposure. The rate
of increase of the $C^-$ and $C_2^-$ ion yields with exposure is initially
very low but gradually increases with increasing exposure. The $C^-$
and $C_2^-$ ion yields subsequently begin to level off after about 5.0
L although they continue to increase at a low rate with additional
exposure. The rates of increase of the negative hydrocarbon ion
yields are also initially very low and gradually increase with
increasing exposure. The hydrocarbon ion yields, however, level
off at about 5.0 L to essentially constant levels with additional
increases in exposures.

The initial low rate of increase of all negative ion yields
with exposure is again indicative of an increasing work function
which generally reduces the ionization probability for negative
ions—an effect opposite to the initial increase in ionization
probability observed for positive ions.

The continuous increase in the $C^-$ and $C_2^-$ ion yields with
exposure above 5.0 L is analogous to the increase in carbon deposition and the $RuC^+/Ru^+$ and $Ru\ C_2^+/Ru^+$ ion ratios shown in Figure
8. This is consistent with observations above indicating that
both positive and negative ions that contain carbon but not hydrogen are related to the same nondesorbable surface carbon. The
similar behavior of all the hydrocarbon ions again indicates these

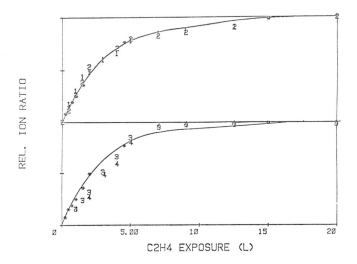

Figure 8. Variation of the $RuC^+/Ru^+$ (<u>1</u>), $RuC_2^+/Ru_2$ (<u>2</u>), $RuC_2H_4^+/Ru^+$ (<u>3</u>) and $RuCH^+/Ru$ (<u>4</u>) ion ratios and the carbon coverage (-*-) with $C_2H_4$ exposure at 323 K. All variables are normalized to their maximum values.

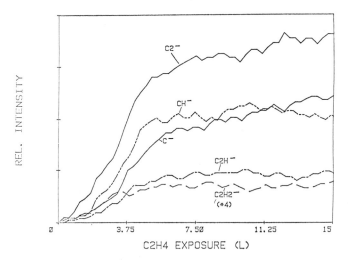

Figure 9. Relative variation of the $C^-$, $C_2^-$, and $C_2H_2^-$ ion yields with $C_2H_4$ exposure at 323 K.

ions are derived from the same source and/or from different species present in the same relative proportions at all exposures.
The results of the next section strongly suggest that the $C_2H_2^-$ ion
is formed by intact emission of acetylenic complexes and possible
fragmentation of molecularly adsorbed ethylene. The other smaller
ions are probably formed by simple fragmentation of acetylenic
complexes and molecularly adsorbed ethylene, as well as by intact
emission and simple fragmentation of other smaller hydrocarbon
complexes.

Temperature Dependence of Secondary Ions. We now consider the
relationship between SIMS spectra and desorbable hydrogen and
hydrocarbon species in more detail by comparing the temperature
dependence of the various hydrocarbon containing ions and $Ru_2C_2^+$
with the thermal desorption spectra of Figure 1.

The data were obtained by exposing a clean Ru(001) surface to
a 15 L dose of ethylene at 323 K, followed by monitoring the mass
spectrometer signal of the desired ion while increasing the sample
temperature at a rate of approximately 65 K/sec. In the case of
the $RuCH^+$, $RuC_2H_2^+$ and the $Ru_2C_2^+$ ions, the mass corresponding to
the most abundant Ru isotope ($M/e = 102$) was monitored. In all
cases, this peak is free from isotope interferences from other
species. The $RuC_2H^+$ ion was not investigated because of isotope
interference with $RuC_2^+$. In order to obtain adequate signal to
noise levels for the less intense ions, the primary ion current
density was increased to $1 \times 10^{-7}$ amps/cm$^2$ in these experiments.
This current density corresponds to a total primary ion dose of
$2 \times 10^{13}$ ions/cm$^2$ during the course of each experiment. The resulting temperature dependences of the various ions are shown in
the lower half of Figure 10, while the previously described thermal desorption spectra are reproduced in the upper half of Figure
10 for comparison. The Ru-containing ions are plotted as ratios
to $Ru^+$ in an attempt to cancel out effects due to changes in ionization probability as previously discussed.

A general examination of Figure 10 shows that the changes in
all ion intensities are highly correlated with the appearance of
the low and high temperature ethylene desorption peaks and hence
with the accompanying ethane and high temperature hydrogen desorption peaks respectively. In particular, a complete disappearance
of the $RuC_2H_4^+/Ru^+$ ion ratio accompanies the low temperature ethylene desorption peak along with partial declines in the intensities of $RuCH^+/Ru^+$, $Ru_2C_2^+/Ru^+$, $CH^-$ and $C_2H^-$. After partially
leveling off in a short plateau region immediately following the
low temperature ethylene desorption peak, the intensity of these
latter ions, together with that of the $C_2H_2^-$ ion, subsequently declines at a rapid rate in the temperature region of the high temperature ethylene and hydrogen desorption peaks. The residual ion
intensities then level off to a much lower rate of decline in the
region of the high temperature ethylene desorption tail.

The complete disappearance of the $RuC_2H_4^+/Ru^+$ ion ratio and

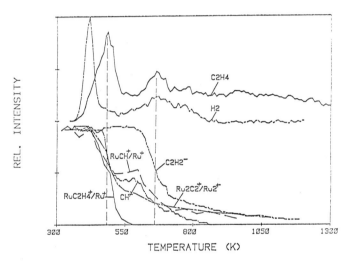

Figure 10. $H_2$ and $C_2H_4$ TDS spectra and temperature dependence of SIMS ions following a 15 L $C_2H_4$ exposure at 323 K.

the partial decline in intensities of $RuCH^+/Ru^+$, $Ru_2C_2^+/Ru_2$, $CH^-$ and $C_2H^-$ accompanying the low temperature ethylene desorption peak strongly suggests a direct relationship of these ions to molecularly adsorbed ethylene. In this interpretation, the $RuC_2H_4^+$ ion is formed by combination of $Ru^+$ with molecular ethylene emitted intact, while a portion of the $RuCH^+$, $Ru_2C_2^+$, $CH^-$ and $C_2H^-$ ions are the result of simple fragmentation of parent ethylene molecules. This assignment is, furthermore, consistent with the appearance of ethane in the TDS spectrum in this temperature region since ethane is likely to be formed by hydrogenation of molecular ethylene as previously discussed.

Assuming that the complete disappearance of the $RuC_2H_4^+/Ru^+$ ion ratio signifies the complete removal of molecular ethylene, then it is clear that the additional desorption of ethylene at higher temperatures must result from the hydrogenation of a more dehydrogenated complex. Thus, the persistence of the $C_2H_2^-$ ion as the largest emitted hydrocarbon ion following the low temperature ethylene desorption peak, together with the rapid loss of this ion in the temperature region of the high temperature ethylene desorption peak, suggests that the $C_2H_2^-$ ion is the result of intact emission of acetylenic complexes. This is, furthermore, consistent with the self-hydrogenation of acetylene observed in this temperature region in preliminary thermal desorption experiments mentioned earlier. The other smaller ion species observed following the low temperature ethylene desorption peak can be formed by fragmentation of parent ions or by intact emission and fragmentation of other possible smaller hydrocarbon complexes. Finally, we note the absence of a decline in intensity of any hydrocarbon-containing secondary ion in the temperature region of the low temperature hydrogen desorption peak. This is consistent with the assignment of this peak to desorption of a hydrogen species bonded to the Ru surface. Furthermore, this suggests that recombination of hydrogen atoms originally bonded to the metal with independently emitted carbon or hydrocarbon-containing ions is of minor importance.

All of these results are consistent with a direct relationship between emitted $Ru_2C_2^+$ and hydrocarbon-containing secondary ions with hydrocarbon species on the surface. The secondary ions are derived by intact emission and/or cationization and simple fragmentation of the parent hydrocarbon species. This picture is completely analogous to the intact emission and simple fragmentation of CO molecules in the formation of secondary ions emitted from a CO-covered Ru(001) surface as described in the introduction (1). It is also consistent with a variety of molecular dynamics model calculations (18,19), including our own preliminary calculations showing $RuC_2H_2^+$ formation by the mechanism cited.

The present results also suggest the following model for the ethylene/Ru(001) interaction. The interaction of ethylene with Ru(001) at 323 K is accompanied by substantial dissociation and subsequent desorption of hydrogen. The resulting adlayer consists primarily of nondesorbable surface carbon in addition to small amounts of dissociated hydrogen bonded to the metal surface and various hydrocarbon species consisting of molecularly adsorbed ethylene, acetylenic complexes and possibly other smaller hydrocarbon complexes. When the adlayer is heated, the hydrogen bonded to the Ru metal desorbs first at about 400 K. As the temperature increases to about 500 K, molecular ethylene desorbs. At the same time, hydrogenation of some of the adsorbed ethylene occurs leading to the simultaneous appearance of ethane with ethylene in the desorption spectrum. The source of hydrogen for this reaction probably comes from the thermal activation of the weakest C-H bonds, although residual hydrogen originally bonded to the metal surface cannot be completely ruled out. As the temperature increases to about 650 K, a relatively large number of C-H bonds associated with various adsorbed hydrocarbon complexes, possibly including some acetylenic complexes as well as others, become thermally activated leading to rapid desorption of hydrogen and to rapid hydrogenation of acetylenic complexes which immediately desorb as ethylene. As the temperature increases further, a relatively wide distribution of a small number of higher enegy C-H bonds are successively activated producing a broad hydrogen desorption peak and the hydrogenation of residual acetylenic complexes, which produce the high energy tail in the ethylene desorption spectrum. A small part of the high temperature ethylene tail may also be due to desorption from the sample holder.

We note finally that in view of the apparent fragmentation of parent species, it seems somewhat surprising that the low temperature ethylene desorption peak is not accompanied by a partial decline in the intensity of the $C_2H_2^-$ ion. In fact, the intensity of this ion appears to increase slightly in this region. We suggest that this is due to the formation of additional acetylenic complexes by decomposition of adsorbed ethylene upon heating.

Conclusions

In this paper we have shown that a direct quantitative relationship can be established between the $RuC^+/Ru^+$ and $Ru_2C^+/Ru_2$ ion ratios in SIMS and the surface carbon formed by the ethylene/Ru(001) interaction at 323 K. We have also shown that a direct relationship can be established between emitted hydrocarbon-containing secondary ions in SIMS and the small amounts of hydrocarbon species formed by the ethylene/Ru(001) interaction. The results are furthermore consistent with an ejection mechanism involving intact emission and simple fragmentation of the parent hydrocarbon species. Finally, the results of this study provide a relatively detailed description of the behavior of the rather small amounts of hydrocarbon species produced by the ethylene/Ru(001) interaction, thus illustrating the potential of SIMS for investigating surface reactions.

## Acknowledgments

We are grateful for support of this work by NSF Grants No. CHE78:
08728, No. CPE-7911597 and No. DMR77:23798, and by the Exxon Education Foundation. We also thank N. Winograd and B. Garrison for valuable discussions regarding this work and access to papers prior to publication.

## Literature Cited

1.  Benninghoven, A.; Sichtermann, W. K. Anal. Chem. 1978, 50, 1180.
2.  Benninghoven, A.; Muller, K. H.; Schemmer, M.; Beckmann, P. Appl. Phys. 1978, 16, 367.
3.  Benninghoven, A.; Beckmann, P.; Muller, K. H.; Schemmer, M. Surf. Sci. 1979, 84, 701.
4.  Muller, K. H.; Beckmann, P.; Schemmer, M.; Benninghoven, A. Surf. Sci. 1979, 325.
5.  Colton, R. J.; Murda, J. S.; Wyatt, J. R.; DeCorpo, J.J. Surf. Sci. 1979, 84, 235.
6.  Grade, H.; Cooks, R. G. J. Am. Chem. Soc. 1978, 100, 5615.
7.  Ray, R. J.; Unger, S. E.; Cooks, R. G. J. Am. Chem. Soc. 1978, 101, 501.
8.  Delgass, W. N.; Lauderback, L. L.; Taylor, D. G. "SIMS of Reactive Surfaces"; Springer-Verlag Series in Chemical Physics 20, 1982.
9.  Karevacki, E.; Winograd, N., to be published, 1982.
10. Lauderback, L. L.; Delgass, W. N. Phys. Rev. B. 1982, 26, 5258.
11. Benninghoven, A.; Lange, W.; Jirikowsky, M.; Holtkamp, D. Surf. Sci. 1982, 123, L721.
12. Benninghoven, A.; Beckmann, P.; Greifendorf, D.; Schemmer, M. Surf. Sci. 1982, 114, L62.
13. Drechsler, M.; Hoinkes, H.; Kaarmann, H.; Wistch, H.; Eith, G.; Weiss, M. Appl. Surf. Sci. 1979, 3, 217.
14. Ko, E. I.; Madix, R. J. Appl. Surf. Sci. 1979, 3, 236.
15. Maday, T. E.; Menzel, D. Japan J. Appl. Physics, Proceedings of 2nd International Conference on Solid Surfaces, Kyoto, 1974.
16. Williams, E. D.; Weinberg, W. H. Surf. Sci. 1979, 82, 93.
17. Lauderback, L. L.; Delgass, W. N., to be published.
18. Garrison, B. J. J. Am. Chem. Soc. 1980, 102, 6553.
19. Winograd, N.; Garrison, B. J.; Harrison, D. E. Jr. J. Chem. Phys. 1980, 73, 3473.

RECEIVED December 20, 1983

# X-Ray Photoelectron Spectroscopy of Cobalt Catalysts

## Correlation with Carbon Monoxide Hydrogenation Activities

D. G. CASTNER and D. S. SANTILLI

Chevron Research Company, Richmond, CA 94802-0627

A series of supported cobalt catalysts ($Co/Al_2O_3$, $Co/K-Al_2O_3$, $Co/SiO_2$, $Co/TiO_2$) have been examined by X-ray photoelectron spectroscopy (XPS) and microreactor studies. A catalyst treatment system attached to the XPS spectrometer was used to prepare in situ treated (air, $H_2$, 1% $H_2S/H_2$) catalysts for determining the identity, concentration, and reducibility of the cobalt species. At least three different types of cobalt species were present on the calcined catalysts. These included large particles of $Co_3O_4$, various $Co^{+2}$ species, and $CoAl_2O_4$. The $Co_3O_4$ particles were more readily reduced to metallic cobalt in $H_2$ than the $Co^{+2}$ species were. After $H_2$ reduction at 480°C, the CO hydrogenation activity in 10 atmospheres of $3H_2{:}1CO$ at 260°C for supported 5 wt % cobalt decreased as $Co/SiO_2 > Co/TiO_2 \geqslant Co/Al_2O_3 > Co/K-Al_2O_3$. The determination of the types of cobalt species present on each support and their reduction properties were used to explain the catalysts' CO hydrogenation activities.

We undertook this investigation in order to examine the relationship of physical structure and composition of cobalt catalysts to catalytic activity. Several different cobalt species have been detected on supported cobalt catalysts (1-7); the type, amount, and reactivity of the cobalt species varied with support, metal loading, and preparation procedures. For this investigation, the supports were varied and the other parameters were held constant. $SiO_2$, $TiO_2$, $Al_2O_3$, and $K-Al_2O_3$ were used as

0097-6156/84/0248-0039$06.00/0

supports. Since cobalt catalysts are known to be good CO hydrogenation catalysts (8) and the CO/H$_2$ reaction has been shown to
be sensitive to metal-support interactions (5, 9, 10), we
selected CO hydrogenation as our reaction probe.

## Experimental

Catalyst Preparation. The supported catalysts were all prepared
by pore fill impregnation with an aqueous solution of Co(NO$_3$)$_2$
to give a nominal Co loading of 5 wt %. The supports were first
calcined at 510°C for one to two hours, then impregnated, dried
at 65°C, and recalcined at 510-540°C for one to two hours. The
supports used were γ-Al$_2$O$_3$ (Catapal), silica gel (Davison), and
TiO$_2$ (Degussa). K-Al$_2$O$_3$ was made from γ-Al$_2$O$_3$ by pore fill
impregnation with K$_2$CO$_3$ followed by drying (65°C, several hours)
and calcining (870°C, four hours). K addition was done prior to
impregnating with Co.

X-Ray Photoelectron Spectroscopy (XPS or ESCA) Analysis. The
catalyst treatment-surface analysis system employed to characterize and treat the cobalt samples has been described previously (4). Briefly, it is a modular system consisting of a
Hewlett-Packard 5950A ESCA spectrometer, two quartz reactors, a
metal evaporator, an Auger electron spectroscopy (AES) and thermal desorption spectroscopy (TDS) station, a sample storage
area, and a rapid sample entry port. A 3.5-m long, 10-cm
diameter transfer tube connects these stations. The samples are
moved between stations by a combination of a chain-driven
trolley and magnetically coupled rods. The transfer tube is
always maintained under ultrahigh vacuum (UHV) with a base pressure of 1 x 10$^{-10}$ torr after bakeout. The reactors and sample
inlet stations are continually cycled between atmospheric pressure and UHV and have working base pressures of 10$^{-9}$ torr,
although pressures in the 10$^{-10}$ torr range can be attained after
a bakeout.
       For catalyst treatments, the sample is transferred into the
quartz reactor in vacuo, the reactor isolated, the gas flow
commenced, and temperature linearly ramped to the desired
value. After the sample has been at temperature for the desired
time period, the sample is cooled to room temperature in the gas
flow, the gas flow is stopped, and the reactor is evacuated.
The sample is then transferred in vacuo to the XPS spectrometer. For ease of handling, a few milligrams of each sample
is pressed into either a 400- or 200-mesh gold grid and then
mounted in a gold sample holder.

The 5950A ESCA spectrometer is interfaced to a desktop computer for data collection and analysis. Six hundred watt monochromatic Al Kα X-rays are used to excite the photoelectrons and an electron gun set at 2 eV and 0.3 mAmp is used to reduce sample charging. Peak areas are numerically integrated and then divided by the theoretical photoionization cross-sections (11) to obtain relative atomic compositions. For the supported catalyst samples, all binding energies (BE) are referenced to the Al 2p peak at 75.0 eV, the Si 2p peak at 103.0 eV, or the Ti $2p_{3/2}$ peak at 458.5 eV.

**Reactor Studies.** The microreactor system consists of a 9.5-mm diameter, 0.9-m long stainless steel reactor tube mounted in a tube furnace and connected to stainless steel feed lines equipped with pressure and flow controllers and an on-line gas chromatograph (GC). The reactor tube is packed with 0.2 g of catalyst in the middle (~6 mm in length) and alundum on both ends. The catalyst bed is separated from the alundum with glass wool plugs. The premixed $CO/H_2$ feed gas (Linde Custom grade) is passed through a -78°C cold trap to remove metal carbonyls before introducing it into the microreactor system. The on-line GC is equipped with an OV 101 capillary column and flame ionization detector (FID) for hydrocarbon product detection. (The hydrocarbon values are not corrected for the small differences in detector efficiencies.) The exiting gases are analyzed for CO, $CO_2$, and $CH_4$ with a gas partitioner equipped with a thermistor detector. The peak areas obtained from the gas partitioner are converted to weight percentages by using the appropriate sensitivity factors (12).

The catalyst runs are conducted as follows. First, the system is flushed with helium, then hydrogen is passed over the catalyst at 150 psi and ca. 50 ml/min. as the reactor is heated from room temperature to 480°C over ~40 min. After 30 min. at 480°C, the reactor is cooled to the desired operating temperature (usually 260°C); and the gas flow is switched from $H_2$ to $3H_2:1CO$. Finally, the flow is adjusted to 20 ml/min.

During each run, the products are analyzed to determine "pseudosteady state conditions" and the temperature is varied to obtain an approximate measure of activity changes. These measurements are made during a span of several hours.

**Temperature Programmed Reduction (TPR) Studies.** In the TPR studies, a gas mixture of 2% $H_2$ in Ar is passed over powdered samples of the calcined catalysts. The catalysts are held in the middle of a 5-mm diameter, 0.4-m long quartz reactor with

glass wool plugs and quartz chips. The samples are heated at a
rate of 4.5°C/min. by a tube furnace mounted around the
reactor. A thermalconductivity detector is used to monitor the
composition of the gas phase after it passes through the reactor
and a water removal trap.

## Results

$Co/Al_2O_3$, $Co/K-Al_2O_3$, $Co/SiO_2$, and $Co/TiO_2$ were all tested for
CO hydrogenation to determine the effect of the supports on the
activity and selectivity of cobalt (Table I). The data in
Table I were taken after the catalysts had been onstream for
several hours. Since catalyst preparation, calcination, reduction, and startup procedures can affect catalyst performance,
procedures were kept as constant as possible to allow meaningful
comparisons. For $Co/K-Al_2O_3$, the reaction temperature was
raised to 315°C because of its low activity at 260°C. All catalysts had to be exposed to $CO/H_2$ for 10-30 min. before CO hydrogenation activity was detected. The onset of reaction was
accompanied by a large increase in the water content of the
product gas.

Table I.  CO Hydrogenation Activities and Selectivities of
Cobalt Catalysts[1]

| Catalyst | Temp. (°C) | $CH_4$[2] HC | $C_5+$[2] HC | Propene Propane | CO Conv. to HC (%) | Total CO Conv. (%)[3] |
|---|---|---|---|---|---|---|
| $Co/SiO_2$ | 260 | 0.51 | 0.19 | 0.2 | 73 | 81 |
| $Co/TiO_2$ | 260 | 0.48 | 0.24 | 1.1 | 23 | 23 |
| $Co/Al_2O_3$ | 260 | 0.54 | 0.18 | 0.5 | 17 | 17 |
| $Co/K-Al_2O_3$ | 260 | – | – | – | <5 | <5 |
|  | 315 | 0.32 | 0.34 | 3 | 53 | 74 |

[1] $3H_2:1CO$, 150 psi, 20 ml/min., 0.2 g catalyst.
[2] Weight ratios, HC = total hydrocarbon product.
[3] Includes conversion to both HC and $CO_2$.

The conversion and selectivity data presented in Table I
were obtained using the same amount of cobalt (5 wt %) on each
support and the same amount of each catalyst (0.2 g) in the

reactor. In this table, we did not correct the catalyst activities for differences in exposed metallic Co surface area for a number of reasons. First, $H_2$ chemisorption, a common technique for measuring metal surface area, was found to be an activated process on cobalt catalysts (13). The degree of activation varied with both support and Co loading (13). Additionally, a variety of cobalt species, dispersions, and reducibilities were observed on our catalysts. Since any of these factors can affect $H_2$ chemisorption experiments, the relationship of the data and the true metal surface area would be difficult to establish. Therefore, to avoid unnecessary complexity, we have ranked the catalysts according to their activity based on total catalyst weight. In the following paragraphs, we present data on the different structural and reduction properties of the catalysts which could account for the activity differences.

The total surface areas determined by the $N_2$ BET method for the calcined, supported catalysts are listed in Table II. The X-ray diffraction (XRD) results showed diffraction peaks from a cubic lattice with a unit cell distance of ~8.1 Å were present on all of the calcined catalysts. Both $Co_3O_4$ and $CoAl_2O_4$ have structures consistent with that lattice spacing, making assignment of the type of crystalline cobalt species present on the alumina supports difficult.

| Table II. Surface Areas for the Calcined Catalyst | |
| --- | --- |
| Catalyst | Surface Area $(m^2/g)$[1] |
| $Co/SiO_2$ | 268 |
| $Co/TiO_2$ | 48 |
| $Co/Al_2O_3$ | 230 |
| $Co/K-Al_2O_3$ | 147 |

[1]Determined by $N_2$ BET measurement.

All the calcined samples were examined by TPR to determine the temperature ranges where reduction of the various cobalt oxide species was occurring. These TPR spectra are shown in Figure 1. All samples were scanned up to 750°C except $Co/Al_2O_3$

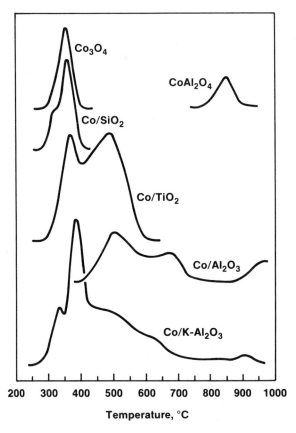

Figure 1. TPR spectra for $Co_3O_4$, $CoAl_2O_4$, 5 wt % $Co/SiO_2$, 5 wt % $Co/TiO_2$, 5 wt % $Co/Al_2O_3$, and 5 wt % $Co/K-Al_2O_3$ taken at heating rate of 4.5°C/min. in 2% $H_2/Ar$

and $Co/K-Al_2O_3$, in which cases the range was extended to 975°C in order to investigate the possible presence of cobalt aluminate species.

The Co 2p XPS spectra from in situ treatments of the catalysts are shown in Figures 2-5.  The Co 2p BE, line shape, and satellite intensity vary noticeably among the samples.  The variation of these three spectral features has been shown to be useful in identifying the different Co species present in a sample (4).  For the Co $2p_{3/2}$ line, the metallic peak is located near 778 eV, the +2 and +3 oxide peaks are near 781 eV, and +2 satellite peak is near 787 eV.  The Co $2p_{1/2}$ features are located at 15-16 eV higher BE than the corresponding Co $2p_{3/2}$ peaks.

Inspection of the Co spectra showed that the fraction of reducible surface cobalt varied significantly with the catalysts and temperature of reduction.  Direct comparison of the catalysts using the fraction of surface Co reduction is not advisable since there are several factors which can modify the observed values.  XPS only detects cobalt that is located within ~30 Å of a sample surface.  Thus, the Co dispersion, the Co distribution among the various Co species, and the support properties can affect the measured Co intensity.  Therefore, the following procedure was employed so direct comparisons among the catalyst could be made.  First, the total Co weight percent is calculated relative to the support from the measured XPS peak areas, correcting for differences among the supports (surface area, density, etc.) using the model and equations given by Kerkhof and Moulijn (14).  Then the percent of surface Co reduction is determined by the contribution the metallic Co peak makes to the Co 2p spectrum.  The metallic cobalt peak widths and shapes on all the supported catalysts were assumed to be the same as that observed for reduced $Co/SiO_2$ (100% reduction).  Table III shows the results for the supported Co catalysts reduced at 480°C in $H_2$.  The large $Co/Al_2O_3$ error limits were due to the uncertainty involved in separating the small metallic peak from the large $Co^{+2}$ peak.  We emphasize that these values represent "surface" cobalt concentrations as measured by XPS and can be substantially different from the bulk values.

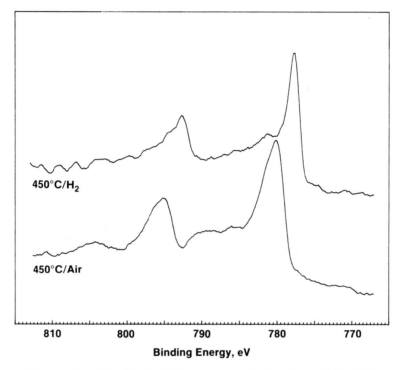

Figure 2. Cobalt 2p XPS spectra of the 5 wt % $Co/SiO_2$ catalyst after a one-hour calcination and a two-hour $H_2$ reduction at 450°C

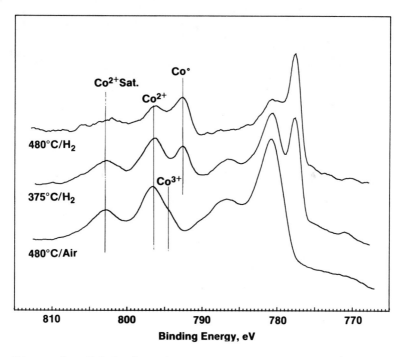

Figure 3. Cobalt 2p XPS spectra of the 5 wt % Co/TiO₂ catalyst after one-hour treatments in air at 480°C and H₂ reduction at 450°C

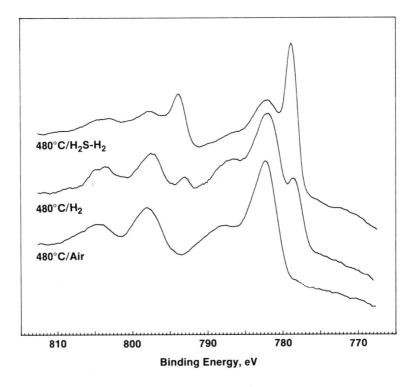

Figure 4. Cobalt 2p XPS spectra of the 5 wt % $Co/Al_2O_3$ catalyst after one-hour treatments in air, $H_2$, and 1% $H_2S/H_2$ at 480°C

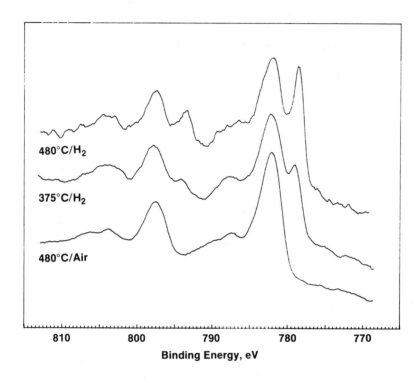

Figure 5.  Cobalt 2p XPS spectra of the 5 wt % Co/K-Al$_2$O$_3$ catalyst after a one-hour calcination at 480°C, a one-hour H$_2$ reduction at 375°C, and a two-hour H$_2$ reduction at 480°C

| Table III. | XPS Measured Surface Cobalt Concentrations After $H_2$ Reduction at 480°C of the Catalysts | | |
|---|---|---|---|
| Catalyst | Total Co (Wt %) | Percent Reduction | Reduced, Co (Wt %) |
| $Co/SiO_2$ | 0.8 | 100 | 0.8 |
| $Co/TiO_2$ | 0.6 | 85 | 0.5 |
| $Co/Al_2O_3$ | 2.4 | 20 ± 10 | 0.5 ± 0.2 |
| $Co/K-Al_2O_3$ | 0.9 | 50 | 0.4 |

## Discussion

Cobalt Species. XRD results indicated crystalline $Co_3O_4$ and/or $CoAl_2O_4$ were present on all the calcined catalysts. For the $SiO_2$ and $TiO_2$ supports, $CoAl_2O_4$ can be eliminated since no alumina is present. For the alumina supports, information obtained in the TPR and XPS experiments was needed to make the assignment. TPR and XPS results establish that unsupported $Co_3O_4$ was completely reduced to metallic Co by 450°C in $H_2$ [see Figure 1 and Reference (4)] with the TPR peak maximum near 350°C. These reduction characteristics and the XPS Co 2p line shape for $Co_3O_4$ can be used to determine the amount of bulklike $Co_3O_4$ present on the supported catalysts.

For the $Co/SiO_2$ catalyst, the TPR profile and Co 2p XPS spectra were nearly indistinguishable from those for the unsupported $Co_3O_4$ sample, indicating that $Co_3O_4$ was the major, and probably only, cobalt species present on the calcined catalyst. The amount of Co detected by XPS for the $Co/SiO_2$ sample was 0.8 wt %, which was significantly lower than the bulk value of 5 wt %. This implied the presence of $Co_3O_4$ particles much larger than the ~30 Å depth probed by XPS. Thus, it appears that the cobalt species present on $SiO_2$ were large bulklike $Co_3O_4$ particles which were completely reduced to metallic cobalt by $H_2$ below 450°C.

The presence of two peaks in the TPR profile for the $Co/TiO_2$ system suggested that two cobalt oxide species were present on the calcined catalyst. Using XPS, these species were identified as $Co_3O_4$ and $Co^{+2}$. The Co $2p_{1/2}$ XPS spectrum of the calcined $Co/TiO_2$ catalyst had a satellite peak due to a $Co^{+2}$

species, a large $Co^{+2}$ peak, and a low BE shoulder that was consistent with $Co^{+3}$ from $Co_3O_4$. After $H_2$ reduction at 375°C, metallic cobalt and the $Co^{+2}$ species were detected by XPS, indicating the 375°C TPR peak is due to reduction of the $Co_3O_4$ particles. This was in agreement with the results for $Co/SiO_2$ and unsupported $Co_3O_4$. After $H_2$ reduction at 480°C, XPS showed that 85% of the cobalt was metallic, with the remainder $Co^{+2}$. Thus, the 480°C TPR peak can be attributed to reduction of the $Co^{+2}$ species to metallic cobalt. Also during the 480°C reduction, the total amount of Co detected by XPS decreased by a factor of two. This could be a consequence of either the cobalt particles sintering, cobalt diffusing into the $TiO_2$ support, or a Ti oxide species diffusing onto the Co particles. Further studies of this system with other techniques would be helpful in determining which process is occurring. In a transmission electron microscopy (TEM) study, Tatarchuk and Dumesic observed by XPS a decrease in iron concentration in a $Fe/TiO_2$ system subjected to $H_2$ reduction near 500°C, presumably due to Fe diffusion into the $TiO_2$ lattice (15, 16). The suppression of $H_2$ chemisorption on $TiO_2$ supported metals has been ascribed to the presence of a $TiO_x$ species on the surface of the metal particles (17). Interestingly, the $Co/TiO_2$ was the only sample in our study which XPS detected a noticeable decrease in Co concentration after treatment with $H_2$ at 480°C.

The $Co/Al_2O_3$ catalyst had three peaks in the TPR profile, none of which occurred at the same temperature as unsupported $Co_3O_4$. On this catalyst, TPR showed >25% of the cobalt reduction occurred above 700°C. Also, $Co^{+2}$ was the major cobalt species detected by XPS after a two-hour $H_2$ reduction at 700°C. In fact, the $Co^{+2}$ features dominated the XPS spectra of both the calcined and 480°C $H_2$ reduced samples. These results suggested the presence of a cobalt aluminate species on the calcined $Co/Al_2O_3$ catalyst. Since no peak characteristic of $Co_3O_4$ was present in the TPR profile, the crystalline cobalt species detected by XRD was probably not $Co_3O_4$. No indication of $Co_3O_4$ was present in the XPS spectrum of calcined $Co/Al_2O_3$, but a small $Co_3O_4$ signal could be masked by the dominant $Co^{+2}$ spectral features. Thus, the crystalline cobalt species is $CoAl_2O_4$ and/or some intermediate between $Co_3O_4$ and $CoAl_2O_4$ with the same crystal structure. The two reduction peaks near 500°C and 675°C in the TPR spectrum could be associated with such an intermediate. Identification of the oxide species reducing in this temperature range was difficult with XPS due to the presence of the aluminate $Co^{+2}$ signal and because the measured cobalt intensity decreased when reduction temperatures near 700°C were

employed. This decrease could be due to the cobalt sintering or
diffusing into the support, analagous to the $Co/TiO_2$ system. In
order to avoid the loss of signal intensity, we treated the
$Co/Al_2O_3$ catalyst in 1% $H_2S/H_2$, which allowed the cobalt oxides
to react at lower temperatures. The results showed a large
decrease in the fraction of cobalt oxide and the appearance of a
S 2p peak at 162.6eV, indicating cobalt sulfide was formed.
Thus on the calcined $Co/Al_2O_3$ there were at least three cobalt
species: one that reduced at 480°C in $H_2$, one that suldifed at
480°C in $H_2S/H_2$, and one that did not reduce or sulfide at
480°C.

Three cobalt oxide species have also been detected on calcined $Co/Al_2O_3$ catalysts in other studies (1, 6, 7). Chin and
Hercules identified these cobalt species as bulklike $Co_3O_4$, $Co^{+2}$
in the octahedral holes of the $\gamma-Al_2O_3$ lattice and $Co^{+2}$ in the
tetrahedral holes of the $\gamma-Al_2O_3$ lattice (1). Chung and Massoth
observed a cobalt species that reduced in $H_2$ at 500°C, one that
sulfided in $H_2S/H_2$ at 400°C, and one that neither reduced or
sulfided (6, 7). The results of Chung and Massoth are consistent with our results on $Co/Al_2O_3$, but evidence for the $Co_3O_4$
species identified by Chin and Hercules was not found in our
study.

The results for the $Co/K-Al_2O_3$ catalyst differed in several
respects from those of $Co/Al_2O_3$. For example, on the $Co/K-Al_2O_3$
system, the amount of cobalt aluminate was significantly lower,
as was demonstrated by XPS which showed a lower $Co^{+2}$ species
concentration, and TPR which showed only a small amount of $H_2$
uptake above 750°C. Additionally, by XPS, we measured a higher
fraction of reduced cobalt on the surface. The $Co/K-Al_2O_3$ catalyst also showed a significant amount of reduction occurring
below 400°C, indicating the presence of $Co_3O_4$ particles.
Furthermore, peaks indicative of $Co_3O_4$ were present in the XPS
spectrum of the calcined catalyst. These peaks disappeared
after $H_2$ reduction at 375°C, and $Co^{+2}$ remained as the major
cobalt oxide species. The $Co/K-Al_2O_3$ sample also had two reduction peaks near 450°C and 650°C in the TPR profile, analogous to
$Co/Al_2O_3$. We interpreted this to mean that the $Co/K-Al_2O_3$ catalyst had the same cobalt species present as those found on
$Co/Al_2O_3$, but that the cobalt aluminate phase concentration was
lower and that bulklike $Co_3O_4$ particles were present.

No determination was made as to whether differences between
$Co/Al_2O_3$ and $Co/K-Al_2O_3$ were due to the presence of K, the
higher support calcination temperature, or the lower surface
area of $K-Al_2O_3$. It should be noted that although the total
amount of bulk cobalt reducible in $H_2$ at 480°C had increased in

$Co/K-Al_2O_3$ compared to $Co/Al_2O_3$, the XPS experiments indicated that the amount of surface metallic cobalt present in the two systems was similar.

## CO Hydrogenation

The catalysts we examined in this study contained several cobalt species with different reduction properties. In this section, we will employ those findings to explain the observed CO hydrogenation activity ranking. The differences observed in catalyst selectivity will not be addressed in this paper.

CO hydrogenation activity was found to depend on both the amount of available surface cobalt metal and the environment of the reduced cobalt. Factors influencing these cobalt properties included the type of support, the type of cobalt species present on a given support, and the presence of additives. Due to the interrelationship of these properties, a detailed examination of the activity and structure of each catalyst is needed in order to determine the importance of each factor for CO hydrogenation activity.

Because of the large difference in cobalt reduction properties on $SiO_2$, $Al_2O_3$, and $TiO_2$, the amount of surface metallic cobalt was the most important variable affecting the relative activities of these catalysts. $Co/SiO_2$ had the highest activity ranking and the highest concentration of reduced surface cobalt, presumably because $Co_3O_4$ was the only species detected on the calcined catalyst, and it was completely reduced at 480°C in $H_2$. $Co/Al_2O_3$ and $Co/TiO_2$ had activities that were similar to each other and noticeably lower than that of $Co/SiO_2$. However, these two catalysts were observed to have different reduction behaviors. For $Co/Al_2O_3$ the activity decrease was probably due to the significant resistance of Co reduction at 480°C as observed by XPS and TPR. On $TiO_2$, the cobalt readily reduced at 480°C (similar to $Co/SiO_2$), but surface cobalt was apparently lost due to sintering or diffusion. Therefore, less cobalt would be available for CO hydrogenation.

$Co/K-Al_2O_3$ had the lowest activity ranking and the lowest concentration of surface metallic cobalt of the catalysts studied. However, the large fall off in CO hydrogenation activity from the $Co/Al_2O_3$ catalyst to the $Co/K-Al_2O_3$ catalyst was not accompanied by a large decrease in the amount of surface metallic cobalt. There was a large change in the relative amounts of the cobalt species present on these catalysts, suggesting the interaction of cobalt with the support was modified by the addition of K. Furthermore, K could affect CO hydrogenation activity by direct interaction with adsorbed CO or hydrogen.

Comparing results of different investigators for cobalt catalysts must be done with care. Preparation, reduction, and reaction conditions can all affect the catalytic activity of a cobalt catalyst. With this caution in mind, we compared our results with those reported by Vannice (10, 18, 19) and Reuel and Bartholomew (5) in Table IV. In all three studies, $Co/SiO_2$ was found to be more active than $Co/Al_2O_3$. However, the relative activity of $Co/TiO_2$ with respect to $Co/SiO_2$ and $Co/Al_2O_3$ varied. We believe the phenomenon responsible for the differing relative $Co/TiO_2$ activities was employment of different calcination and reduction conditions by the different groups. The range of reduction conditions would tend to generate samples with varying amounts of reduced surface Co. The different reduction behavior of the various cobalt oxide species present on $TiO_2$ would lead to varying fractions of reduced cobalt with treatment conditions. Additionally, the loss of surface metallic cobalt could have varied with the different reduction temperatures employed. Our catalysts were calcined near 525°C, while in the other studies the catalysts were only dried at 100-120°C. This difference could affect both the type and amount of cobalt species present on $TiO_2$.

Catalyst characterization studies and catalyst activity measurements were found to be a powerful combination for determining the important catalytic properties of the cobalt catalysts. By using XPS and TPR, it was possible to characterize types of surface cobalt species present on $SiO_2$, $TiO_2$, $Al_2O_3$, and $K-Al_2O_3$ supports. In situ treatment methods allowed us to determine the temperature dependent reduction behavior of each cobalt species. This data was used to explain the observed CO hydrogenation activities of each catalyst and to determine the important catalytic properties of a given system.

| Table IV. Activity Rankings of Cobalt Catalysts per gram Co for CO Hydrogenation | | | | |
|---|---|---|---|---|
| Reduction Conditions h/°C | Reaction Conditions | Cobalt Loading (Wt %) | Activity Ranking by, Support | Reference |
| 0.5/480 | $3H_2:1CO$, 10 Atm, 260°C | 5 | $SiO_2 > TiO_2 \geqslant Al_2O_3$ | This Study |
| 1/450 | $3H_2:1CO$, 1 Atm, 275°C | 1.5-4 | $SiO_2 > Al_2O_3 \geqslant TiO_2$ | (10, 18, 19) |
| 16/400 | $2H_2:1CO$, 1 Atm, 225°C | 3 | $TiO_2 \geqslant SiO_2 > Al_2O_3$ | (5) |

## Conclusions

1. XPS and TPR results showed that cobalt was present in different forms on $SiO_2$, $TiO_2$, $Al_2O_3$, and $K-Al_2O_3$.

2. The activity per gram of catalyst decreased as $Co/SiO_2 >$ $Co/TiO_2 \geq Co/Al_2O_3 > Co/K-Al_2O_3$.

3. The amount, type, and reducibility of the cobalt species detected on each catalyst was important in determining its CO hydrogenation activity.

4. The different cobalt species detected on the calcined, supported catalysts included large $Co_3O_4$ particles, $Co^{+2}$, and $CoAl_2O_4$.

5. Reduction to metallic cobalt in $H_2$ occurred between 300-425°C for the $Co_3O_4$ particles, between 425-700°C for some $Co^{+2}$ species, and above 850°C for $CoAl_2O_4$.

6. The amount of surface cobalt present on $TiO_2$ decreased at temperatures of 480°C in $H_2$. This also occurred for $Co/Al_2O_3$ at 700°C in $H_2$.

## Acknowledgments

The contributions and assistance of R. O. Billman and B. Lee in obtaining the experimental results are gratefully acknowledged. J. N. Ziemer is thanked for discussing and providing the TPR results. The XRD and surface area measurements were done by the Analytical Services Division at Chevron Research Company. Informative discussions with R. A. Van Nordstrand and R. T. Lewis are gratefully acknowledged.

## Literature Cited

1. Chin, R. L.; Hercules, D. M., *J. Phys. Chem.* 1982, 86, 360-7.
2. Lycourghiotis, A.; Defosse, C.; Delannay, F.; Lemaitre, J.; Delmon, B., *J.C.S. Faraday I* 1980, 76, 1677-88.
3. Lycourghiotis, A.; Tsiatsios, A.; Katsanos, N. A., *Z. Phys. Chem. Neue Folge* 1981, 126, 85-93.
4. Castner, D. G. *Proc. Adv. Catal. Chem. II* 1982, submitted.
5. Reuel, R. C.; Bartholomew, C. H. *J. Catal.*, submitted.

6.  Chung, K.S.; Massoth, F. E., J. Catal. 1980, 64, 320-331.

7.  Chung, K. S.; Massoth, F. E., J. Catal. 1980, 64, 332-345.

8.  Anderson, R. B. in "Catalysis"; Emmett, P. H.; Ed.;
    Reinhold: New York, 1956; Vol IV, Chap. 1-3.

9.  Tauster, S. J.; Fung, S. C.; Baker, R. T. K.;
    Horsley, J. A., Sci. 1981, 211, 1121-1125.

10. Vannice, M. A.,  J. Catal. 1982, 74, 199-202.

11. Scofield, J. H. J. Electron Spectrosc. Relat. Phenom. 1976,
    8, 129-137.

12. McNair, H. M.; Bonelli, E. J. "Basic Gas Chromatography";
    Consolidated Printers: Berkeley, 1969; Chap. 7.

13. Zowtiak, J. M.; Weatherbee, G. D.; Bartholomew, C. H.,
    J. Catal. 1983, 82, 230-235.

14. Kerkhof, F. P. J. M.; Moulijn, J. A., J. Phys. Chem. 1979,
    83, 1612-1619.

15. Tatarchuk, B. J.; Dumesic, J. A., J. Catal. 1981, 70,
    323-334.

16. Tatarchuk, B. J.; Dumesic, J. A., J. Catal. 1981, 70,
    335-346.

17. Jiang, X. Z.; Hayden, T. F.; Dumesic, J. A., J. Catal.
    1983, 83, 168-181.

18. Vannice, M. A., J. Catal. 1977, 50, 228-236.

19. Vannice, M. A., J. Catal. 1975, 37, 449-461.

RECEIVED December 5, 1983

# Modifications of Surface Reactivity by Structured Overlayers on Metals

ROBERT J. MADIX

Department of Chemical Engineering, Stanford University, Stanford, CA 94305

The dehydrogenation of alcohols has been studied on both copper and nickel surfaces. On clean copper, excepting $CH_3OH$, alcohols dehydrogenated to their respective aldehydes via the surface alkoxyl species. Methanol required the presence of surface oxygen to remove hydrogen as water. With the alkoxide identified as the reaction intermediate from these experiments, the dehydrogenation of $CH_3OH$ was examined on Ni(100) with adsorbed sulfur acting as a reaction modifier. A complete change from total to partial dehydrogenation was noted at 0.38 monolayer of sulfur. This change in selectivity was due to stabilization of the methoxyl group by the sulfur, and to an increase in the activation barrier for C–H bond cleavage. Since the methoxyl group formed $H_2CO$ at a higher temperature, and the barrier for C–H bond rupture increased, formaldehyde desorbed rather than dehydrogenated. Using vibrational spectroscopy and CO as a probe of the surface, the site responsible for $CH_3O$ formation was identified with the four-fold hollow.

Selective poisoning is widely employed in catalytic processes. This selective poisoning is achieved by treating the catalysts with carbon, chlorine, sulfur and/or oxygen-containing compounds to modify the reactivity of the catalysts. The methods of surface science offer the opportunity to clarify the chemistry responsible for these effects. Metal surfaces with well-defined overlayers of reaction modifiers can be prepared, and the kinetics and mechanism of reactions can be definitively studied, elucidating both the elementary steps and their rate constants. In this paper the effects of modifying nickel single-crystal surfaces with sulfur on the dehydrogenation of methanol is discussed. The results clearly

0097–6156/84/0248–0057$06.00/0

illustrate that surface modifiers can alter the stability of reaction intermediates and thereby change selectivity.

## Experimental

The tools used for the experiments described below have been described in several books and review articles (1-3). Surface structure is determined by low energy electron diffraction (LEED), surface composition by Auger electron spectroscopy (AES), and reaction kinetics and mechanism by temperature programmed reaction spectroscopy (TPRS). Standard ultra-high vacuum technology is used to maintain the surface in a well-defined state. As this article is a consolidation of previously published work, details of the experiments are not discussed here.

## Results and Discussion

In order to recognize the pattern of reactivity of alcohols on modified nickel surfaces, it is essential to know the reaction pathways exhibited by less reactive surfaces. Initially the dehydrogenation of $CH_3OH$ was studied on copper (4) and silver (5) single crystal surfaces. On Cu(110), following the preadsorption of submonolayer quantities of atomic oxygen, methanol reacted via the following sequence (4,6):

$$CH_3OH_{(g)} \longrightarrow CH_3OH_{(a)} \tag{1}$$

$$CH_3OH_{(a)} + O_{(a)} \longrightarrow CH_3O_{(a)} + OH_{(a)} \tag{2}$$

$$CH_3OH_{(a)} + OH_{(a)} \longrightarrow CH_3O_{(a)} + H_2O_{(g)} \tag{3}$$

$$CH_3O_{(a)} \longrightarrow H_2CO_{(g)} + H_{(a)} \tag{4}$$

$$2H_{(a)} \longrightarrow H_{2(g)} \tag{5}$$

This reaction sequence was definitively shown by use of temperature programmed reaction spectroscopy (7). The key to the success of this method was that reaction (4) was the rate-limiting step, allowing positive identification of the $CH_3O_{(a)}$ intermediate by TPRS. Isotopic substitution with $^{18}O$ and deuterium was used to identify steps (2) and (3).

With the stability of this intermediate established, its spectral features in photoemission and high resolution electron energy loss (vibrational) spectroscopy (EELS) could be determined. Indeed, with ultraviolet photoelectron spectra (UPS) it was shown that methanol reacted with the preadsorbed oxygen to form $CH_3O_{(a)}$ on $Cu(110)$ with the C-O axis perpendicular to the surface(6,8). On clean copper (in the absence of oxygen), subsequent to methanol adsorption only methanol was observed to desorb. With the spectral fingerprints of $CH_3O_{(a)}$ established, however, it was determined that on clean copper

$$CH_3OH \xleftrightarrow{\text{reversible}} CH_3O_{(a)} + H_{(a)} \tag{6}$$

Thus, upon heating, $CH_3O_{(a)}$ and $H_{(a)}$ recombination occurred, and no hydrogen atom recombination was observed.

Parenthetically, it is interesting to note that no other alcohols show this completely reversible adsorption on copper. Ethanol, n-propanol, i-propanol, n-butanol and ethylene glycol all dehydrogenate to their respective aldehydes on clean $Cu(110)$ (9,10). Temperature programmed reaction spectra for those alcohols are shown in Figure 1. In general the alcohols show several peaks. The lowest two TPRS peaks for the alcohol itself correspond to desorption of the alcohols from multilayer and monolayer coverages, respectively, while the peak observed at the highest temperature is due to recombination of adsorbed alkoxyl species with adsorbed hydrogen atoms (peaks at 175, 220, 300 K for isopropanol m/e45, for example). In the case of ethylene glycol the dehydrogenation product is the dialdehyde $(CHO)_2$. A summary of the results with the alcohols is given in Table I. With the exception of ethylene glycol which bonds via both functional groups, the stability of the alkoxyl group can be correlated with the C-H bond strength on the carbon in the $\alpha$-position with respect to the hydroxyl oxygen (9); this bond must be broken to form the aldehyde.

On nickel, a more reactive surface than copper, the reactivity of methanol is somewhat different. Temperature programmed reaction spectroscopy produces only CO and hydrogen products subsequent to $CH_3OH$ adsorption on $Ni(100)$ or $Ni(111)$ surfaces, each product being evolved from the surface in a desorption-limiting step(11). EELS results on both $Ni(111)(12)$ and $Ni(100)(13)$ clearly show the formation of $CH_3O_{(a)}$ at low temperatures (see Table II). However, the dehydrogenation activity is high enough to break the C-H bonds below 300 K, producing adsorbed CO and atomic hydrogen which desorb as CO and $H_2$ at 355 K and 445 K, respectively. In other words the C-H bond is easily cleaved, and the reaction proceeds toward adsorbed atomic hydrogen.

As sulfur is added to the $Ni(100)$ surface, the pattern of reactivity changes. The temperature programmed reaction spectra, a portion of which is shown in Figure 2, clearly reflects the formation of $H_2CO$, $H_2$ and CO in a common rate-limiting step (rls). This step, the dehydrogenation of the methoxyl species, becomes apparent as the sulfur coverage ($\theta_s$) approaches 0.25 monolayer. With increasing sulfur coverage the peak temperature of the products formed by the reaction sequence

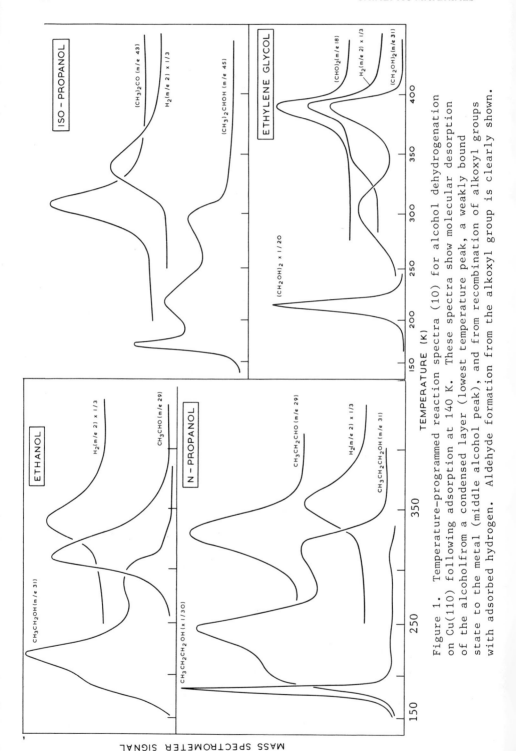

Figure 1.  Temperature-programmed reaction spectra (10) for alcohol dehydrogenation on Cu(110) following adsorption at 140 K.  These spectra show molecular desorption of the alcohol from a condensed layer (lowest temperature peak, a weakly bound state to the metal (middle alcohol peak), and from recombination of alkoxyl groups with adsorbed hydrogen.  Aldehyde formation from the alkoxyl group is clearly shown.

Table I.  Summary of Reactions of Alcohols on Cu(110)

| Adsorbed Species | TPRS Products | Coverage [a] (species $cm^{-2}$) |
|---|---|---|
| $CH_3OH$[b] | Some desorbs, rest dissociates at 200 K | $11 \times 10^{14}$ |
| $CH_3O$[b] | Decomposes to form $H_2CO$ + $H_2$ + $CH_3OH$ at 355 K | $6 \times 10^{14}$ |
| $C_2H_5OH$ | Some desorbs, rest dissociates at 200 K | $10 \times 10^{14}$ |
| $C_2H_5O$ | Decomposes to $CH_3CHO$, $H_2$ and $CH_3CH_2OH$ at 348 K | $6 \times 10^{14}$ [c] |
| $n-C_3H_7OH$ | Desorption of physisorbed material at 180 K | $\sim 50 \times 10^{14}$[d] |
| $n-C_3H_7OH$ | Desorption of chemisorbed material at 245 K, with some dissociation | $10 \times 10^{14}$ |
| $n-C_3H_7O$ | Decomposes to $CH_3CH_2CHO$, $H_2$ and $CH_3CH_2CH_2OH$ at 340 K | $5 \times 10^{14}$ [c] |
| $iso-C_3H_7OH$ | Some desorbs, rest dissociates at 200 K | $8 \times 10^{14}$ |
| $C_2H_5O$ | Decomposes to $CH_3CHO$, $H_2$ and $CH_3CH_2OH$ at 348 K | $6 \times 10^{14}$ |

Continued on next page

Table I. Continued

| | | |
|---|---|---|
| $n\text{-}C_3H_7OH$ | Desorption of physisorbed material at 180 K | $\sim\!50\!\times\!10^{14}$ (d) |
| $n\text{-}C_3H_7OH$ | Desorption of chemisorbed material at 245 K, with some dissociation | $10\!\times\!10^{14}$ |
| $n\text{-}C_3H_7O$ | Decomposes to $CH_3CH_2CHO$, $H_2$ and $CH_3CH_2CH_2OH$ at 340 K | $5\!\times\!10^{14}$ (c) |
| $iso\text{-}C_3H_7OH$ | Some desorbs, rest dissociates at 220 K | $8\!\times\!10^{14}$ |
| $iso\text{-}C_3H_7O$ | Decomposes to $(CH_3)_2CO$, $H_2$ and $(CH_3)_2CHOH$ at 335 K | $5\!\times\!10^{14}$ (c) |
| $(CH_2OH)_2$ | Desorption of physisorbed material at 220 K | $40\!\times\!10^{14}$ (d) |
| $(CH_2OH)_2$ | Desorption of chemisorbed material at 300 K, with dissociation | $6\!\times\!10^{14}$ |
| $CH_2OH$ | Decomposes to $(CHO)_2$, $H_2$ and $(CH_2OH)_2$ at 390 K | $4\!\times\!10^{14}$ (c) |
| $CH_2O$ | | |

(a) Estimated from the XPS O(1s) area relative to p(2X1)O as half a monolayer ($5.5\!\times\!10^{14}$ cm$^{-2}$); uncertainty in estimate 20%.

(b) From ref. (6).

(c) Values for the maximum allkoxy coverage obtained with 1/4 monolayer pre-dosed oxygen.

(d) Multilayer desorption.

$$CH_3O_{(a)} \xrightarrow{\text{rls}} H_2CO_{(a)} + H_{(a)} \qquad (7)$$

$$H_2CO_{(a)} \longrightarrow H_2CO_{(g)} \qquad (8)$$

$$H_2CO_{(a)} \longrightarrow 2H_{(a)} + CO_{(a)} \qquad (9)$$

$$CO_{(a)} \longrightarrow CO_{(g)} \qquad (10)$$

$$2H_{(a)} \longrightarrow H_{2(g)} \qquad (11)$$

shifts to higher temperature, revealing the increased stability of the methoxy group in the presence of surface sulfur. The activation energy for reaction (7) increases by 6 kcal/gmol for sulfur coverages between 0.20 and 0.46 monolayer. Note that the rate-limiting step is the C-H bond cleavage.

The dramatic change from total to partial dehydrogenation above $\Theta_s = 0.20$ is illustrated in Figure 3. Above $\Theta_s = 0.32$ the CO produced goes to zero. The residual mass spectrometer signal at m/e = 28 above this sulfur coverage can be accounted for entirely as a cracking fraction of $H_2CO$. The buildup of $H_2CO$ at

---

Table II. Vibrational Spectra of Methoxyl
Species on Metal Surfaces

---

frequency in $cm^{-1}$

| mode | Cu(100)[16] | Ni(111) | Ni(100) |
|------|-------------|---------|---------|
| $\nu$(M-O) | 290(s) | 405(m) | -- |
| $\nu$(C-O) | 1010(s) | 1040(s) | 980(s) |
| $\delta$(CH$_3$) | 1450(w) | 1440(m) | 1455(m) |
| $\nu$(C-H) | 2910(m) | 2955(m) | 2966(m) |

---

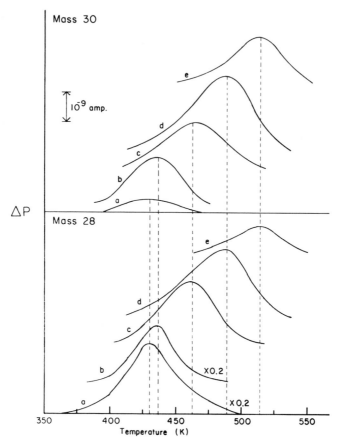

Figure 2.          Temperature-programmed reaction spectra for $H_2CO$
                   and CO from $CH_3OH$ as a function of preadsorbed
                   sulfur.
                   (a)  $n_s$ = 3.2X10$^{14}$,   (b)  4.2X10$^{14}$,   (c)
                   4.8X10$^{14}$,   (d)  5.1X10$^{14}$,  and  (e) 7.4X10$^{14}$
                   atoms/cm$^2$.  Hydrogen also desorbs in a peak at
                   the same temperature as $H_2CO$ and CO.  The
                   simultaneous evolution of $H_2CO$ (m/e = 30), $H_2$,
                   and CO(m/e = 28) is indicative of a common rate-
                   limiting step identified as $CH_3O_{(a)}$ dehydro-
                   genation.

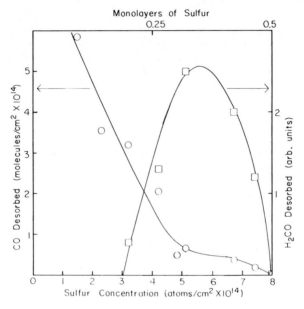

Figure 3.      Variation of the amounts of $H_2CO$ and CO formed in $CH_3OH$ dehydrogenation as a function of pre-adsorbed sulfur.

the expense of CO is clear. This trade-off occurs as the methoxide becomes more stable and the activation barrier for C-H bond
cleavage increases. This increase is the result of a decrease in
the metal-hydrogen bond strength with increasing sulfur
coverage. If the binding energy (BE) of $H_2CO$ were constant as $\theta_s$
increased, it is apparent why the selectivity must change. Since
reaction (7) occurs at increasing temperatures the surface
lifetime for $H_2CO_{(a)}$ must decrease markedly; thus it has less
time to react to CO and $H_2$. Furthermore, the activation barrier
for C-H bond cleavage (step (9)) increases also. Consequently
step (8) dominates step (9) with increasing $\theta_s$. In fact this
selectivity change is further enhanced as the binding energy of
$H_2CO$ to the surface actually <u>decreases</u> with increasing $\theta_s$ (11).
     The change in selectivity and reactivity can be related to
the structure of sulfur on the surface. The two-dimensional
structure on Ni(100) is well known (14). Two ordered structures
form under the circumstances of these experiments; they are shown
in Figure 4. As the sulfur coverage approaches 0.25 monolayer the
p(2X2) structure (Figure 4a) develops. Between 0.25 and 0.50
monolayer the four-fold hollow positions in the center of the
p(2X2) sulfur unit cell are filled until the c(2X2) structure
forms (Figure 4b). Between $\theta_s$ = 0.25 and 0.50 the stability of
$CH_3O$ (and consequently the selectivity to $H_2CO$) increases, but the
total number of binding sites for the methoxyl species
decreases. Thus the total reactivity for $H_2CO$ formation goes
through a maximum. Note that the "spectator" sulfur in the
c(2X2) position blocks a site for $CH_3O$ formation, but stabilizes
$CH_3O$ toward further dehydrogenation.
     The site responsible for $CH_3O$ formation can be identified
with the use of CO as a probe molecule. As increasing amounts of
sulfur are added to Ni(100), the desorption state characteristic
of CO on the clean surface disappears, and two new states appear
at 315 and 380 K, respectively. These states persist from about
$\theta_s$ = 0.15 to $\theta_s$ 0.46 (see Figure 5). In this coverage range one
sulfur atom blocks adsorption of one CO molecule (15).
     In order to determine the nature of the site for $CH_3O_{(a)}$
formation, CO was adsorbed in varying amounts in the $\beta_1$ state
prior to exposure to methanol at $\theta_s$ = 0.38 to see if this binding
of CO to this site would block methanol dissociation (15). The
results showed a linear decrease in $H_2CO$ production with CO
precoverage (15). Saturation of the $CO(\beta_1)$ state <u>only</u> completely
blocked methanol dehydrogenation. The binding of this CO was
examined further with EELS. The vibrational loss spectra are
shown in Figure 6 for CO adsorbed to saturation coverage at 90 K
on Ni(100) with varying degrees of presulfiding. Corresponding
temperature programmed desorption spectra are shown on the right
hand side of the figure. At $\theta_s$ = 0.37 three CO binding states
were identified. These correspond to CO in four-fold nickel
hollows in the p(2X2) structure (1750 $cm^{-1}$), two-fold bridging
sites on p(2X2) (1935 $cm^{-1}$) and four-fold nickel hollows on the
c(2X2) structure (2115 $cm^{-1}$). Partial desorption experiments

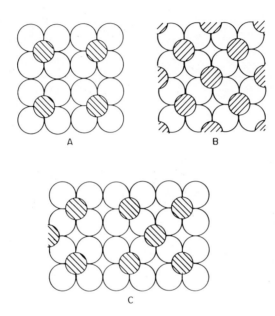

Figure 4.        Real space structures for  (a) Ni(100)p(2X2)S,
                 (b)  Ni(100)c(2X2)S  and  (c) one intermediate
                 coverage (see the text).

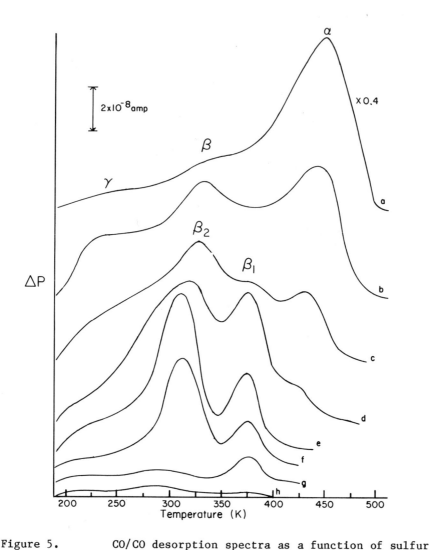

Figure 5.        CO/CO desorption spectra as a function of sulfur
                 concentration.  CO was adsorbed to saturation in
                 each case.  (a) $n_s$ = 0,  (b)  1.5 X $10^{14}$,
                 (c) 2.3X$10^{14}$,  (d)  3.2 X $10^{14}$,  (e)  4.8 X $10^{14}$,
                 (f)  5.1X$10^{14}$,  (g)  7.4X$10^{14}$, and
                 (h)  8.0X$10^{14}$ atoms/cm$^2$.  The $\alpha$-peak charac-
                 teristic of binding of CO to clean Ni(100) is
                 displaced by sulfur.  In its place two desorption
                 peaks,  $\beta_1$  and  $\beta_2$, appear.

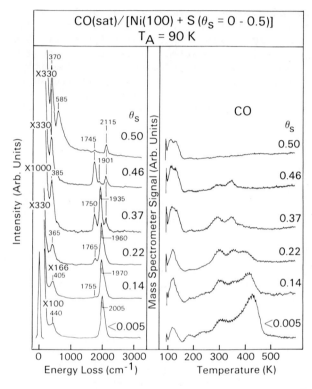

Figure 6.    Vibrational loss spectra for CO adsorbed on Ni(100) with sulfur coverages between zero and 0.50 monolayers. For $\theta_s$ = 0.37 three binding states are evident (see text).

showed that the vibrational loss 1750 $cm^{-1}$ originated from the
$CO(\beta_1)$ state. Clearly then formation of the $CH_3O_{(a)}$ is
associated with the four-fold hollow sites on the partially
sulfided surface. This explains the decrease in overall activity
with sulfur adsorption.

## Summary

Model studies of methanol dehydrogenation on Ni(100) show that
dramatic changes in selectivity can be achieved by preadsorbing
sulfur in four-fold hollow sites. The selectivity changes result
from increases in the activation barrier for C-H bond cleavage.
This change causes a sharp differentiation in competing steps
subsequent to the initial C-H bond rupture of the surface methoxyl
species. Desorption of $H_2CO$ is overwhelmingly favored over
further dehydrogenation with increasing sulfur concentration.

## Literature Cited

1.  Ertl, G.; Kupers, J. "Low Energy Electrons and Surface
    Chemistry"; Verlag Chemie: Weinheim, 1974.
2.  Madix, R.J. Adv. in Catalysis 1980, 29, 1.
3.  Roberts, M.W. Adv. in Catalysis 1980, 29, 55.
4.  Wachs, I.E.; Madix, R.J. J. Catalysis 1978, 53, 208.
5.  Wachs, I.E.; Madix, R.J. Surf. Sci. 1978, 76, 531.
6.  Bowker, M.; Madix, R.J. Surf. Sci. 1980, 95, 190.
7.  Madix, R.J. in "Chemistry and Physics of Solid Surfaces";
    Vol. II, Vanselow R., Ed., p.63.
8.  Hofmann P.; Mariani C.; Horn, K.; Bradshaw A.M. "Proceedings
    of the 3rd European Conference on Surface Science"; Degras,
    D.A.; Costa, M. Eds., Vol. 1, 1980, 720.
9.  Bowker, M.; Madix, R.J. Surf. Sci. 1982, 116, 549.
10. Jorgensen, S.; Madix, R.J. Surf. Sci. 1983 130, L291.
11. Johnson, S.W.; Madix, R.J. Surf. Sci. 1981 103, 361.
12. Demuth, J.E.; Ibach, J. Chem. Phys. Lett. 1979, 60, 395.
13. Madix, R.J.; Gland, J.L. To be published.
14. Perereau, M.; Oudar, J. Surf. Sci. 1970, 20, 80.
15. Madix, R.J.; Lee, S.B.; Thornburg, M. J. Vac. Sci. Tech. 1983,
    A1, 1254.
16. Sexton, B.A. Surf. Sci. 19979, 88, 299.

RECEIVED December 19, 1983

# X-Ray Absorption Fine Structure, Mössbauer, and Reactivity Studies of Unsupported Cobalt–Molybdenum Hydrotreating Catalysts

BJERNE S. CLAUSEN, HENRIK TOPSØE, and ROBERTO CANDIA

Haldor Topsøe Research Laboratories, DK-2800 Lyngby, Denmark

BRUNO LENGELER

IFF, Kernforshungsanlage Jülich, D-5170 Jülich, Germany

Catalytic and structural information has been obtained for unsupported Co-Mo hydrotreating (HDS) catalysts. The structural information has been provided by means of in situ Mössbauer emission spectroscopy (MES) and in situ EXAFS (for both the Mo and the Co K-edges). By comparing these results with the thiophene HDS rate and the rate of secondary hydrogenation of butenes the nature of the active sites for these reactions has been elucidated. The results suggest that the Co-Mo-S phase can be regarded as a $MoS_2$ structure with Co atoms located at the edges. These Co atoms are found to strongly promote the HDS reactions but have a much less influence on the hydrogenation rate.

The increasing need for efficient treatment of various fossil fuel feedstocks has resulted in many studies (for a recent review, see e.g., Ref. (1)) devoted to the understanding of the catalytic properties of hydroprocessing catalysts (e.g., Mo or W based catalysts promoted by Co or Ni). The efforts have been directed towards an understanding of the structural form in which the different atoms are present, and to establish connections between the structural information and the various catalytic functions (hydrodesulfurization (HDS), hydrogenation, hydrodenitrogenation (HDN), etc). It has, however, been very difficult to make progress since for a long time direct information regarding the structural state of the active elements has been almost impossible to obtain. This is probably the reason why greatly diverging views on the structure exist (2-5). Recently, it has been shown that two techniques, Mössbauer emission spectroscopy (MES) (6-11) and extended X-ray absorption fine structure (EXAFS) (12, 13), can provide some of the needed structural information. This has not only resulted in a better description of the structural state in such catalysts but it has also allowed one to understand some of the catalytic implications of the different structural features. In this connection,

0097-6156/84/0248-0071$06.00/0

it is important that both of the above techniques conveniently allow studies to be carried out in situ.

The MES investigations showed that part of the promoter atoms in Co-Mo catalysts is generally present in a structure also containing molybdenum and sulfur atoms (6). This structure was termed the Co-Mo-S structure (8) and since the promotion of the HDS activity was found to be associated with this structure (9, 11) much work has been initiated in order to characterize further the properties of this Co-Mo-S structure.

All of the results obtained so far, including the recent Mo EXAFS studies of Co-Mo/$Al_2O_3$ catalysts (12), indicate that the Co-Mo-S structure has a $MoS_2$-like structure (see e.g., 1, 10). With respect to the location of the Co atoms in the highly dispersed $MoS_2$ structure, the early MES studies (6, 8) showed that the Co atoms are located at surface positions but it was not possible to definitively conclude whether these positions are at Mo sites in the $MoS_2$ structure or on the basal or edge planes. Recent studies (1, 10, 11, 14, 15) seem to favor the latter positions and since the Co-Mo-S structure is formed in systems where single $MoS_2$ slabs (or layers) dominate (16, 17), the Co positions are most likely edge substitutional or interstitial positions and not edge intercalation positions.

The previous EXAFS studies were restricted to supported catalysts. Furthermore, the structural properties determined by MES and EXAFS were mainly related to the HDS activity and not to the other catalytic functions. Presently, we will report EXAFS (both Mo and Co), MES, HDS and hydrogenation activity studies of unsupported Co-Mo catalysts. These catalysts have been prepared by the homogeneous sulfide precipitation method (18) which permits large amounts of Co to be present as Co-Mo-S. The choice of unsupported catalysts allows one to avoid some of the effects which inherently will be present in alumina supported catalysts, where support interactions may result in both structural and catalytic complexities.

Experimental

Sample Preparation. The preparation of the unsupported Co-Mo catalysts has been carried out using the homogeneous sulfide precipitation (HSP) method as described earlier (18) and only few details will be given here. A hot (335-345 K) solution of a mixture of cobalt nitrate and ammonium heptamolybdate with a predetermined Co/Mo ratio is poured into a hot (335-345 K) solution of 20% ammonium sulfide under vigorous stirring. The hot slurry formed is continuously stirred until all the water has evaporated and a dry product remains. This product is finally heated in a flow of 2% $H_2S$ in $H_2$ at 675 K and kept at this temperature for at least 4 hr. Catalysts with the following Co/Mo atomic ratios were prepared: 0.0, 0.0625, 0.125, 0.25, 0.50, 0.75, and 1.0.

EXAFS Measurements. The sulfided catalysts were studied in situ by placing self-supporting wafers (1.125" in diameter) of pressed catalyst powder in specially designed cells, equipped with X-ray transparent windows (13). After sulfiding of the catalysts in 2% $H_2S$ in $H_2$ at 675 K, the cells were sealed off prior to the measurements. EXAFS studies of the model compounds and the passivated catalysts were carried out by placing appropriate amounts of the sample in thin aluminum frames equipped with Kapton windows. In order to ensure a homogeneous sample thickness, boron nitride was used as a low absorption filler. By partly immersing the aluminum frames into liquid nitrogen, EXAFS spectra of these samples could also be recorded at 77 K. The absorber thickness, x, of all the samples was chosen such that $\mu x \sim 1$ ($\mu$ is the linear absorption coefficient) on the high absorption side of the edge, and great care was taken in order to make the samples of homogeneous thickness. Indeed, the jump height at the absorption edge corresponded nicely (in most cases within 10%) to the theoretical value (based on the amount of sample used).

The EXAFS experiments were conducted at DESY in Hamburg, using the synchrotron radiation from the DORIS storage ring and the EXAFS-setup at HASYLAB. This spectrometer is somewhat different from that used in our earlier studies (12). The X-rays, which were emitted by electrons in the storage ring with an energy of 3.3 GeV and a typical current of 60 mA, were monochromatized by two Si (111) single crystals when EXAFS above the Co K-edge was measured and by two Si (220) single crystals in the case of the Mo K-edge measurements. The beam intensity was measured, before ($I_o$) and after ($I$) passing through the sample, by use of two ionization chambers filled with one atmosphere of $N_2$ (Co K-edge) or one atmosphere of Ar (Mo K-edge). In order to eliminate changes in intensity due to sample inhomogeneities, the sample table is moved simultaneously with the beam during the scan to ensure that the beam is hitting the sample at the same place at all times.

In the present study we have extracted the EXAFS from the experimentally recorded X-ray absorption spectra following the method described in detail in Ref. (19, 20). In this procedure, a value for the energy threshold of the absorption edge is chosen to convert the energy scale into k-space. Then a smooth background described by a set of cubic splines is subtracted from the EXAFS in order to separate the non-oscillatory part in $\ln(I_o/I)$ and, finally, the EXAFS is multiplied by a factor k and divided by a function characteristic of the atomic absorption cross section (20). The reason for multiplying with a k weighting factor is to compensate for the decrease of the EXAFS amplitudes at high k values due to the Debye-Waller factor, the backscattering amplitude, and the $k^{-1}$ dependence of the EXAFS (see, e.g., Ref. (21)).

In order to interpret an EXAFS spectrum quantitatively, the phase shifts for the absorber and backscatterer and the backscattering amplitude function must be known. Empirical phase shifts and amplitude functions can be obtained from studies of known structures which are chemically similar to that under investigation (22). Calculated phase shifts and amplitude functions have, however, recently been tabulated for a large number of elements (23).

By Fourier transforming the EXAFS oscillations, a radial structure function is obtained (24). The peaks in the Fourier transform correspond to the different coordination shells and the position of these peaks gives the absorber-scatterer distances, but shifted to lower values due to the effect of the phase shift. The height of the peaks is related to the coordination number and to thermal (Debye-Waller smearing), as well as static disorder, and for systems, which contain only one kind of atoms at a given distance, the Fourier transform method may give reliable information on the local environment. However, for more accurate determinations of the coordination number N and the bond distance R, a more sophisticated curve-fitting analysis is required.

In the present study we have used the phase and amplitude functions of absorber-scatterer pairs in known model compounds to fit the EXAFS of the catalysts. By use of Fourier filtering, the contribution from a single coordination shell is isolated and the resulting filtered EXAFS is then non-linear least squares fitted as described in Ref. (19, 20).

Mössbauer Measurements. Co-Mo catalysts cannot be studied directly in absorption experiments since neither cobalt nor molybdenum has suitable Mössbauer isotopes. However, by doping with $^{57}Co$ the catalysts can be studied by carrying out Mössbauer emission spectroscopy (MES) experiments. In this case information about the cobalt atoms is obtained by studying the $^{57}Fe$ atoms produced by the decay of $^{57}Co$. The possibilities and limitations on the use of the MES technique for the study of Co-Mo catalysts have recently been discussed (8, 25).

The MES experiments were performed using a constant-acceleration spectrometer with a moving single-line absorber of $K_4Fe(CN)_6 \cdot 3H_2O$ enriched in $^{57}Fe$. Zero velocity is defined as the centroid of a spectrum obtained at room temperature with a source of $^{57}Co$ in metallic iron. Positive velocity corresponds to the absorber moving away from the source. The in situ MES spectra were recorded with the catalysts placed in a Pyrex cell (8) connected to a gas handling system allowing the catalysts to be studied in a $H_2S/H_2$ or in a thiophene/$H_2$ gas mixture. The use of $H_2S$ instead of thiophene did not have any noticeable influence on the MES spectra (6).

Catalyst Activity Measurements. Activity measurements for thiophene HDS and the consecutive hydrogenation of butene were carried out in a Pyrex-glass, fixed-bed reactor at 625 K and at atmospheric pressure as described in Ref. (9). Before the measurements the catalysts were presulfided in 2% $H_2S$ in $H_2$ at 675 K. For each catalyst conversions were measured at different space velocities of the thiophene/$H_2$ mixture (2.5% thiophene) and the catalytic activities are here expressed as pseudo first-order rate constants assuming that the HDS reaction is first order in thiophene and that the hydrogenation of butene can be considered as a first order consecutive reaction.

## Results

Mössbauer Spectroscopy. Figure 1 shows room temperature Mössbauer emission spectra of two of the unsupported Co-Mo catalysts which we have studied in the present investigation. It is observed that the MES spectra of the two catalysts are quite different. For the catalyst with the low Co/Mo ratio (0.0625), a quadrupole doublet with an isomer shift of $\delta=0.33$ mm/s and a quadrupole splitting of $\Delta E_Q=1.12$ mm/s are observed (spectrum a). These parameters are very similar to those observed previously for the Co-Mo-S phase in other catalysts (6-9). Furthermore, the spectrum of an unsupported catalyst with Co/Mo = 0.15 is found to be essentially identical to spectrum (a). The MES spectrum (b) of the catalyst with Co/Mo = 0.50 shows the presence of a broad single line with apparent "shoulders" near the background absorption line. The single broad line can be identified as originating from $Co_9S_8$ in the catalyst, whereas the spectral component, which shows up as the "shoulders" in the spectrum, is typical of the spectrum of the Co-Mo-S structure. Thus, it is observed that for the present unsupported catalysts, the Co-Mo-S structure is the only Co phase present at low Co concentrations, whereas $Co_9S_8$ is also formed at higher Co/Mo ratios, and at very high Co content this phase may be the dominating Co phase. The distribution of Co atoms among the two phases as obtained by computer analyzing the Mössbauer emission spectra is given in Table I for the different unsupported catalysts. A more detailed analysis of the MES data for the unsupported catalysts has been given in Ref. (8, 11).

Table I. The distribution of Co atoms among the two phases present in the unsupported catalysts as determined by MES.

| Co/Mo | Co as Co-Mo-S (%) | Co as $Co_9S_8$ (%) |
|---|---|---|
| 0.0625[1] | 100 | 0 |
| 0.15[2] | 100 | 0 |
| 0.25[2] | 80 | 20 |
| 0.50[1] | 23 | 77 |

[1] From Ref. (26).  [2] From Ref. (11).

Mo EXAFS. In Figure 2a we have shown an X-ray absorption spectrum near the Mo K-edge of the unsupported catalyst with Co/Mo = 0.125. The spectrum has been obtained in situ and at room temperature. After background subtraction, multiplication by k and normalization,

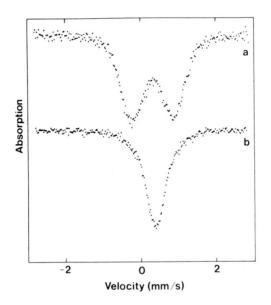

Figure 1. Examples of in situ Mössbauer emission spectra of unsupported Co-Mo catalysts. a) Co/Mo = 0.0625; b) Co/Mo = 0.50.

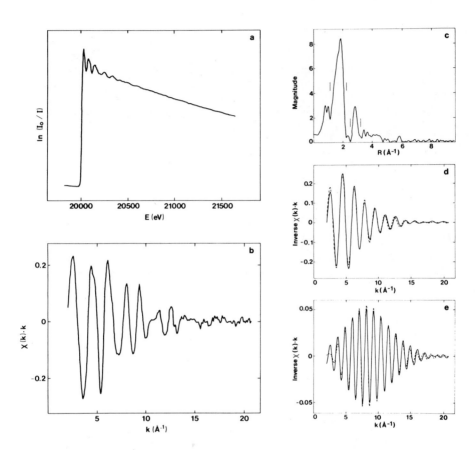

Figure 2. a) X-ray absorption spectrum near the Mo K-edge of the Co/Mo = 0.125 unsupported Co-Mo catalyst recorded *in situ* at room temperature; b) normalized Mo EXAFS spectrum; c) absolute magnitude of the Fourier transform; d) fit of the first shell; e) fit of the second shell. The solid line in d) and e) is the filtered EXAFS, and the dashed line is the least squares fit.

we obtain the EXAFS oscillations as shown in Figure 2b. The EXAFS is now Fourier transformed from k = 2 Å$^{-1}$ to k = 21 Å$^{-1}$ and the resulting radial structure function (Figure 2c) shows the presence of two distinct peaks, one located at about 1.90 Å and the other at about 2.86 Å. It should be noted here that the Fourier transformed EXAFS of the model compound, $MoS_2$, recorded in the present study is essentially identical to that of $MoS_2$ recorded by us on another EXAFS-setup (12). Furthermore, the locations of the two main peaks in the transform for $MoS_2$ are very close to those of the two peaks in Figure 2c. Also the heights of the first shell peak are similar and only the second shell peak for the catalyst is reduced in height. Therefore, the phase and amplitude functions of the absorber-scatterer pair Mo-S (first shell) and Mo-Mo (second shell) in the model compound $MoS_2$ recorded at room temperature have been used to fit the Fourier filtered EXAFS in order to obtain the interatomic distances and the coordination number of the first and second neighbor shells in the catalysts. The Fourier filtered EXAFS and the corresponding fit of the first shell contribution are shown in Figure 2d and those of the second shell are shown in Figure 2e. The calculated bond lengths and coordinations numbers for the two shells are given in Table II.

Table II. Bond lengths and coordinations numbers obtained by fitting the Fourier filtered Mo EXAFS of the Co-Mo unsupported catalyst recorded in situ at room temperature.

| Co/Mo | 1. shell | | 2. shell | |
|-------|----------|-----|----------|-----|
|       | R(Å)     | N   | R(Å)     | N   |
| 0.125 | 2.42     | 6.1 | 3.16     | 3.7 |

   In order to obtain data with reduced temperature smearing, experiments were also carried out at 77 K. However, such experiments could not be carried out in situ and the catalysts were thus exposed to air before the measurements. EXAFS data of three catalysts with Co/Mo atomic ratios of 0.0., 0.25, and 0.50 were obtained. The results show many similarities with the data recorded in situ and were fitted in a similar fashion using phase and amplitude functions of the well-crystallized model compound $MoS_2$ recorded at 77 K. The results, which are given in Table III, show that the bond lengths for the first and second coordination shell are the same for all the catalysts and identical to the values obtained for the catalyst recorded in situ (Table II). The coordination numbers for both shells appear, however, to be somewhat smaller. Although coordination numbers determined by EXAFS cannot be expected to be determined with an accuracy better than ± 20%, the observed reduction

Table III. Structural parameters obtained by fitting the Mo EXAFS
of various unsupported catalysts recorded after exposure to air.

| | 1. shell | | 2. shell | |
|---|---|---|---|---|
| Co/Mo | R(Å) | N | R(Å) | N |
| 0.00 | 2.41 | 5.6 | 3.15 | 3.4 |
| 0.25 | 2.42 | 4.5 | 3.16 | 2.6 |
| 0.50 | 2.42 | 4.8 | 3.15 | 2.9 |

in the first shell coordination numbers is probably due to the
fact that these catalysts were measured after exposure to air.
Several authors (27–29) have reported that sulfided Mo based HDS
catalysts have a considerable $O_2$ uptake and thus we tentatively
explain the reduced first shell coordination number by an influence of oxygen atoms in the local surroundings of the Mo atoms. The
possible reasons for the smaller coordination number for the second shell will be discussed below.

Co EXAFS.   X-ray absorption spectra near the Co K-edge have also
been recorded for the Co/Mo = 0.125 unsupported catalyst in order
to get information about the local surroundings of the Co atoms.
Figures 3a–c show the X-ray absorption spectrum, the normalized
EXAFS, and the Fourier transform, respectively. Only one strong
backscatterer peak is observed in the Fourier transform indicating
highly disordered surroundings outside the first shell. However,
it should be noted here that the Co EXAFS results are associated
with greater uncertainty due to the much smaller signal-to-noise
ratio compared to the Mo EXAFS, and contributions from backscatterer atoms outside the first shell – if present – may escape detection in the transform. A region surrounding the observed backscatterer peak was transformed back into k-space (Figure 3d) and
fitted by use of the phase and amplitude functions of the Co-S absorber-scatterer pair in the $CoS_2$ model compound. The values obtained for the interatomic distance and number of atoms for the
coordination shell around the Co atoms in the catalyst are listed
in Table IV. The relevance of using a Co-S absorber-scatterer pair
as in $CoS_2$ is justified by the in situ MES results which show that
all the Co atoms in the unsupported catalysts are surrounded by
sulfur. In order to get further information on the location of Co
in the catalysts we have recorded an X-ray absorption spectrum near
the Co K-edge of the Co/Mo = 0.125 unsupported catalyst after exposure to air at room temperature (Figure 4). This spectrum is different from the corresponding one recorded in situ. This is most
easily seen at the absorption edge which shows a peak at apex for

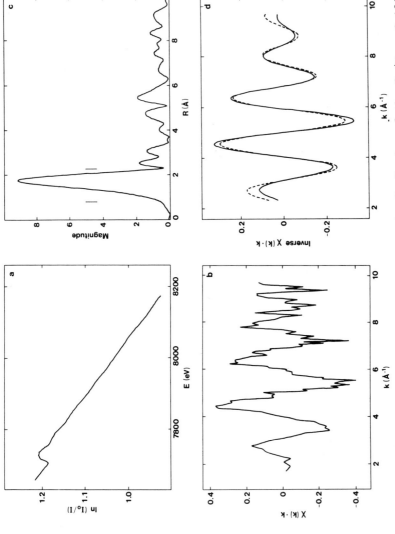

Figure 3. (a) X-ray absorption spectrum near the Co K-edge of the Co/Mo = 0.125 catalyst recorded in situ at room temperature; (b) normalized Co EXAFS; (c) absolute magnitude of the Fourier transform; (d) fit of the Fourier filtered EXAFS. The solid line is the filtered EXAFS and the dashed line is the fit.

Table IV. Structural parameters obtained by fitting the Fourier
filtered Co EXAFS of the Co-Mo unsupported catalyst
recorded in situ at 300 K.

| Co/Mo | R(Å) | N |
|-------|------|---|
| 0.125 | 2.27 | 4.6 |

the air exposed catalyst (Figure 4), whereas this is not present
for the catalyst measured in situ (Figure 3a).

Reactivity Studies. In Figure 5A the ratio between the hydrogenation and the hydrodesulfurization rate constants is shown as a
function of the Co/(Co+Mo) atomic ratio of the unsupported catalysts. This selectivity ratio is observed to be very dependent on
the Co/(Co+Mo) ratio with a relative high selectivity for hydrogenation of butane over HDS for the unpromoted $MoS_2$ catalysts,
whereas it is much lower for the whole series of Co promoted catalysts. In the plot, we have also included the relative selectivity
for an unsupported $Co_9S_8$ catalyst (i.e. the value at Co/(Co+Mo) =
1.0). This catalyst shows a somewhat higher selectivity ratio than
the promoted catalysts. Also the observed dependence of the butane/
butene ratio on the conversion for $Co_9S_8$ was different from that
of all the promoted catalysts (30) indicative of different kinds
of kinetics. In Figure 5B, we have plotted separately the HDS and
the hydrogenation rate constants as a function of the Co/(Co+Mo)
atomic ratio. It is seen that while the promotion with Co has a
large effect on the HDS rate parameter, the hydrogenation activity
is only slightly influenced.

Discussion

It has previously been found (3, 11, 18, 31-34) that unsupported
catalysts exhibit a HDS activity behavior quite similar to that of
supported catalysts. This suggests that although the support is of
importance, it does not have an essential role for creation of the
active phase. Thus, it is very relevant to study unsupported catalysts, both in their own right and also as models for the more e-
lusive supported catalysts. Many different explanations have been
proposed to explain the similarity in behavior of unsupported and
supported catalysts (3, 31-34). Recently, we have observed that
for both types of catalysts the HDS activity behavior can be related to the fraction of cobalt atoms present as Co-Mo-S (9-11, 35).
    In the present study, MES was used to establish the cobalt
phase distribution. In analogy with previous results (6, 8, 11,

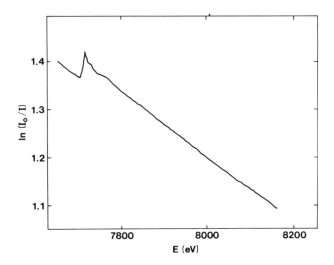

Figure 4. X-ray absorption spectrum near the Co K-edge of the Co/Mo = 0.125 catalyst after exposure to air at room temperature.

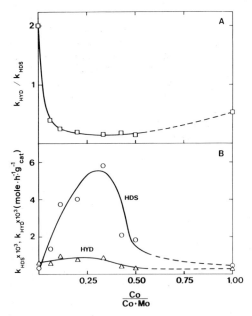

Figure 5. (A) Relative selectivities for hydrogenation of butenes for the unsupported Co-Mo catalysts; (B) first-order rate parameters for hydrogenation ($k_{HYD}$) and HDS ($k_{HDS}$).

18) this is seen to be very dependent on the Co/Mo ratio. Table I
shows that even for quite large Co concentrations (Co/Mo ~ 0.20),
the only Co phase observed in the present catalysts is the Co-Mo-S
structure. At higher Co/Mo ratios $Co_9S_8$ is also identified in the
MES spectra. From previous analyses of the Mössbauer spectra of
the cobalt atoms, which give rise to the Co-Mo-S structure, quite
detailed insight into the nature of these cobalt atoms could be
obtained (6, 8). However, such studies do not, of course, directly
provide information about the nature of the molybdenum atoms in
Co-Mo-S.

The Co/Mo = 0.125 catalyst has all the cobalt atoms present
as Co-Mo-S and, therefore, the EXAFS studies of this catalyst can
give information about the molybdenum atoms in the Co-Mo-S structure. The Fourier transform (Figure 2c) of the Mo EXAFS of the
above catalyst shows the presence of two distinct backscatterer
peaks. A fit of the Fourier filtered EXAFS data using the phase
and amplitude functions obtained for well-crystallized $MoS_2$ shows
(Table II) that the Mo-S and Mo-Mo bond lengths in the catalyst
are identical (within 0.01 Å) to those present in $MoS_2$ ($R_{Mo-S}$ =
2.41 Å and $R_{Mo-Mo}$ = 3.16 Å). These results therefore reveal that
the Co-Mo-S structure is $MoS_2$-like. The data given in Table III
indicate that cobalt does not have any significant influence on
the average bond length observed in the catalysts.

An analysis of the coordination numbers gives further understanding of the nature of the $MoS_2$-like structure. For the catalyst recorded in situ it is seen that the first coordination shell
contains about 6 atoms (Table II) which are in agreement with the
6 nearest neighbor sulfur atoms in well-crystallized $MoS_2$. A similar result was previously obtained for alumina supported catalysts
(12).

For the catalyst, the average coordination number for the
second shell is about 3.5 (Table II) which is significantly lower
than the 6 molybdenum atoms present in the second shell in $MoS_2$.
There may be several reasons for this. Some degree of disorder
outside the first coordination shell may give rise to an apparent
lower average coordination number in the second shell. However,
the presence of very small crystallites (having a large fraction
of surface atoms) may also give rise to a reduction in the coordination number for the second shell. As shown in Figure 6, the
transmission electron micrographs obtained on these catalysts reveal that the layers are highly disordered which may well be the
main reason for the reduced second shell contribution. The reduction in the second shell contribution corresponds to a "domain"
size of about 15 Å (i.e., the dimension parallel to the basal-
plane direction over which order similar to that of well-ordered
$MoS_2$ exists). The presence of the disordered structure where the
domain size is smaller than the particle size can explain the relatively low BET surface area (15 $m^2/g$ (11)) since large amounts
of "domain" boundaries may not be accessible to the $N_2$ molecules.

In alumina supported catalysts, a reduced second shell contribution was also seen ($\underline{12}$) but in this case neither X-ray diffraction nor TEM investigations can resolve the $MoS_2$ structure. Therefore, for the supported catalysts there may be a much closer relationship between the domain size estimated from EXAFS and the crystallite size (i.e., an estimate of the edge dispersion in these systems may be obtained from the EXAFS data).

The EXAFS data recorded after exposure to air of the unsupported Co-Mo catalysts with different cobalt content allow one to examine the effect of cobalt. In spite of a great uncertainty in the coordination numbers, the promoted catalysts seem to have a somewhat smaller domain size than the unpromoted catalyst as indicated both by the smaller second shell coordination numbers and by the larger effect of air exposure (i.e., reduced sulfur coordination number in first shell). This influence of cobalt on the domain size may be related to the possibility that cobalt atoms located at edges of $MoS_2$ stabilize the domains towards growth in the basal plane direction. Recent results on $Co-Mo/Al_2O_3$ catalysts indicate that Co may also have a similar stabilizing effect in supported catalysts ($\underline{36}$).

In a previous EXAFS study ($\underline{12}$) no difference in the domain size of the $MoS_2$ phase in a $Mo/Al_2O_3$ and in a $Co-Mo/Al_2O_3$ catalyst was observed. This may be related to the fact that the $Co-Mo/Al_2O_3$ catalyst was prepared from the same $Mo/Al_2O_3$ catalyst. Also, in general the dispersion of the $MoS_2$-like phase present in supported catalysts may be less susceptible to changes due to support interactions.

The Co EXAFS data show that Co occupies some type of surface location. First of all, it is observed from the study of the Co K-edge itself that the Co atoms in unsupported catalysts are affected by air exposure in quite an analogous way as the previously observed behavior for supported catalysts ($\underline{13}$) (previous MES studies were only able to reveal such effects for supported catalysts thus causing some ambiguities with respect to the possible Co locations (see Ref. ($\underline{8}$, $\underline{10}$) for a discussion)). Secondly, an analysis of the EXAFS data gives, in spite of the poor quality, further support for such a Co location since both the absence of a strong second coordination shell and the Co-S bond distance are incompatible with cobalt substituting for molybdenum (or intercalating) inside the $MoS_2$ structure. Also the apparent low number of sulfur atoms coordinated to cobalt is in accord with a "surface" location.

From the Co EXAFS results alone one cannot conclude whether the Co atoms are located at edges or basal planes but a comparison of the Co EXAFS data with the above Mo EXAFS results indicates that the edge position is the most likely one. This Co location is illustrated in Figure 7. For the unsupported catalysts, many of these "surface" positions may be present at internal edges (i.e., at the "domain" boundaries). Recently, direct evidence confirming the edge position has been obtained by combining MES results (to ensure that Co is present as Co-Mo-S in the samples studied) with ir spectroscopy ($\underline{14}$) or with analytical electron microscopy ($\underline{15}$).

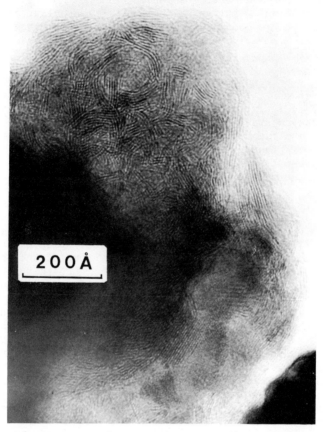

Figure 6. Electron micrograph of the Co/Mo = 0.25 unsupported catalyst.

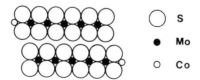

○ S

● Mo

○ Co

Figure 7. Schematic picture indicating the edge location of Co in MoS$_2$.

Therefore, the data indicate that Co-Mo-S can be considered as a
MoS$_2$ structure with Co atoms located in edge positions. As discussed below, these Co atoms play a direct role in the catalysis.
Furthermore, it is generally accepted that the HDS reaction involves adsorption on sulfur vacancies. The low sulfur coordination number (large coordinative unsaturation) estimated from the
Co EXAFS may, in fact, reflect that active sites (vacancies) are
associated with the Co atoms.

Relationship between Structure and Catalytic Activity. Nature of
Active Sites. By combining the above structural information with
the catalytic activity data it is possible to elucidate the location and the nature of the active sites in both promoted and unpromoted Co-Mo catalysts. Figure 5B shows that the HDS activity
increases by a factor of about 20 upon cobalt promotion. The present EXAFS results show that the promotion only leads to small
changes in the edge dispersion. Also, previous results show that
the BET surface area was quite constant for all these catalysts
(11). The results, therefore, confirm our previous proposal that
the primary role of the promoter atoms is not to increase the dispersion of MoS$_2$ and thereby creating more molybdenum sites, but
rather, the promoter atoms play a more direct role in the catalytic activity. In fact, an analysis (11) of some of the data shown
in Table I and Figure 5B shows that the increase in the HDS activity is proportional to the number of cobalt atoms present as
Co-Mo-S. Therefore, it appears that the promotion of the HDS activity is linked to the creation of active sites associated with the
Co atoms located at the edges of MoS$_2$. Figure 5 shows that these
sites behave catalytically different from those present in Co$_9$S$_8$.
This is not unreasonable since atomically dispersed Co atoms will
presumably have both structural and electronic properties which
are strongly influenced by the MoS$_2$ "support". Apart from this influence of Mo it is also possible, in view of the above structural
information, that the active sites involve vacancies created from
the removal of sulfur atoms coordinated to both Co and neighboring
Mo atoms.

    A comparison of the data for the unsupported and the support-
ed catalysts reveals that the activity per Co atom present as Co-
Mo-S is much lower for the former catalyst system. This is probably related to the fact that in these catalysts many of the Co
atoms are either inaccessible to the reactants or are subjected to
diffusion restrictions.

    Figure 5A shows that the selectivity towards butane formation
(i.e. the rate of formation of butane relative to that of the butenes) decreases as the Co/Mo ratio increases in the unsupported
catalysts. Similar results have previously been reported for alumina supported Co-Mo catalysts (37, 38) and this behavior does
therefore appear to be a quite general feature of Co-Mo catalysts.
The large change in the selectivity is observed (Figure 5B) to be
related to a greater promotion of the HDS reaction rate compared

to that of the hydrogenation rate of butenes. In fact, it is seen
that the unpromoted catalyst has almost the same hydrogenation
activity as the promoted catalysts. Therefore, it is possible
that for the promoted catalysts, the hydrogenation reaction associates mainly with unpromoted sites, i.e. for these catalysts the
HDS and the hydrogenation reactions may occur on different sites
(see Ref. (39) for a review of other data supporting this view).
The small increase in the hydrogenation rate upon promotion with
Co could then be related to the slight increase in the edge dispersion as indicated by EXAFS. The present result, however, does
not allow one to rule out that the hydrogenation may take place
on the sites associated with the Co atoms present as Co-Mo-S. If
the latter is the case, the hydrogenation activity of these promoted sites cannot be much different from that of the unpromoted
sites (i.e. a behavior different from that of the HDS activity).

## Conclusion

In the present study it has been demonstrated how the combined
use of two *in situ* techniques, EXAFS and MES, can provide quite
detailed structural information about unsupported Co-Mo hydrotreating catalysts. The Co and Mo EXAFS results show that the Co-
Mo-S structure can be regarded as $MoS_2$ with cobalt atomically
dispersed along the edges of the $MoS_2$ "support". The active sites
responsible for the large promotion of the HDS activity is seen
to be associated with these Co atoms. The results furthermore indicate that the $MoS_2$ "support" plays an essential role in determining the structural, electronic, and catalytic properties of the
Co atoms present at the edges. Besides providing information on
structural parameters, the EXAFS technique may also be used to
estimate the edge dispersion of these $MoS_2$ domains. This parameter
is particularly important since the maximum promotional effect,
which can be obtained, will depend on the number of the cobalt
atoms present as Co-Mo-S. Therefore, the promotional effect will
depend on the edge dispersion. While these promoted sites are important for HDS, they seem to have no or only a slight influence
on the hydrogenation activity of such catalysts. Consequently, the
hydrogenation to HDS selectivity ratio will depend on the relative
abundance of unpromoted sites and sites associated with the Co
atoms at the $MoS_2$ edges.

## Acknowledgments

We are grateful to HASYLAB for offering beam time on the synchrotron radiation facility of DESY and for the access to the EXAFS
spectrometer. We also thank J.W. Ørnbo for help in the preparation of the samples, N.-Y. Topsøe for helpful discussions, and O.
Sørensen for carrying out the electron microscopy investigations.

Literature Cited

1.   Topsøe, H. in "Surface Properties and Catalysis by Non-Metals:
     Oxides, Sulfides, and other Transition Metal Compounds",
     Bonnelle, J.P., et al., Ed.; D. Reidel Publishing Company,
     1983, p. 326.
2.   Schuit, G.C.A.; Gates, B.C. AIChE J. 19, 1973, p. 417.
3.   Farragher, A.L.; Cossee, P. Proc. 5th Int. Congr. on Cataly-
     sis, Palm Beach, 1972"; Hightower, J.W., Ed.; 1973, p. 1301.
4.   Massoth, F.E. in "Advances in Catalysis"; Eley, D.D., Pines,
     H.; Weizs, P.B., Eds.; Academic Press, New York, 27, 1978,
     p. 265.
5.   Delmon, B. Proc. of the Climax Third Intern. Conf. on Chemi-
     stry and Uses of Molybdenum; Barry, H.F.; Mitchell, P.C.H.,
     Eds.; Climax Molybdenum Co., Ann Arbor, Michigan, 1979, p. 73.
6.   Clausen, B.S.; Mørup, S.; Topsøe, H.; Candia, R. J. Phys.
     Colloq. C6, 37, 1976, p. C6-249.
7.   Topsøe, H.; Clausen, B.S.; Burriesci, N.; Candia, R.; Mørup,
     S. in "Preparation of Catalysts II"; Delmon, B.; Grange, P;
     Jacobs, P.A.; Poncelet, G., Eds.; Elsevier Scientific Pub-
     lishing Company, Amsterdam, 1979, p. 479.
8.   Topsøe, H.; Clausen, B.S.; Candia, R.; Wivel, C.; Mørup, S.
     J. Catal. 68, 1981, p. 435.
9.   Wivel, C.; Candia, R.; Clausen, B.S.; Mørup, S.; Topsøe, H.
     J. Catal. 68, 1981, p. 453.
10.  Topsøe, H.; Clausen, B.S.; Candia, R.; Wivel, C.; Mørup, S.
     Bull. Soc. Chim. Belg. 90, 1981, p. 1187.
11.  Candia, R.; Clausen, B.S.; Topsøe, H. J. Catal. 77, 1982,
     p. 564.
12.  Clausen, B.S.; Topsøe, H.; Candia, R.; Villadsen, J.; Lenge-
     ler, B.; Als-Nielsen, J.; Christensen, F. J. Phys. Chem. 85,
     1981, p. 3868.
13.  Clausen, B.S.; Lengeler, B.; Candia, R.; Als-Nielsen, J.;
     Topsøe, H. Bull. Soc. Chim. Belg. 90, 1981, p. 1249.
14.  Topsøe, N.-Y.; Topsøe, H. J. Catal. in press.
15.  Topsøe, H.; Topsøe, N.-Y.; Sørensen, O.; Candia, R.; Clausen,
     B.S.; Kallesøe, S.; Pedersen, E. Am. Chem. Soc., Div. Petrol
     Chem., Preprints, 28, 5, 1983, p. 1252.
16.  Topsøe, N.-Y. J. Catal. 64, 1980, p. 235.
17.  Topsøe, N.-Y.; Topsøe, H. Bull. Soc. Chim. Belg. 90, 1981,
     p. 1311.
18.  Candia, R.; Clausen, B.S.; Topsøe, H. Bull. Soc. Chim. Belg.
     90, 1981, p. 1225.
19.  Lengeler, B.; Eisenberger, P. Phys. Rev. B21, 1980, p. 4507.
20.  Eisenberger, P.; Lengeler, B. Phys. Rev. B22, 1980, p. 3551.
21.  Sayers, D.E.; Stern, E.A.; Lytle, F.W.; Phys. Rev. Lettr. 27,
     1971, p. 1204.
22.  Citrin, P.H.; Eisenberger, P.; Kincaid, B.M. Phys. Rev.
     Lettr. 36, 1976, p. 1346.
23.  Teo, B.-K.; Lee, P.A. J. Am. Chem. Soc. 101, 1979, p. 2815.

24. Stern, E.A.; Sayers, D.E.; Lytle, F.W.; Phys. Rev. B11, 1975, p. 4836.
25. Mørup, S.; Clausen, B.S.; Topsøe, H. J. Phys. Colloq. C2, 1979, p. C2-87.
26. Alstrup, I.; Chorkendorff, I.; Candia, R.; Clausen, B.S.; Topsøe, H. J. Catal. 77, 1982, p. 397.
27. Tauster, S.J.; Percoraro, T.A.; Chianelli, R.R. J. Catal. 63 1980, p. 515.
28. Bachelier, J.; Duchet, J.C.; Cornet, D.; Bull. Soc. Chim. Belg. 90, 1981, p. 1301.
29. Zmierczak, W.; MuraliDhar, G.; Massoth, F.E. J. Catal. 77, 1982, p. 432.
30. Topsøe, H.; Bartholdy, J.; Clausen, B.S.; Candia, R. results presented at the ACS symposium on Structure and Activity of Sulfided Hydroprocessing Catalysts, The Division of Petroleum Chemistry, Inc. Kansas City Meeting, Sept. 12-17, 1982.
31. Hagenbach, G.; Courty, Ph.; Delmon, B. J. Catal. 23, 1971, p. 295.
32. Hagenbach, G.; Courty, Ph.; Delmon, B. J. Catal. 31, 1973, p. 264.
33. Delmon, B. Am. Chem. Soc., Div. Petrol. Chem., Reprints, 22, 2, 1977, p. 503.
34. Furimsky, E.; Amberg, C.H. Can. J. Chem. 53, 1975, p. 2542.
35. Candia, R.; Topsøe, N.-Y.; Clausen, B.S.; Wivel, C.; Nevald, R.; Mørup, S.; Topsøe, H. Proc. of the Climax Fourth Int. Conf. on the Chemistry and Uses of Molybdenum; Barry, H.F.; Mitchell, P.C.H., Eds.; Climax Molybdenum Co., Ann Arbor, Michigan, 1982, p. 374.
36. Candia, R.; Clausen, B.S.; Sørensen, O.; Topsøe, H.; Topsøe, N.-Y.; Villadsen, J. to be submitted for publication.
37. Hargreaves, A.E.; Ross, J.R.H. Proc. 6th Int. Congr. Catal., Bond, G.C.; Wells, P.B.; Tompkins, F.C., Eds; Chem. Soc. London 2, 1977, p. 937.
38. Massoth, F.E.; Chung, K.S. Proc. 7th Int. Congr. Catal., Tokyo; Seiyama, T.; Tanabe, K., Eds.; Elsevier, New York, 1980, p. 629.
39. Massoth, F.E.; MuraliDhar, G. Proc. of the Climax Fourth Int. Conf. on the Chemistry and Uses of Molybdenum; Barry, H.F.; Mitchell, P.C.H., Eds.; Climax Molybdenum Co., Ann Arbor, Michigan, 1982, p. 343.

RECEIVED October 31, 1983

# Magnetic Resonance Studies of Metal Deposition on Hydrotreating Catalysts and Removal with Heteropolyacids

B. G. SILBERNAGEL, R. R. MOHAN, and G. H. SINGHAL[1]

Corporate Research, Science Laboratories, Exxon Research and Engineering Company, Linden, NJ 07036

Nuclear magnetic resonance and electron spin resonance techniques have been used to trace the deposition of vanadium on Co-Mo/alumina catalysts employed to hydrotreat heavy, highly metalized petroleum feedstocks. The vanadium is deposited on the catalyst in a sequence of chemical forms which vary with the level of metals loading: initially a paramagnetic $VO^{2+}$ form dominates, followed by a diamagnetic vanadium form believed to be associated with the alumina surface. At levels in excess of several percent, the dominant form is a vanadium sulfide of approximate composition $V_2S_3$. Molybdophosphoric acid exposure facilely removes the sulfides, with the surface and paramagnetic forms being much more refractory. A scheme for enhancing metals removal by complete conversion to the sulfide is discussed.

The present paper will describe the application of magnetic resonance spectroscopy, both nuclear magnetic resonance (NMR) and electron spin resonance (ESR), for analyzing the process of metals deposition on catalysts during the treatment of heavily metalized petroleum feedstocks and for studying the removal of these metals during extraction with heteropolyacids. We begin with a brief description of the hydrotreating process and the systematics of metals deposition, as traced by elemental analysis and microprobe techniques. We then compare NMR and ESR studies of model systems synthesized in the laboratory with analyses of discharged catalysts, illustrating the different vanadium forms present on the catalyst. Finally, we describe the heteropolyacid extraction process and demonstrate the role that magnetic resonance plays in its analysis.

As the world's known crude oil reserves diminish, we are confronted with the prospect of treating progressively less desirable crude oils. These materials contain high levels (typically several percent) of organic sulfur and nitrogen, as well as organically complexed vanadium and nickel at the level of

[1]Current address: Baytown Research and Development Laboratories, Exxon Research and Engineering Company, Baytown, TX 77520.

0097-6156/84/0248-0091$06.00/0

hundreds of parts per million. These heavier feedstocks also contain a significant high-molecular-weight fraction which is difficult to refine. A preliminary step in refining such materials is called hydrotreating, a process intended to demetalize the feed, significantly reduce the levels of sulfur and nitrogen, and to reduce the molecular weight of the petroleum. Such a process must contend with the high levels of sulfur generated as a by-product. Treating 25,000 barrels/day of a feed containing 5 wt% sulfur produces 200 tons/ day of sulfur! The catalysts must thus be sulfur-tolerant, and it has been the practice to use sulfide catalysts.(1)  A high-surface-area alumina support is impregnated with metals such as cobalt and molybdenum (or possibly nickel and tungsten), calcined, and then exposed to an environment which "sulfides" the catalyst.  The precise form of the catalytic metals after such a treatment is a subject of present research interest.(2,3)

    We will focus the present discussion on the topic of metals, mostly vanadium and nickel, which are deposited from the feed onto the catalyst during the hydrotreating process. This is not an insigificant amount of material, since 25,000 barrels of a feed containing 100 ppm. metals will deposit one-half ton of vanadium and nickel on the catalyst being employed to treat it. This metals chemistry has been the subject of considerable recent research.(4,5)  Profiles of the metals distribution in individual catalyst pellets using energy-dispersive x-ray techniques show substantial metals accumulation near the pellet surface, the form of the profile depending on the size distribution of the catalyst pores and the character of feedstock.(5,6) As we will demonstrate, it is very useful to know the amount and chemical form of metals deposited throughout the reactor, changes in deposit chemistry as a function of time, and the influence of deposited metals on catalyst activity and selectivity.

    Vanadium and nickel in the starting petroleum fractions occur largely as organically complexed species. Figure 1 shows a typical ESR absorption from a crude oil, revealing the well-articulated, sixteen-component spectrum associated with a vanadyl ($VO^{2+}$) species. ESR g-values and hyperfine coupling constants, deduced from the magnetic field positions and splittings of these components, indicate that this is a vanadyl porphyrin species.(7)  The integrated intensity of the spectrum indicates that more than 80% of the vanadium in the sample is in the porphyrin form.  Figure 2 shows the ESR spectrum of a discharged hydrotreating catalyst.  Two other paramagnetic species are also prominent in this spectrum; a very intense, narrow signal associated with carbon radicals in the catalyst coke and an absorption associated with an oxygen-coordinated form of $Mo^{5+}$. However, the dominant ESR features are again those of a vanadyl species.  The spectral components are significantly broader, and a detailed analysis of the g-value and hyperfine coupling con-

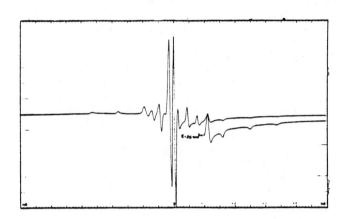

Figure 1: ESR Spectrum of Vanadyl Porphyrins in Crude Oil ($\nu_0$ = 9.1 GHz, $H_0$ = 3240G, field scan = 2 KG)

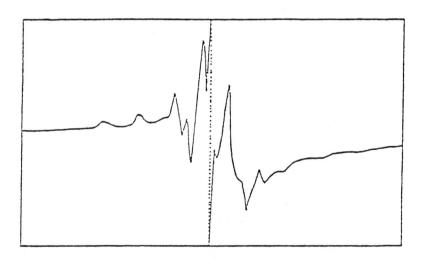

Figure 2: ESR spectrum of Carbon Radicals, $Mo^{5+}$ Species and Vanadyl Ions on a Discharged Catalyst ($\nu_0$ = 9.1 GHz, $H_0$ = 3240G, field scan = 2 KG).

stants reveals that this $VO^{2+}$ is not a vanadyl porphyrin. The signature observed here appears associated with a $VO^{2+}$ ion associated with defect sites on the alumina support and has no relationship to the starting organic species. This was demonstrated by exposing pure alumina support material and alumina pre-impregnated with cobalt and molybdenum to aqueous solutions of $VOSO_4$. Upon reduction in an $H_2/H_2S$ environment at 350°C for 45 minutes, a paramagnetic signal was observed which was identical to that shown in Figure 2, independent of the presence of the active metals.(7) The number of paramagnetic vanadyl species as determined by integrating the ESR spectrum is found to be proportional to the surface area of the alumina support employed, suggesting that specific surface sites on the alumina serve as the residence for these $VO^{2+}$ ions.

The intensity data also suggest that the maximum number of vanadium atoms that can be accommodated in this form is on the order of a fraction of a percent (by weight) of the catalyst. The balance of the vanadium must occur in other forms, which we

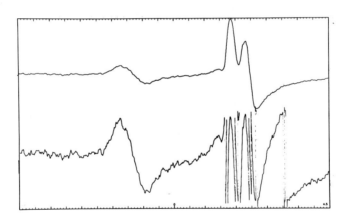

Figure 3:    NMR Spectrum of Vanadium and Aluminum Forms on a Discharged Catalyst ($\nu_o$ = 15 MHz, $H_o$ = 13401G, field scan = 400G).

have traced by NMR techniques. Figure 3 shows an NMR absorption spectrum for a discharged catalyst. This spectrum exhibits an NMR absorption from the alumina support, a second component from the aluminum metal of the NMR bridge and two absorptions associated with vanadium species. The well-defined signal at lower field is identified on the basis of its shape and resonance position as a vanadium sulfide of approximate composition $V_2S_3$.(7) The more poorly defined signal at somewhat higher field values is related to a diamagnetic vanadium species. The lack of definition of the signal suggests an amorphous vanadium form; the interaction of the vanadium quadrupole moment with a disordered environment causes the kind of "smearing" observed here. The amount of material associated with this diamagnetic vanadium phase (several weight percent) would be consistent with a thin surface layer on the catalyst.

These data taken together suggest that vanadium is deposited on the catalyst in three successive forms. The initial vanadium which appears on the catalyst is primarily an isolated $VO^{2+}$ species, presumably associated with alumina defect sites. This is followed by the diamagnetic vanadium surface phase and finally by the vanadium sulfides. This progression is illustrated by the analysis of catalyst samples taken from different positions in a reactor which had been employed in a pilot-plant treatment of a petroleum residuum (Figure 4). Note that all of

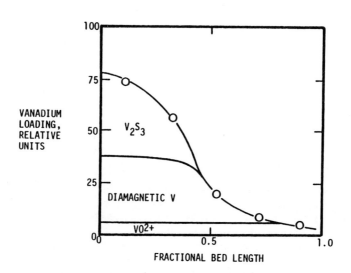

Figure 4: Vanadium Distribution and Chemical Forms at Different Positions in a Reactor Used to Treat a Heavy Feed.

the samples show the $VO^{2+}$ ESR absorption, while four samples show the diamagnetic vanadium species. This indicates that these vanadium species, once deposited, do not convert as the catalyst ages. A recent conjecture that the different forms of vanadium encountered on the catalyst result from different organic host molecules(8) is not consistent with these systematics.

There are several incentives to regenerate such discharged catalysts by removing the deposited metals and carbon, which will become increasingly greater when larger amounts of heavy, metals-bearing petroleum must be refined. The high metals levels of heavy feeds typically contemplated for RESIDFINING will result in a high metals deposition rate. The amount of deposited metals will equal the starting weight of the catalyst for treatment of feeds in the 300-500-ppm range after a run of several months' duration. This discharged catalyst must be either regenerated or replaced. The increased expense of the catalyst, coupled with the substantial ecological problems associated with the disposal of large volumes of used catalysts, suggests that regeneration merits careful consideration. Furthermore, these catalysts could prove to be valuable sources of vanadium and nickel metal, since discharged catalysts may have a higher content of these metals than many native ores.

For regeneration to be technically viable, it must be able to remove deposited vanadium and nickel quantitatively as well as the carbonaceous coke which was co-deposited. The catalytically active metals should remain unaffected in amount, chemistry, and state of dispersion. The alumina support should remain intact, with the surface area, pore-size distribution and crush strength after treatment comparable to that of the original. To be economically viable, the process should be accomplished in a minimum of steps at nearly ambient temperatures and preferably in aqueous solution. The ultimate proof of any such scheme is for the catalytic activity of the regenerated catalyst to be equal to that of a fresh one.

The simultaneous deposition of metals and coke during hydrotreating provides a particular complication. In the absence of the deposited metals, these catalysts can be regenerated by burning the coke off the catalyst surface.(9) However, the presence of vanadium and nickel on the catalyst surface leads to substantial sintering of the catalyst support. Similarly, acid extraction of metals from such catalysts has been proposed, but such techniques nonselectively remove metals from the catalyst surface and often do severe damage to the support.(10) One class of acids in which these problems are minimized is the heterpolyacids,(11) one example being molybdophosphoric acid (MPA): $H_3PMo_{12}O_{40} \cdot xH_2O$. As discussed below, these acids are easily synthesized in aqueous solution and stable under the reaction conditions. They can be made to provide a sufficiently low pH in water solution to solubilize vanadium and nickel sulfides. The reaction associated with their formation is general-ly described by:

$$\alpha\ M_oO_3 + H_3PO_4 + H_2O \xrightarrow{\Delta} H_3PMo_yO\cdot xH_2O \qquad (1)$$

The phase behavior of these molybophosphoric acid complexes has been studied as a function of pH and a variety of complexes are observed.(10) For pH values below 4, the dominant complex is $H_3PMo_{12}O_{40}\cdot xH_2O$. Most of the experiments described here have been performed with pH≈2.

Batch extraction studies were employed with 0.4 wt% MPA solutions, at temperatures of ~45°C. The extraction process was continuously monitored by examining the metals in the extract. The extraction process was essentially complete in a time scale of several days, although the catalyst type and pore geometry influenced the extraction rate. The amounts of metals extracted after such a treatment are shown in Table I, for a relatively small-pore (~50Å diameter) catalyst and a larger-pore (~150Å) one. The nickel is nearly quantitatively extracted, while no changes in either molybdenum or aluminum level were observed. Conversely, only three-quarters of the deposited vanadium was removed. A conspicuous negative feature of this extraction procedure is that nearly half of the cobalt was lost during the treatment.

NMR and ESR provide some reasons for the lack of complete vanadium removal during the treatment. NMR shows that the va-nadium sulfide species has been completely removed to within the precision of the observation, while loss of the diamagnetic surface phase is considerably less pronounced. ESR observations reveal little if any change in the number of paramagnetic vana-dyl species on the catalyst. Thus it appears that the sulfide forms are particularly susceptible to extraction by MPA, while the other forms are not. NMR studies of the extract solution suggest that the vanadium removed from the catalyst is incor-porated into the MPA cluster, giving it a composition of $H_3PV$ $Mo_{11}O_{40}\cdot xH_2O$.

The cobalt loss is a very great concern for the practice of this technology. One means of remedying the difficulty is to restore the cobalt by a separate impregnation after the MPA treatment. This has been done in two ways: by impregnating the cobalt in a single step and then calcining, and by doing two successive impregnations with half the total amount of cobalt added in each case. As Table II indicates, this process leads to a nearly complete recovery of surface area and pore volume of the support. The successive increase of ~50% in both quantities following the MPA treatment and again after the calcination stages produces surface-area and pore-volume values comparable to the original catalyst. Perhaps as significant is the fact that the crush strength of the treated pellets is comparable to the starting catalyst support. Hydrodesulfurization tests on

Table I:  Metals Extraction Effects

| SPENT | PERCENT EXTRACTED | | | | |
|---|---|---|---|---|---|
| CATALYSTS | VANADIUM | NICKEL | COBALT | MOLYBDENUM | ALUMINUM |
| SMALL PORE (d~50Å) | 70-80 | 98 | ~45 | 0 | 0 |
| LARGE PORE (d~150Å) | 70-80 | 93 | ~40 | 0 | 0 |

Table II:  Surface Area and Pore Volume Recovery

| TREATMENT | SURFACE AREA $(m^2/g)$ | PORE VOLUME $(cm^3/g)$ |
|---|---|---|
| DISCHARGED CATALYST | 136 | 0.24 |
| AFTER MPA TREATMENT | 200 | 0.34 |
| SINGLE CO IMPREGNATION | 225 | 0.43 |
| DOUBLE CO IMPREGNATION | 275 | 0.55 |
| FRESH CATALYST | ~275 | ~0.50 |

these regenerated catalysts yield activities of 75% of fresh catalyst for the single calcination sequence and ~100% for the dual calcination.

In reviewing these results, we would like to emphasize the information they provide about metals deposition chemistry as well as the potential utility of the proposed regeneration process. The metals extraction studies confirm the fact that three distinct phases of vanadium occur on these catalysts and that they are of varying reactivity. By contrast, the nickel removal is essentially complete under similar conditions. We suggest that the chemical stability of the nickel sulfide phase causes it to form under a broader range of temperatures and sulfur partial pressures than the vanadium sulfides, and that the nickel sulfide is readily soluble in MPA. All vanadium species can be converted to the sulfide by treatments in $H_2S$ at temperatures of 500°C-600°C for durations of several hours.(10,13) Samples receiving that pretreatment can be quantitatively demetalized by MPA extraction.(13)

The economic practice of an extraction technology based on MPA hinges on several factors. The extraction rate can be enhanced dramatically by the addition of $H_2O_2$ to the MPA solution.(14) The removal of cobalt during extraction is a cause for concern because recovering the cobalt from the extract and reimpregnation of the catalyst adds steps to the treatment process. The details of such a regeneration process in actual practice would be defined by the specific nature of the catalyst to be treated.

## Literature Cited

1.  For an introduction, see Weisser, O.; Landa, S. "Sulfide Catalysts, Their Properties and Applications"; Pergamon: New York, 1973.
2.  Silbernagel, B. G.; Pecoraro, T. A. Chianelli, R. R., J. Catalysis 1982, 78, 380-88.
3.  Clausen, B. S.; Topsøe, H.; Candia, R; Lengeler, B. This volume.
4.  Sie, S. in "Catalyst Deactivation, vol. 6", Delmon, B.; Froment, G., Eds.; Elsevier: Amsterdam, 1980, p. 545-69.
5.  Silbernagel, B. G.; Riley, K. L. in "Catalyst Deactivation, vol. 6," Delmon, B.; Froment, G., Eds.; Elsevier: Amsterdam, 1980, pp. 313-21.
6.  Tamm, P. W.; Harnsberger, H. F.; Bridge, A. G. Ind. Eng. Chem. Process Des. Dev. 1981, 20, 262-73.
7.  Silbernagel, B. G. J. Catalysis, 1979, 56, 315-20.
8.  Mitchell, P.C. H.; Valero, B., 1983, Inorganica Chimica Acta 71, 179-84.
9.  See, e.g. Satterfield, C. N. "Heterogeneous Catalysis in Practice", McGraw-Hill: New York, 1980, pp. 272-7.
10. Gamble, F. R.; Levy, R. B., U.S. Patent 4,014,815, 1977.
11. See, e.g. Filowitz, M.; Ho, R. K. C.; Klemperer, W. G.; Shum, W. Inorg. Chem, 1979, 18, 93-103.
12. Mohan, R. R.; Singhal, G. H., U.S. Patent, 4,272,401, 1981.
13. Silbernagel, B. G.; Mohan, R. R.; Singhal, G. H. U.S. Patent 4,272,400, 1981.
14. Mohan, R. R.; Silbernagel, B. G.; Singhal, G. H., U.S. Patent 4,268,415, 1981.

RECEIVED December 5, 1983

# Applications of High-Resolution ¹³C-NMR and Magic-Angle Spinning NMR to Reactions on Zeolites and Oxides

ERIC G. DEROUANE[1] and JANOS B. NAGY

Facultés Universitaires de Namur, Laboratoire de Catalyse, Rue de Bruxelles 61, B-5000 Namur, Belgium

Conventional high resolution $^{13}$C-NMR is shown to provide quantitative information on the conformation of 1-and 2-butene molecules adsorbed on zeolites and on mixed tin-antimony oxides. The kinetics of isomerization of 1-butene into the two 2-butene isomers is readily determined and enables the proposition of a reaction mechanism involving an intermediate cyclic complex. Rate constants can be related to the acid-base properties of the mixed oxide surfaces. An in-situ characterization of reaction products is readily achieved in the conversion of methanol, ethanol and ethylene on the highly acidic and shape selective ZSM-5 zeolite. In addition, $^{13}$C isotopic labeling proves to be a powerful technique to discriminate between possible reaction pathways of ethylene. High resolution magic angle spinning $^{13}$C-NMR appears as a superior technique to investigate the nature of carbon-containing residues in which molecular motion is highly reduced. There exists a relationship between the nature of these species and the acidic and molecular shape selective properties of the zeolites which are considered (ZSM-5 and mordenite). This technique allows a distinction between hydrocarbon molecules which are trapped inside the zeolite framework as a result of pore plugging and strongly chemisorbed or surface alkoxide species.
Large molecules or ions can also be trapped inside the zeolitic framework during cristallization. Tetrapropylammonium ions in ZSM-5 zeolite and tetrabutylammonium ions in ZSM-11 zeolite (intact in their respective frameworks) occupy the channel intersections and their alkyl chain extend in the linear and zig-zag channels in ZSM-5 zeolite or in the perpendicular linear channels of ZSM-11 zeolite.

1

[1]Current address: Central Research Division, Mobil Technical Center, Princeton, NJ 08540.

0097–6156/84/0248–0101$07.00/0
© 1984 American Chemical Society

High resolution [13]C-NMR spectra of systems presenting a catalytic
interest can be obtained in two different modes. The first one,
which is the approach used over the last ten years, consists in a
straightforward application of conventional high resolution Fourier
transform [13]C-NMR to studies of reacting adsorbates, looking into
product formation and reaction kinetics. A satisfactory resolution is achieved in this case by considering a statistical number
of reactant monolayers such that the molecules will retain sufficient mobility in the adsorbed phase. The second one, more recent,
is referred to as high resolution magic angle spinning (HRMAS),
that is a rotation, at high frequency (ca. 3kHz) and at a given
and fixed angle, of the sample placed in the external magnetic
field. This procedure averages out the dipolar interaction between nuclei and leads to sharp NMR spectra from solid samples in
which molecular motion is reduced. The use of both techniques is
illustrated in the present work.

The first mode of the high resolution [13]C-NMR of adsorbed molecules was recently reviewed (1-3) and the NMR parameters were
thoroughly discussed. In this work we emphasize the study of the
state of adsorbed molecules, their mobility on the surface, the
identification of the surface active sites in presence of adsorbed
molecules and finally the study of catalytic transformations. As
an illustration we report the study of 1- and 2-butene molecules
adsorbed on zeolites and on mixed tin-antimony oxides (4,5). Another application of this technique consists in the in-situ identification of products when a complex reaction such as the conversion of methanol, of ethanol (6,7) or of ethylene (8) is run on a
highly acidic and shape-selective zeolite. When the conversion of
methanol-ethylene mixtures (9) is considered, [13]C isotopic labeling
proves to be a powerful technique to discriminate between the possible reaction pathways of ethylene.

HRMAS [13]C-NMR is a promising method to investigate the nature
of carbon-containing residues trapped in zeolitic structures when
running hydrocarbons and oxygenates conversion reactions (10).
These deposits can be heavy molecular weight molecules or groups
attached to the zeolite surface. Large molecules can also be
trapped inside the zeolite framework during cristallization.
Examples are found in the syntheses of zeolites ZSM-5 and ZSM-11
in which tetrapropylammonium and tetrabutylammonium cations have
been shown to act as templates (11). High resolution solid state
magic angle spinning [13]C-NMR enables one to show that these occluded molecules are intact in their respective frameworks and provides information on their configurations.

Future applications will also be considered, in particular
with respect to the use of these techniques combined with the NMR
of other nuclei (Na, Al, Si) for the investigation of zeolite synthesis mechanisms (12-14).

Experimental

Materials    NaGeX zeolite was kindly supplied by Dr. G. Poncelet
(Université Catholique de Louvain) and the mixed tin-antimony
oxide catalysts (SnSbO) by I.C.I. Ltd. The H-Z is the acidified
form of commercially available Norton mordenite. The ZSM-5 and
ZSM-11 zeolites were synthesized following the patent literature
(15,16). 1-Butene (Prochem) was a natural abundance compound,
while methanol (95 % $^{13}$C, British Oxygen Corporation (B.O.C.)),
ethanol (95 % $^{13}$C, B.O.C.) and ethylene ($\sim$ 90 % $^{13}$C, Prochem) were
$^{13}$C-enriched compounds. For the latter a 30 % v/v dilution was
realized prior to adsorption.

All the catalysts (ca.0.6 g) were progressively dehydrated
and activated at temperatures 573-673 K and at a final pressure
of 2.10$^{-6}$Torr. The adsorption for all the reactants was made at
room temperature. The mixed tin-antimony oxide samples were rapidly cooled down to 77 K in order to avoid isomerization of 1-butene
following adsorption.

The kinetic measurements were performed by successive heating
cycles at the reaction temperature, the $^{13}$C-NMR spectra being recorded at a lower temperature where the reaction was quenched.

$^{13}$C-NMR measurements    Conventional high resolution $^{13}$C-NMR spectra
were recorded on a Bruker WP-60 spectrometer working in the Fourier transform mode, using a ca.$\pi$/6 pulse length under proton
broad band decoupling. The gated sequence 4s-1s-0.1 s was used to
record the nuclear Overhauser effect (NOE) - suppressed decoupled
spectra in order to obtain quantitative values for the relative
intensities. The chemical shifts were determined using benzene
as an external reference.

T$_1$ measurements were performed at 250, 275 and 300 K by inversion-recovery ($\pi$-$\tau$-$\pi$/2-5T$_1$) sequences on a JEOL-FX-100 and a
Bruker WP-80 spectrometers. On this latter the "repetitive frequency shift" method of Brevard et al. (18) was used, where two
systematic instrumental errors (drift, round off errors in FT pro-
cessing...) are uniformly distributed through all data points. The
NOE measurements are reproducible within 10-20 %, while the average standard error on the T$_1$ values is of about 5 %.

High resolution magic angle spinning cross polarization (CP/
MAS) $^{13}$C-NMR spectra were recorded at room temperature on a Bruker
CXP-200 spectrometer. The $^{13}$C (50.3 MHz) and $^1$H (200.0 MHz) rf-
fields are 39.0 G and 9.8 G respectively satisfying the Hartmann-
Hahn condition. The CP-MAS spectra were obtained using a single
contact sequence (17). A contact time of 5.0 ms and a recycle
time of 4.0 s were used in these experiments. A $\pi$/2 $^1$H pulse is
first applied, followed by a $\pi$/2 phase shift, after which the
$^1$H-$^{13}$C cross-polarization is allowed and $^1$H decoupling is maintained during data acquisition (4 K data points). Delrin or poly-
methylmethacrylate rotors were span at 3.1 kHz at the magic
angle. One thousand spectra were accumulated prior to the Fourier
transformation.

## Results and Discussion

### State of the adsorbed molecules and nature of the active sites

The state of the adsorbed molecules will be characterized by the usual NMR parameters, i.e. chemical shift, linewidth and relaxation times.

The chemical shifts of liquid, gaseous and adsorbed 1-butene and trans 2-butene are reported in Table I.

The volume magnetic susceptibility corrections are usually made following the procedure of Fraissard et al. (19) where the chemical shifts of the adsorbed species are extrapolated to zero surface coverage. The chemical shift corrections were computed to be $-0.5$ ppm for NaGeX and NaY, taking $\chi_v = -0.626 \times 10^{-6}$ for $C_6H_6$ and $\chi_v = -0.594 \times 10^{-6}$ for ethanol (21) ($\chi_v = 0.39 \times 10^{-6}$ for NaGeX and NaY zeolites (20)). The values for the zeolites were determined by the Faraday method and are in good agreement with theoretical calculations using ionic susceptibilities for isolated solid components (20).

A fundamental question concerns the state of the adsorbed gas, namely whether it is closer to the gaseous or the liquid state. At 301 K, the solvent shift is mainly observed on the terminal carbon atoms which are more exposed to intermolecular interactions (22). The carbon $C_1$ and $C_4$ of 1-butene experience a small low field shift with respect to the gas, the $C_3$ carbon a small high field shift, while the methinic $C_2$ carbon atom is much more influenced than the other carbon atoms (low field shift) suggesting a specific interaction at this site of the molecule. In trans 2-butene, the chemical shift variations are almost identical, which could correspond to a general intermolecular interaction. However, if the comparison is made with liquid trans 2-butene, specific interactions are clearly shown again by the greater change in the methinic $C_2$ carbon atom.

From our experimental results and different models used in theoretical calculations using either CNDO/2 (23-25, 37,38) and PCILO methods (26,27), or the electric field effect by INDO finite perturbation theory (28), the following models can be supposed :

$\pi$-complexes

Table I.   $^{13}$C-NMR chemical shifts of gaseous 1- and trans 2-butene, $\delta^{TMS}$ (ppm), and their variations upon adsorption[a].

| Compound | Support | θ | T(K) | $C_1$ | $C_2$ | $C_3$ | $C_4$ |
|---|---|---|---|---|---|---|---|
| 1-butene | gas[c] | – | 301 | 110.9 | 139.6 | 26.4 | 11.2 |
| | p = 2.5 atm NaGeX[d] | 0.5 | 298 | 0.1 [-2.3][b] | 5.8 [5.3] | -1.0 [-1.7] | 0.5 [-1.6] |
| | | | 245 | 0.9 [-1.5] | 5.9 [5.4] | -1.1 [-1.8] | 0.5 [-1.6] |
| | NaY[d] | 0.8 | 298 | 0.4 [-2.0] | 6.0 [5.5] | -0.5 [-1.2] | 0.5 [-1.6] |
| | | 0.8 | 300 | -2.2 [-4.6] | 4.9 [4.4] | -2.9 [-3.6] | -2.3 [-4.4] |
| trans 2-butene | gas[c] | – | 301 | 14.2 | 124.1 | | |
| | p = 2.0 atm NaGeX[d] | 0.8 | 298 | 2.8 [0.6] | 2.7 [2.1] | | |

a) $\Delta\delta = \delta_{ads} - \delta_{gas}$

b) $[\Delta\delta = \delta_{ads} - \delta_{liq}]$

c) ref. 22

d) $\chi_v = -0.39 \times 10^{-6}$   (-0.5 ppm)

These results obtained at room temperature favor the specific
interactions of a "π-complex" type. Similar conclusions were
drawn from 1-and trans 2-butene adsorption studies on NaY zeolites
as well as on HY zeolite activated at 750 K (29).

Nevertheless, the proposed models are very qualitative as not
only the magnitude, but even the sign of interaction can differ
from one method of calculation to the other. Some other effects
should be included in order to obtain satisfactory agreement with
experiment.

The specific interaction of the admolecule with the surface
is then rather well established, while the geometry of the adsorbed species is only tentative. One important conclusion to be
drawn from the study of the chemical shifts, is that they cannot
by themselves indicate unambiguously the exact geometry of a "con-
tact-type complex". Nevertheless the π-complex nature of the adsorbed species was also suggested by the dependence of the adsorption coefficient of n-butenes on their energy of ionization (4).

The nature of the active site in zeolites is still contro-
versial (30, 31). However, there are several reports about a specific interaction with sodium cations. One is the $T_1$ measurements
of $^1H$ in benzene adsorbed on NaX zeolite (31) where a distribution
of correlation times were seen due to the presence of fixed and
mobile $Na^+$ ions as potential adsorption sites. In NaY zeolite,
only one correlation time was necessary to explain the relaxation
data, hence only one type of sodium ion at the $S_{II}$ site interacts
with the adsorbed molecule. Very recently the mobility of benzene
adsorbed on NaY zeolite was reported, and it was shown unambiguously that four benzene molecules saturated the $S_{II}$ cation sites in
the supercage. At this surface coverage, the mobility was the
lowest, while at smaller or greater surface coverages greater mobility due to exchange between different sites (lower coverage) or
inside the cavity (higher coverage) was observed (32).

## Mobility of the adsorbed molecules on the surface

Linewidths    Figure 1 shows the variation with temperature of the
$^{13}C$-NMR spectrum of 1-butene adsorbed on NaGeX zeolite at a surface coverage of θ = 0.5 while Table II reports the linewidth (ΔH
at half intensity) variations for gaseous, liquid and adsorbed
1-and trans 2-butene molecules.

The mobility of 1-butene molecules decreases from the gas
(ΔH = 5 Hz) to the adsorbed state (ΔH = 35 Hz) at room temperature.
The latter depends on surface coverage as previously shown for
benzene (33). At the low surface coverage ( θ = 0.5) where not all
the available $S_{II}$ sites are occupied - there are only 3 molecules
for 4 different positions (4) - the mobility is smaller than in
the higher surface coverage sample (θ = 0.9) and this effect is
even more important at lower temperaure (250 K). In the latter,
characterized by 4.6 molecules for 4 $S_{II}$ sites - the molecules
are exchanged between the strong adsorption sites and the inner
cavity, hence the smaller linewidth.

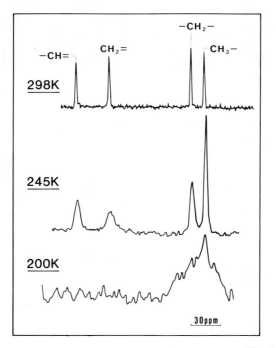

Figure 1.   Variation with temperature of the $^{13}$C-NMR spectra of 1-butene adsorbed on NaGeX zeolite ($\theta$ = 0.5).

Table II.  $^{13}$C-NMR linewidth ($\Delta$H in Hz) of gaseous, liquid and adsorbed 1- and trans 2-butenes.

| Compounds | θ | T(K) | $C_1$ | $C_2$ | $C_3$ | $C_4$ | Ref |
|---|---|---|---|---|---|---|---|
| 1-butene gas | p = 2.5 atm | 301 | 5 | 5 | 5 | 5 | 22 |
| liquid | | 293 | 0.5 | 0.5 | 0.5 | 0.5 | 38 |
| NaGeX | 0.5 | 298 | 35 | 35 | 35 | 35 | this work |
| | | 250 | 220 | 140 | 110 | 60 | " |
| | 0.9 | 303 | 30 | 30 | 25 | 20 | " |
| | | 275 | 30 | 30 | 25 | 25 | " |
| | | 250 | 90 | 80 | 45 | 30 | " |
| trans 2-butene gas | p = 2.0 atm | 301 | 5 | 5 | | | 22 |
| liquid | | 293 | 0.5 | 0.5 | | | 38 |
| NaGeX | 0.8 | 302 | 30 | 40 | | | this work |
| | | 258 | 65 | 90 | | | " |
| | | 247 | 65 | 120 | | | " |
| | | 235 | 100 | 700 | | | " |
| | | 214 | 300 | 1300 | | | " |

Figure 1 shows that decreasing temperature (250 K) results in a lower mobility of the adsorbed molecules. If the temperature is further decreased (200 K), the translational motion of the molecules is almost quenched, while the internal rotation of the methyl groups remains. Let us emphasize that in the linewidth and the relative intensity variations, there exists a greater ressemblance between carbons 1,2 and 3, which denote a greater interaction of these carbons with the surface.

Nevertheless, the linewidth variations are in the order C1> C2>C3>C4 for both coverages, a greater change being observed for the sample θ = 0.5. At high surface coverage, a phase transition occurs which will be discussed in a separate paper (34).

Similar conclusions can be drawn from the behavior of 2-butene on the surface. The main difference stems from the greater interaction of trans 2-butene with the surface ($\Delta H \approx 40$ Hz). At low temperature the difference in mobility is even more striking : $\Delta H$ = 120 Hz (247 K) for $C_2$ of trans 2-butene and only 80 Hz for $C_2$ of 1-butene (250 K). The adsorption coefficients determined from adsorption isotherms and kinetic data which are $5 \times 10^{-2}$ torr$^{-1}$ for 1-butene and $8.2 \times 10^{-2}$ torr$^{-1}$ for trans 2-butene (4) support entirely this difference in mobility.

Longitudinal relaxation times ($T_1$) and NOE measurements   The relevant data are reported in Table III. The $T_1$ values were measured by the inversion-recovery method.

The first comparison is based on the $T_1$ values of gaseous, liquid and adsorbed molecules. Unfortunately, no measurements are available for butenes in the gas or liquid phase. Nevertheless a reasonable parallel can be drawn with propylene where the three different phases were investigated (35) at 295 K : for the gas (1 atm) $T_1$ of $C_2$ is 0.095 s ; in the liquid state (2.6 M in CDCl$_3$) 59.9, 58.7 and 65.2 s for $C_1$, $C_2$ and $C_3$ respectively ; adsorbed on NaY zeolite 0.81, 1.6 and 0.81 s. The shortest relaxation times characterize the gas phase where the spin-rotation mechanism (NOE factor : $\eta$ = 0) is very effective (30,35). In the liquid, dipole-dipole and spin-rotation mechanisms both play a role and the total relaxation rate is about three orders of magnitude lower than in the gas phase. The adsorbed molecules show therefore an intermediate behaviour between gas and liquid, as it was also suggested by chemical shift data.

The relaxation times of the carbon atoms of 1-butene (θ = 0.5) vary in the order $C_4$ > $C_3$ > $C_2$ > $C_1$ at 300 K, while they are similar for carbons $C_1$ and $C_3$, and $C_2$ and $C_4$ at 250 K. This rapid glance already suggests the π-complex nature of the adsorbed species at 300 K, where $C_1$ is more influenced than $C_2$ and this influence decreases in the chain. At low temperature, the different carbon atoms are close to the surface and the carbon atoms 1 and 3 and the carbon atoms 2 and 4 are relaxed at comparable rates respectively. These important results show that both carbons $C_1$

Table III.  Longitudinal relaxation times $T_1$ (in s) and NOE's[a] of $^{13}$C-NMR lines of 1- and trans 2-butene adsorbed on NaGeX zeolite.

| 1-butene | θ | T(K) | ν(MHz) | $C_1$ | $C_2$ | $C_3$ | $C_4$ |
|---|---|---|---|---|---|---|---|
| | 0.5 | 300 | 25.1 | 0.14 [1.2] | 0.24 [1.0] | 0.38 [1.2] | 1.0 [1.2] |
| | | | 20.1 | 0.11 [1.2] | 0.22 [1.0] | 0.33 [1.2] | 1.1 [1.2] |
| | | 250 | 25.1 | 0.13 [0.8] | 0.32 [0.6] | 0.17 [0.9] | 0.27 [1.6] |
| | 0.9 | 303 | 20.1 | 0.10 [1.2] | 0.20 [1.0] | 0.41 [1.2] | 1.08 [1.2] |
| | | 275 | 20.1 | 0.10 [1.0] | 0.12 [0.8] | 0.17 [1.0] | 0.65 [1.4] |
| | | 250 | 20.1 | – [0.7] | 0.2 [0.5] | 0.14 [0.7] | 0.21 [1.6] |
| | | | | 0.18$^b$ | 0.71$^b$ | 0.35$^b$ | 1.32$^b$ |
| trans 2-butene | 0.8 | 300 | 20.1 | 0.53 [1.0] | 0.24 [1.0] | | |

a) NOE factors in parenthesis (η) : ratio of signals with and without NOE – 1

b) Relaxation time of the pseudo-liquid (34).

and $C_3$ interact with the surface. This was called the low temperature cyclic model (29) :

In this complex the molecule is quasi parallel to the surface and the two adjacent interacting sites could be the $Na^+$ ion linked to $C_1$ and the oxygen atom interacting with one proton of $C_3$.

The relaxation times at $\theta = 0.9$, are not strongly temperature dependent. The $T_1$ value of carbon 1 is almost constant, that of carbon 2 shows a minimum at 275 K, while the relaxation times of carbons 3 and 4 decrease with decreasing temperature. Taking the value of $T_1$ of carbon 2 at this approximate minimum (275 K), 0.12 s and the corresponding value of the apparent transverse relaxation time $T_2^*$ determined from the linewidth, 0.01 s the corresponding ratio $T_1/T_2^* \approx 10$ is rather high. It is characteristic of a distribution of correlation times (36,37). From the estimated minimum for carbon 2, a mean correlation time $\bar{\tau}_c \approx 10$ ns can be calculated at 275 K, assuming $\omega_0 \bar{\tau}_c \approx 1$.

The NOE factors are close to unity in all carbon atoms, except the methyl group at 250 K, where it is 1.6. The dipole-dipole carbon-hydrogen relaxation mechanism accounts for about 50 % of the total rate. The NOE factor is smaller at lower temperature for carbons $C_1$, $C_2$ and $C_3$, which are closer to the surface than for $C_4$ which is always less influenced and remains quasi free.

At 20.1 MHz and 300 K, relaxation times are independent on surface coverage ($\theta = 0.5$ and 0.9). On the contrary, they do depend on surface coverage at lower temperature (250 K) where a two exponential behavior was observed for carbons 2, 3 and 4. The low $T_1$ values at $\theta = 0.9$ are similar to those determined at $\theta = 0.5$ and are characteristic of the adsorbed phase. The higher $T_1$ values show the formation of a new phase in the adsorbed layer such as a pseudo-liquid for which the interaction with the surface, although in a lesser extent, is still significant (34).

The systematic study of the $^{13}$C-NMR parameters can lead to
very interesting conclusions.
(1)   The chemical shift variations, the linewidths and the lon-
      gitudinal relaxation times are all in agreement in favor of
      two different adsorbed species, at high and low temperature,
      respectively. At high temperature the 1- and 2-butene mo-
      lecules are more stabilized as π-complexes on the NaGeX
      zeolite surface, while the cyclic type adsorbed species is
      preferred at low temperature.
(2)   The adsorbed species behaves like a gas in a polar medium
      or like a non polar solute in a polar solvent. This inter-
      mediate behavior between gas and liquid is well suggested
      by all the parameters studied.
(3)   A deep insight in the mechanism of interaction between the
      adsorbed molecule and the surface is gained by measuring
      $^{13}$C relaxation times. The active sites of a sodium ion and
      of an oxide ion can be clearly suggested on this zeolite
      surface.

The study of catalytic transformations

        Usually the concentration of desorbed reactants or products
are determined to monitor heterogeneous catalytic reactions.
$^{13}$C-NMR spectroscopy of the adsorbed molecules has, in addition,
the advantage of differentiating between the different carbon
atoms and or identifying the specific behavior of the adsorbed
state.

Isomerization of 1-butene   Figure 2 shows the evolution of $^{13}$C-NMR
spectra of the reactants and products adsorbed on NaGeX zeolite
at various time intervals. Similar spectra were reported for the
reaction of 1-butene on SnSbO catalysts. From the variation of
the line intensities as a function of time the double-bond shift
kinetic constant, $k_{1-2}$ and the geometric isomerization constant
$k_{c-t}$ can be determined (4). The data for the SnSbO catalysts are
reported in Table IV. These apparent kinetic constants are strictly first order as evidenced by the good linearity of the loga-
rithmic plots. The $k_{1-2}$ constants for the disappearance of 1-
butene and the formation of 2-butenes are equal, showing clearly
that no side reaction takes place simultaneously.
        A common feature of both NaGeX and SnSbO catalysts is the
higher cis/trans equilibrium constants obtained in the adsorbed
state in comparison with that of the gas phase value. This difference reflects the greater tendency for adsorption of the cis
2-butene with respect to the trans isomer (4,5,39). On most catalysts the difference in the apparent heats of adsorption is roughly 0.5 kcal mol $^{-1}$ (40).
        The second similarity between the two types of catalyst is
the rather high initial cis/trans ratio which is characteristic
of an anionic transition state (4,5,41) : indeed, the cis-π-allyl

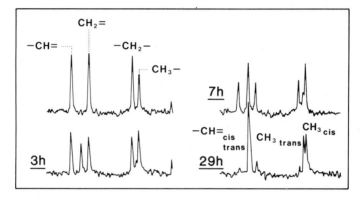

Figure 2.    $^{13}$C-NMR spectra of 1-butene ($B_1$) and cis ($B_{2c}$) and trans 2-butene ($B_{2t}$) adsorbed on NaGeX zeolite, at various time intervals during the isomerization reaction. "Reproduced with permission from Ref. 4 Copyright 1981, Academic Press".

Table IV. Surface basicity ($\theta_B$), positional ($k_{1-2}$) and geometric ($k_{c-t}$) isomerization constant of 1- and 2-Butene at 297 K and energy of activation for diffusion of 1-Butene adsorbed on SnSbO catalysts.

| Catalyst | $\theta_B$ [a] | $k_{1-2} \times 10^5$ (min$^{-1}$.m$^{-2}$) | $k_{c-t} \times 10^5$ (min$^{-1}$.m$^{-2}$) | cis/trans [b] (equilibrium) | $E_{B1}$ [c] (kcal.mol$^{-1}$) |
|---|---|---|---|---|---|
| SnO$_2$ | 0.66 | 0 | 0 | - | 1.2 |
| Sb(at.%) | | | | | |
| 5.3 | 0.36 | 23 | 4.7 | 0.67 | 1.0 |
| 10.4 | 0.41 | 1.1 | 0.13 | 0.53 | 0.4 |
| 19.0 | 0.50 | 6.7 | 2.1 | 0.57 | 1.0 |
| 30.0 | 0.77 | $\geqslant$ 150 [d] | 22 | - | 1.5 |
| 40.0 | 0.89 | 63 | 9 | 0.67 | 1.7 |
| 75.2 | 0.79 | 27 | 6.2 | 0.40 | 0.9 |
| 87.0 | 0.62 | 2.7 | 2.0 | 0.76 | 0.9 |
| Sb$_2$O$_4$ | 0.45 | 0 | 0 | - | - |

a) Obtained from the amount of adsorbed acetic acid and standardized to unity for the value of 0.55 at.% Sb catalyst (Ref. 52).

b) (cis/trans) gas phase = 0.26.

c) Energy of activation for diffusion obtained from the study of linewidth as a function of temperature.

d) Estimated value from the ln $k_{1-2}$ vs ln $k_{c-t}$ relationship.

anion is known to be more stable than its trans conformation (42-44). The variation of the double-bond shift rate constant $k_{1-2}$ with the surface basicity of the various SnSbO catalysts also supports the hypothesis of an anionic-type transition state (Table IV).

The presence of π- and cyclic-complexes on the catalyst surface and the anionic nature of the transition state suggest the following mechanism where π- and σ-allyl anions are intermediate species on the reaction path (Scheme I).

The acidic site A and the basic site B are only indicative because different types of acid-base-pairs can intervene on these catalysts.

Conversion of methanol and ethanol on H-ZSM-5 zeolite    The first in situ characterization of the adsorbed species on this catalyst has been reported for the conversion of methanol and ethanol to hydrocarbons (6,7).

The strong interaction of methanol with the surface hydroxyl groups leads to a methylated surface which is clearly shown by Figure 3. The CP/MAS $^{13}C$-NMR line of 59.9 ppm is attributed to surface methyl groups (10). On a NaGeX zeolite, the amount of these groups was computed from the comparison of the relative amounts of $CH_3OH$ and of $(CH_3)_2O$ obtained from either mass spectrometry data or $^{13}C$-NMR measurements (45).

At higher temperatures dimethylether is formed on the H-ZSM-5 catalyst (Figure 4). Further heating leads to aliphatic and aromatic hydrocarbons. The product distribution can only be explained by a remarkable shape selectivity due to the dimension of the channel system. The variation of the relative intensity of reactant and products as determined by either gas chromatography measurements or $^{13}C$-NMR results show a great similarity (7). The transformation of ethanol leads to a distribution of aliphatic and aromatic products (7), which is quite close to that obtained from methanol. As the H-ZSM-5 zeolite is highly acidic and it contains cages and channels the structures of which are characterized by high electrostatic fields and gradients, carbenium ions can certainly be formed and stabilized in these zeolites. Therefore a mechanism for dehydration-polymerization of methanol and ethanol was proposed, which involved carbenium ions as basic intermediates (Scheme II).

After the first dehydration step, the reaction propagates by successive dehydration-methanolation steps, competing with polymerization-cyclization-aromatization processes. The existence of dehydration-methanolation mechanism is inferred from the constant presence of a small amount of methanol (from in situ $^{13}C$-NMR observation) on the catalyst. Further evidence has been acquired in favor of the carbenium ion chain-growth mechanism from the $^{13}C$-NMR study of CO incorporation into the products during the conversion of methanol (46).

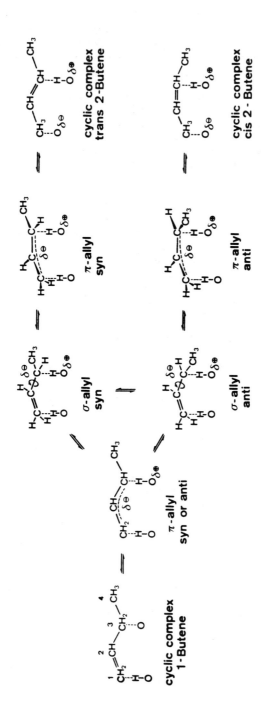

Scheme I. Reproduced with permission from Ref. 5 Copyright 1981, Academic Press

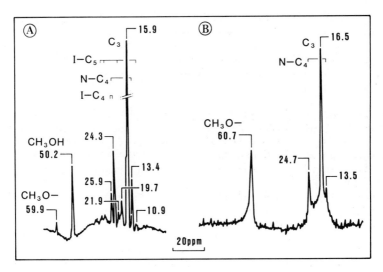

Figure 3.  CP/MAS-[13]C-NMR spectrum of the carbonaceous deposit from the catalyzed methanol conversion to hydrocarbons on zeolite H-ZSM-5 (A) and on H-Mordenite (B).  "Reproduced with permission from Ref. 10 Copyright 1981, Butterworth and Co (Publishers) Ldt c ".

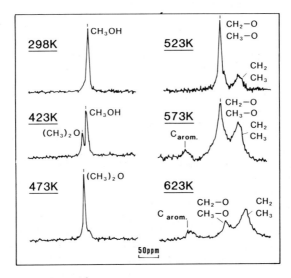

Figure 4.  Typical [13]C-NMR spectra observed during the transformation of methanol to hydrocarbons on ZSM-5 zeolite. "Reproduced with permission from Ref. 7 Copyright 1981, Academic Press".

The small molecules entrapped in the H-ZSM-5 zeolite channels because of coke deposition which occured at the outer surface of the particles, have been examined by CP/MAS-$^{13}$C-NMR spectroscopy (10). Figure 3 shows that isoparaffins are more abundant than linear chains, which agrees with other classical data on methanol conversion (47,48).

Conversion of Ethylene on H-ZSM-5 zeolite  Ethylene was reacted in three distinct ways. First, it was oligomerized at room temperature on an activated (at 673 K) zeolite. Secondly, water was then adsorbed at 295 K and the sample steam-cracked at 573 K. The third reacting way was carried out in presence of water at 573 K.

The polymerization of ethylene in absence of water leads essentially to a linear polymer of 10-12 carbon atoms (10), in agreement with previous data (49) (Figure 5).

Upon steam-cracking the higher molecular weight oligomers are broken down and $C_5$ and $C_6$ aliphatic products are formed pointing out the importance of primary cracking steps (Figure 6A). The end products yield paraffins, olefins and aromatics (8).

When ethylene is reacted at 573 K in the presence of water in static conditions, oligomerization and conjunct polymerization give rise to paraffinic, olefinic and aromatic products (8). Nevertheless, the distribution of the aliphatics and aromatics is quite different from that of the steam-cracking products. In the former a great variety of products is formed : they include propane, n-butane, isobutane and isopentane as aliphatics, and toluene, xylenes and ethylbenzene as aromatics (Figure 6B).

Conversion of methanol-ethylene mixtures  The $^{13}$C isotopic labeling is a powerful technique to discriminate between the possible reaction pathways of ethylene (9,50).

Four different experiments were realized by labeling either the methanol or the ethylene molecules (Figure 7). The reactions were studied in static conditions adsorbing either methanol (A and B) or ethylene (C and D) prior to the second reactants. The $^{13}$C-NMR spectra of Figure 7 reveals that the order of adsorption of the reactants is very important for the reactivities : at first, surface alkylation occurs and is followed by separated reaction pathways for $CH_3OH$ and $C_2H_4$.

When $CH_3OH$ is adsorbed first, the strongly adsorbed $CH_3OH$ is transformed into ethers, olefins, paraffins and aromatics, in a similar way, when it was the only reactant present (7) (A). $C_2H_4$ remains inactive below 623 K. At this temperature, it begins to react as it is shown by the NMR signals in the paraffinic region (B). It can be assumed that in such conditions, ethylene alkylates aromatics obtained from methanol.

When $C_2H_4$ is adsorbed first, the slightly adsorbed methanol molecules cannot replace ethylene on the surface. $C_2H_4$ undergoes a polymerization (C). At higher temperature, the cracking products

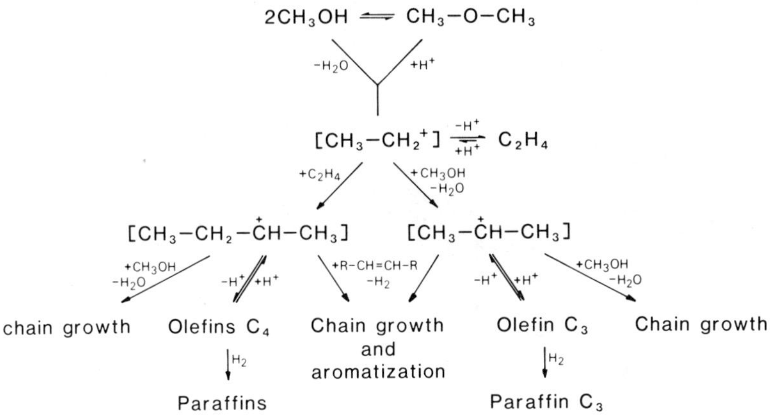

$$2CH_3OH \rightleftharpoons CH_3-O-CH_3$$

$-H_2O$ \\ $+H^+$

$$[CH_3-CH_2^+] \underset{+H^+}{\overset{-H^+}{\rightleftharpoons}} C_2H_4$$

$+C_2H_4$ \\ $+CH_3OH$ \\ $-H_2O$

$$[CH_3-CH_2-\overset{+}{C}H-CH_3] \qquad [CH_3-\overset{+}{C}H-CH_3]$$

$+CH_3OH$ / $-H_2O$    $-H^+$ / $+H^+$    $+R-CH=CH-R$ / $-H_2$    $-H^+$ / $+H^+$    $+CH_3OH$ / $-H_2O$

chain growth    Olefins $C_4$    Chain growth and aromatization    Olefin $C_3$    Chain growth

$\downarrow H_2$                 $\downarrow H_2$

Paraffins              Paraffin $C_3$

Scheme II. Reproduced with permission from Ref. 48 Copyright 1981, Academic Press

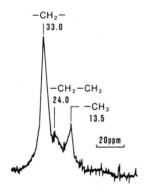

$-CH_2-$
$|33.0$

$-CH_2-CH_3$
$24.0$
$-CH_3$
$13.5$

$20ppm$

Figure 5. CP/MAS $^{13}$C-NMR spectrum of oligomerized ethylene on H–ZSM–5 zeolite at 295 K. (Reproduced with permission from Ref. 10. Copyright, 1981, Butterworth and Co.)

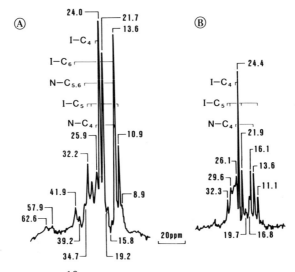

Figure 6. CP/MAS-$^{13}$C-NMR spectrum of the steamcracking products of the ethylene oligomer at 573 K (A) and of the reaction products of ethylene in presence of water at 573 K (B).
"Reproduced with permission from Ref. 10 Copyright 1981,
Butterworth and Co (Publishers) Ltd c ".

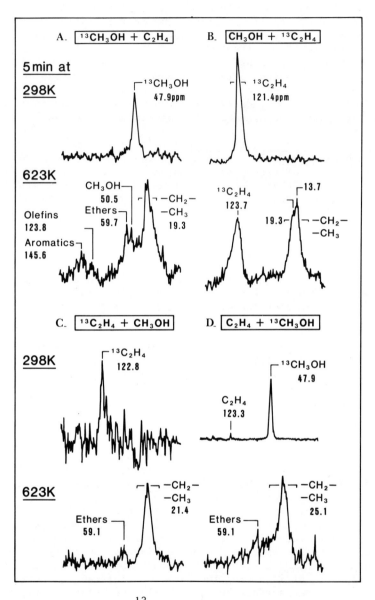

Figure 7. Typical $^{13}$C-NMR spectra observed during the transformation of methanol in presence of ethylene on H-ZSM-5 zeolite.

can react with methanol yielding essentially paraffins and isoparaffins (C and D), as it was emphasized earlier (51). No aromatics were detected by $^{13}$C-NMR in these conditions, probably because of the low mobility of eventually formed aromatic molecules.

One important conclusion concerns the role of ethylene in the methanol conversion. Ethylene cannot accumulate during this transformation because, as it was shown above, the gaseous ethylene gives rise to oligomerization products and it can only react with intermediates obtained in the conversion of methanol.

CP/MAS-$^{13}$C-NMR identification of occluded tetrapropyl- and tetrabutylammonium ions in ZSM-5 and ZSM-11 zeolites  The CP/MAS $^{13}$C-NMR spectrum of tetrepropylammonium (TPA) ions occluded in ZSM-5 zeolite and that of tetrabutylammonium (TBA) ions in ZSM-11 zeolite are reported in Figure 8. They show clearly that the occluded entities are chemically intact in the zeolitic channels.

One important observation is the two chemically different methyl groups in both compounds (11). They are probably not due to crystallographic effects, because two quite different crystallographic structures induce a similar splitting in the NMR spectrum. They are attributed therefore to environmental effects. The line splittings together with the relative line intensities allowed us to position the organic cations in their respective frameworks.

In ZSM-5 zeolite, almost every channel intersection is occupied by a tetrapropylammonium ion. Oppositely, tetrabutylammonium ions occupy preferentially the large cavities in ZSM-11 zeolite, the small cavities being only partially occupied. Their alkyl chains extend in the channel system in order to fill completely the available channel length (11).

Conclusions

$^{13}$C-NMR spectroscopy proves to be a powerful tool to characterize adsorbed species on various zeolites.

The nature of the adsorbed species can be inferred from the usual chemical parameters, i.e. chemical shifts, linewidths and relaxation times. These latter allow the study of the mobility on the surfaces. As an analytical tool, $^{13}$C-NMR spectroscopy can also be used to determine the concentration of reactants or products as a function of time and hence kinetic constants can easily be determined. As a conclusion, a rather complete kinetic study can be carried out involving the nature of interaction between the admolecule and the surface and eventually the nature of the surface active centers. One can finally arrive at the proposition of a reaction mechanism.

As an illustration, the isomerization of 1-butene adsorbed on NaGeX or mixed tin-antimony oxides has been carried out. In the methanol to hydrocarbon conversion on the shape selective H-ZSM-5 zeolite, the surface methylation could be observed, the role of

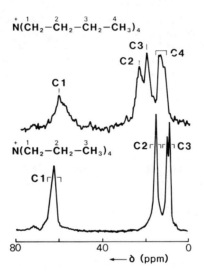

Figure 8.  High resolution CP/MAS solid state $^{13}$C-NMR spectra of tetrapropyl- and tetrabutylammonium ions in ZSM-5 and ZSM-11 zeolites respectively. "Reproduced with permission from Ref. 11 Copyright 1981, Butterworth and Co (Publishers) Ltd c ".

methanol as autocatalytic agent was established and the importance
of carbenium ions was assessed as intermediates in the dehydration-
polymerization process. The ethylene-methanol conjunct reaction
was carefully monitored by selective $^{13}$C-labeling on one of the
reactant at a time. It is concluded that ethylene and methanol
are reacting quite independently. The former essentially reacts
as an alkylation agent with oligomerized species obtained from
methanol.

High resolution magic angle spinning $^{13}$C-NMR using cross polarization enabled us to detect strongly adsorbed species such as
surface alkyl groups or aromatics and aliphatics entrapped in the
zeolite channels during coke deposition. An interesting study
concerns the characterization of tetraalkylammonium ions which
were introduced into the channels of the pentasil zeolites during
their synthesis processes. Moreover, the progressive incorporation of these ions under synthetic conditions opens new ways of
studying the nucleation and crystal growth mechanisms (12,13).
This technique complements quite well the systematic study of intermediate gels by $^{29}$Si- or $^{27}$Al-NMR spectroscopy (12,14).

## Literature cited

1.  Pfeifer, H. Physics Reports (Section C), 1976, 26, 293.
2.  Tabony, J. Progr. Nucl. Magn. Reson. 1980, 14, 1.
3.  B.Nagy, J. in Proceedings of III International Symposium
    on Magnetic Resonance in Colloid and Interface Science,
    Torun, Poland, June 1983.
4.  B.Nagy, J.; Guelton, M.; Derouane, E.G. J. Catal. 1978,
    55, 43.
5.  B.Nagy, J.; Abou-Kaïs, A.; Guelton, M.; Harmel, J.;
    Derouane, E.G. J. Catal. 1982, 73, 1.
6.  Derouane, E.G.; Dejaifve, P.; B.Nagy, J.; van Hooff,J.H.C.;
    Spekman, B.P.; Naccache, C.; Védrine, J.C. C.R. Acad. Sci.
    Paris C, 1977, 284, 945.
7.  Derouane, E.G.; B.Nagy, J.; Dejaifve, P.; van Hooff, J.H.C.;
    Spekman, B.P.; Védrine, J.C.; Naccache, C. J. Catal, 1978,
    53, 40.
8.  Derouane, E.G.; Gilson, J.-P.; B.Nagy, J. J. Mol. Catal.
    1981, 10, 331.
9.  Gilson, J.-P. Ph.D. Thesis, Facultés Universitaires de
    Namur, Belgium, 1982.
10. Derouane, E.G.; Gilson, J.-P.; B.Nagy, J. Zeolites, 1982,
    2, 42.
11. B.Nagy, J.; Gabelica, Z.; Derouane, E.G. Zeolites, 1983,
    3, 43.
12. Gabelica, Z.; B.Nagy, J.; Debras, G.; Derouane, E.G. Sixth
    International Zeolite Conference, Reno, Nevada, July, 1983.
13. Gabelica, Z.; B.Nagy, J.; Debras, G. J. Catal., in press.

14. Bodart, P.; Gabelica, Z.; B.Nagy, J.; Debras, G. in Advanced Study Institute on Zeolites : Science and Technology, Alcabideche, Portugal, May, 1983, Ed.; F.R. Ribeiro.
15. Argauer, R.J.; Landolt, G.R. U.S. Patent, 3,899,544.
16. Derouane, E.G.; Detremmerie, S.; Gabelica, Z.; Blom, N. Appl. Catal. 1981, 1, 201.
17. Mehring, M. "High Resolution NMR spectroscopy in Solids", Springer Verlag, Berlin, 1976.
18. Granger, P.; Chapelle, S.; Brevard, C. 4th European Experimental NMR Conference, Grenoble/Austrans, France, 1979, p. 66.
19. Bonardet, J.L.; Snobbert, A.; Fraissard, J. C.R. Acad. Sci. Paris, Serie C, 1971, 272, 1836.
20. Michel, D.; Meiler, W.; Gutsze, A.; Wronkowski, A. Z. Phys. Chemie, Leipzig, 1980, 261, 953.
21. Pople, J.A.; Schneider, W.G.; Bernstein, H.J. High-resolution Nuclear Magnetic Resonance, McGraw Hill, New-York, 1959.
22. Gay, I.D.; Kriz, J.F. J. Phys. Chem. 1978, 82, 319.
23. Michel, D. Surf. Sci. 1974, 42, 453.
24. Schwind, H.; Deininger, D.; Geschke, D. Z. Phys. Chemie, Leipzig, 1974, 255, 149.
25. Geschke, D.; Hoffmann, W.D.; Deininger, D. Surf. Sci. 1976, 57, 559.
26. Lochmann, R.; Meiler, W. Z. Phys. Chemie, Leipzig, 1977, 258, 1059.
27. Salzer, K. Z. Phys. Chemie, Leipzig, 1978, 259, 795.
28. Seidman, K.; Maciel, G.E. J. Am. Chem. Soc. 1977, 99, 3254.
29. B.Nagy, J.; Gigot, M.; Gourgue, A.; Derouane, E.G. J. Mol. Catal. 1977, 2, 265.
30. Clague, A.D.H.; Maxwell, I.E.; Van Dongen, J.P.C.M.; Binsma, J. Applications of Surf. Sci. 1978, 1, 288.
31. Pfeifer, H. in "Magnetic Resonance in Colloid and Interface Science"; Resing, H.A.; Wade, C.G., Eds.; ACS SYMPOSIUM SERIES 34, Washington, 1976, p. 36.
32. Borovkow, V.Yu.; Hall, W.K.; Kazanski, V.B. J. Catal. 1978, 51, 437.
33. Gay, I.D., Kriz, J.F. J. Phys. Chem. 1975, 79, 2145.
34. B.Nagy, J.; Michel, A.; Guelton, M. to be published.
35. Denney, D.; Mastikhin, V.M.; Namba, S.; Turkevich, J. J. Phys. Chem. 1978, 82, 1752.
36. Pfeifer, H. in "NMR Basic Principles and Progress", Eds.; Diehl, P.; Fluck, E.; Kosfeld, K. Springer, Berlin, 1972, p. 53.
37. Torchia, D.A.; Vanderhart, D.L. in "Topics in Carbon-13 NMR Spectroscopy", vol. 3, Ed.; Levy, G.D. Wiley, New-York, 1979, p. 325.
38. Levy, G.C.; Nelson, G.L., Carbon-13 Nuclear Magnetic Resonance for Organic Chemists, Wiley-Interscience, New-York, 1972.
39. B.Nagy, O.; B.Nagy, J. Ind. Chim. Belge, 1971, 36, 829 and 929.

40. Harlfinger, R.; Hoppach, P.; Hofmann, H.P.; Quitzsch, K.
    Z. Phys. Chem., Leipzig, 1979, 260, 905.
41. Gáti, Gy.; Knözinger, H. in "Proceedings of 5th International
    Congress on Catalysis, Palm Beach, 1972"; Hightower, J.W.
    Ed.; North-Holland, Amsterdam, 1973, p. 819.
42. Grabowski, W.; Misono, M.; Yoneda, Y. J. Catal. 1977, 47, 55.
43. Eyring, H.; Stewart, G.H.; Smith, R.P. Proc. Natl. Acad. Sci.
    U.S.A., 1958, 44, 259.
44. Bingham, R.C. J. Am. Chem. Soc. 1976, 98, 535.
45. Derouane, E.G.; Dejaifve, P.; B.Nagy, J. J. Mol. Catal.
    1977/78, 3, 453.
46. B.Nagy, J.; Gilson, J.-P.; Derouane, E.G. J. Mol. Catal.
    1979, 5, 393.
47. Chang, C.D., Silvestri, A.J. J. Catal. 1977, 47, 249.
48. Dejaifve, P.; Védrine, J.C.; Bolis, V.; Derouane, E.G.
    J. Catal. 1980, 63, 331.
49. Wolthuizen, J.P.; Van de Berg, J.P.; Van Hooff, J.H.C.,
    in "Catalysis by Zeolites", Eds.; Imelik, B. et al.,
    Elsevier, Amsterdam, 1980, p. 85.
50. Gilson, J.-P.; Gabelica, Z.; B.Nagy, J.; Derouane, E.G.
    in "Proceedings of 5th International Conference on Zeolites"
    Eds.; Sersale, R. et al., Giannini, Napoli, 1981, p. 189.
51. Derouane, E.G.; Dejaifve, P.; Gilson, J.-P.; Védrine, J.C.;
    Ducarme, V. 7th North American Meeting of the Catalysis
    Society, Boston, 1981.
52. McAteer, J.C. J. Chem. Soc. Faraday Trans. 1, 1979, 75, 2762.

RECEIVED December 20, 1983

# The Role of Oxygen Ions in the Partial Oxidation of Hydrocarbons

## Electron-Proton Resonance and Activity Measurements

JACK H. LUNSFORD

Department of Chemistry, Texas A&M University, College Station, TX 77843

The paramagnetic oxygen ions $O^-$, $O_2^-$, and $O_3^-$ have been formed on magnesium oxide and studied by EPR spectroscopy. The reactivity of these ions with hydrocarbons follows the sequence $O^- >> O_3^- >> O_2^-$. Both with alkanes and alkenes the initial reaction is thought to be hydrogen atom abstraction. The resulting radicals are not usually observed, but thermal desorption products indicate the nature of the surface intermediates. Molybdenum(V) dispersed on silica also gives rise to $O^-$ and $O_2^-$ ions when exposed to $N_2O$ and $O_2$, respectively. The $O^-$ ion on this surface may be used to activate methane and ethane in a catalytic cycle which leads to their partial oxidation.

Unlike diamagnetic $O^{2-}$ ions, the oxygen ions $O^-$, $O_2^-$, and $O_3^-$ are paramagnetic and may be observed on certain metal oxide surfaces by EPR spectroscopy. These ions have characteristic spectra which allow one to follow their formation, thermal stability, and reactivity with hydrocarbons. Although one might expect that organic radicals would be formed as intermediates, such radicals in general are highly reactive, and only in a few cases have their spectra been observed (1-4).

The importance of these radicals in catalytic processes may be evaluated by studying their behavior in stoichiometric reactions and by extrapolating this information to catalytic conditions. In following the stoichiometric reactions, magnesium oxide has been an excellent model surface since the three types of oxygen ions may be selectively formed and are stable at temperatures where most hydrocarbons of interest will react. Magnesium oxide, on the other hand, is basic and reactive itself; therefore intermediates may react differently on this surface than on silica, for example.

The high sensitivity of EPR spectroscopy is essential for this work since the maximum concentration of paramagnetic species

0097-6156/84/0248-0127$06.00/0

is on the order of $10^{-6}$-$10^{-7}$ mol/g catalyst. The concentrations are quite adequate for EPR, but are marginal for other types of spectroscopy such as infrared. Unfortunately, because of line broadening and other factors, EPR is not very effective at temperatures greater than $100^\circ$C; therefore, in situ spectroscopy on reactions which occur at elevated temperatures cannot be carried out.

## Formation and Identification of the Oxygen Ions

In this paper we will focus on two materials: magnesium oxide and molybdenum highly dispersed on silica. The formation of oxygen anions requires electrons at a high potential. On magnesium oxide these electrons are present as $F_s$ centers, which are formed by irradiation ( $\lambda$ = 254 nm) of a degassed sample in the presence of $H_2$ (5). On Mo/SiO₂ the electrons are derived by the oxidation of Mo(V) to Mo(VI) (6,7).

Since $N_2O^-$ is an unstable anion, electron transfer to $N_2O$ results in the formation of $O^-$ via the reaction (5)

$$N_2O + e \rightarrow N_2 + O^- \tag{1}$$

The EPR spectrum of $^{17}O^-$ on MgO is characterized by the g value and hyperfine coupling constants listed in Table I. From the values of the coupling constants one may conclude that the unpaired electron, and presumably the charge, is localized in a 2p orbital on the oxygen atom. The value of $g_{2,3}$ reflects the splitting of the $p_z$ orbital from the $p_x$ and $p_y$ orbitals by the crystal field at the surface. A similar spectrum has been obtained for $^{17}O^-$ on Mo/SiO₂, for which the magnetic parameters are given in Table I (6). In addition, the spectrum also exhibits superhyperfine structure due to $^{95}Mo$ and $^{97}Mo$ isotopes which confirms that the $O^-$ is localized on molybdenum, rather than on the silica. With proper pretreatment it is possible to prepare a sample on which the Mo(V) EPR signal almost completely disappears as the $O^-$ signal is being formed (8).

Superoxide ions, $O_2^-$, are readily formed by the transfer of electrons from $F_s$ centers on MgO or from Mo(V) on Mo/SiO₂ to molecular oxygen (7,9). The value of $g_3$ for $O_2^-$ is particularly sensitive to the crystal field gradient at the surface and thus varies from one metal oxide to another (10). In fact, the spectrum of $O_2^-$ on MgO indicates that the ions are held at four distinctly different sites (11,12). The oxygen-17 hyperfine splitting (Table I) for $^{17}O^{17}O^-$ on MgO confirms that both oxygen atoms are equivalent, on supported molybdenum the atoms are nonequivalent, suggesting a peroxy-type bond to the metal (7,13).

The ozonide ion, $O_3^-$, may be formed on MgO by the following reactions (14,15):

$$O^- + O_2 \rightarrow O_3^- \tag{2}$$

Table I. Magnetic Parameters for Oxygen Ions[a]

| Ion/Oxide | g values | | | hyperfine splittings | | |
|---|---|---|---|---|---|---|
| | $g_1$ | $g_2$ | $g_3$ | $a_1$ | $a_2$ | $a_3$ |
| $O^-$/MgO | 2.0013 | 2.042 | 2.042 | 103 | 19.5 | 19.5 |
| $O^-$/Mo/SiO$_2$ | 2.004 | 2.019 | 2.019 | 95 | – | – |
| $O_2^-$/MgO | 2.001 | 2.007 | 2.070–2.090 | 77 | 0 | 15 |
| $O_2^-$/Mo/SiO$_2$ | 2.004 | 2.010 | 2.018 | 69,82 | – | – |
| $O_3^-$/MgO | 2.001 | 2.010 | 2.017 | 26,65,82 | – | – |

[a] The hyperfine values, expressed in gauss, are for $^{17}O$; with $O^-$/Mo/SiO$_2$ for $^{95,97}Mo$ $a_\parallel$ = 7.6 G and $a_\perp$ = 8 G.

$$O_2^- + N_2O \xrightarrow{} O_3^- + O \qquad\qquad (3)$$
$$O_2^- + O_2 \xrightarrow{h\nu} O_3^- + O \qquad\qquad (4)$$

Reaction 2 occurs readily at low temperatures and is a problem in the formation of $O^-$ if trace amounts of $O_2$ are present as an impurity in $N_2O$. Reaction 3 is an activated process which requires temperatures on the order of $100°C$. Reaction 4 is promoted by 254-nm ultraviolet light. A mechanism has been proposed in which $O_2^-$ is photodissociated to $O^-$ and $O$; the $O^-$ subsequently reacts with $O_2$ to form the $O_3^-$ (15). Maximum concetrations of $O_3^-$ may be achieved by irradiation of MgO in the presence of $N_2O$. This process probably involves reactions 1 and 2, as well as the photo-dissociation of $N_2O$ to $N_2$ and $O_2$.

The ozonide ion has widely spaced energy levels, and to first order the g values are not influenced by the host lattice or the surface. Thus, the absolute values of the g values are useful in the identification of the ion. These g values, along with the hyperfine coupling constants, are given in Table I. The three sets of hyperfine constants indicate that the oxygen atoms are not equivalent, at least when the ozonide ion is formed according to reaction 2. The geometry of the ion on MgO is believed to be

$$
\begin{array}{c}
O^c \\
\backslash \\
\quad O^b - \\
\diagup \\
\boxed{\, O^a \,}
\end{array}
$$

with the unpaired electron being predominantly on atoms $O^b$ and $O^c$.

There is no clear evidence that $O_3^-$ exists on $Mo/SiO_2$. If one first forms $O^-$ and then adds $O_2$, an intense spectrum results which is characteristic of $O_2^-$. These results, however, are ambiguous because the g values of $O_2^-$ on $Mo/SiO_2$ and those expected for $O_3^-$ are similar (Table I). The oxygen-17 hyperfine structure favors the $O_2^-$ ion, but because of the complexity of the spectrum it is difficult to exclude the presence of some $O_3^-$.

## Stoichiometric Reactions

Reactions of $O^-$ with Alkanes. Bohme and Fehsenfeld (16), working in the gas phase, have shown that $O^-$ reacts with simple alkanes by abstracting a hydrogen atom, and except for methane the efficiency of this reaction is quite high. Even in the case of methane, the reaction occurs with a reaction probability of 0.08. Surface reactions between alkanes and surface $O^-$ ions follow a similar pathway. As indicated by a disappearance of the $O^-$ signal, stoichiometric reactions occur at temperatures sufficient to allow the diffusion of the alkane across the surface (17). For example, most of the $O^-$ on $Mo/SiO_2$ reacts with $CH_4$ in a period of 10 min at 77K (2). Thus, hydrogen atom abstraction does not appear to be an activated process.

With one exception the resulting alkyl radicals have not been detected by EPR because of their short lifetime. The exception is the detection of $CH_3\cdot$ during the reaction of $CH_4$ with $O^-$ on $Mo/SiO_2$ (2). The partial spectrum of the $CH_3\cdot$ radicals, together with the remaining $O^-$ spectrum, is depicted in Figure 1a. The complete methyl radical spectrum of Figure 1b was obtained following UV irradiation of $Mo^{VI}/SiO_2$ in the presence of $CH_4$. The methyl radicals are formed by the reactions

$$Mo^{VI}O^{2-} \xrightarrow{h\nu} Mo^{V}O^{-} \qquad (5)$$

$$Mo^{V}O^{-} + CH_4 \rightarrow Mo^{V}OH^{-} + CH_3\cdot \qquad (6)$$

The concentration of methyl radicals on the surface is considerably less than the concentration of $Mo^V$ on the surface, which suggests that many of the radicals have reacted, perhaps to form methoxide ions, even at 77K. The formation of methoxide ions is a reductive addition reaction which would transform $Mo^{VI}$ ions into $Mo^V$ ions.

Following the initial reactions, more stable intermediates which require elevated temperatures for decomposition are formed on the surface of MgO. The thermal desorption pattern, shown in Figure 2 for the reaction of ethane with $O^-$, reveals that the corresponding alkene appears in the gas phase up to temperatures of 300°C (17). The selectivity for the dehydrogenation reactions was 30 to 45%. Similar experiments carried out over $Mo/SiO_2$ yielded ethylene at 150–300°C. Most of the hydrocarbon products appear at that temperature. The surface intermediates which give rise to these products have been determined indirectly using model compounds. For example, ethoxide ions on MgO decompose at 300°C, giving ethylene as the gas-phase product.

Based upon such evidence the following mechanism has been proposed for ethane:

$$C_2H_6 + (O^-)_s \rightarrow \cdot C_2H_5 + (OH^-)_s \qquad (7)$$

$$\boxed{\cdot} + \cdot C_2H_5 + (O^{2-})_s \rightarrow (OC_2H_5^-)_s + \boxed{\cdot} \qquad (8)$$

where $\boxed{\cdot}$ denotes an electron trapped at an oxide ion vacancy at the surface. At elevated temperatures the ethoxide ions decompose according to

$$(OC_2H_5^-)_s \rightarrow C_2H_4 + (OH^-)_s \qquad (9)$$

Since some ethylene appears in the gas phase even at 25°C, it is reasonable to expect that the following competing reactions may also occur.

$$\cdot C_2H_5 + \boxed{\cdot} \rightarrow C_2H_5^+ + \boxed{\cdot} \qquad (10)$$

$$C_2H_5^+ + (O^{2-})_s \rightarrow C_2H_4 + (OH^-)_s \qquad (11)$$

These results suggested that $O^-$ may be an important intermediate in the oxidative dehydrogenation of alkanes.

A particularly interesting product distribution is observed

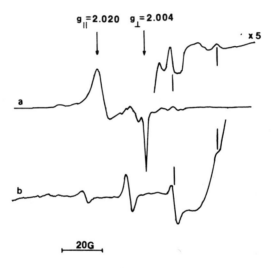

Figure 1. EPR spectra of methyl radicals: (a) after reaction of CH₄ with O⁻ on Mo/SiO₂, (b) after UV irradiation of
oxidized Mo/SiO₂ in the presence of CH₄. Reactions were
carried out and spectra recorded with the sample at 77K.

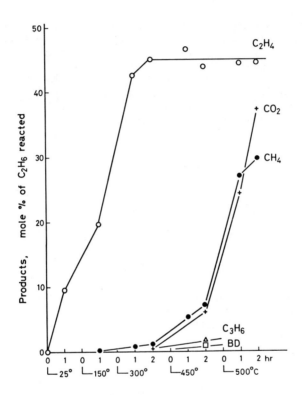

Figure 2. Variations of products as a function of the thermal desorption program following the reaction of ethane with $O^-$ on MgO at 25°C. (BD indicates butadiene) (17).

following the reaction of $CH_4$ with $O^-$ on a $Mo/SiO_2$ catalyst. Infrared bands appear at 2959, 2928, and 2857 $cm^{-1}$, which indicate
the formation of methoxide ions ($\underline{8}$). Since the methoxide ion cannot decompose to form a corresponding alkene, one might expect
that other decomposition products such as ethane, from methyl radical coupling, or formaldehyde might be observed. In fact, the
principal product is methanol, which probably results from the
reaction of methoxide ions with small amounts of water which are
present in the system ($\underline{18}$). In a separate experiment it has been
shown that $MoO(OCH_3)_4$ readily reacts with water at room temperature to form methanol ($\underline{8}$).

Reactions of $O^-$ with Alkenes. Surface reactions between $C_2$ to $C_4$
alkenes and $O^-$ ions on MgO are also rapid at 25°C, with a stoichiometry of one alkene reacted per one $O^-$ ion ($\underline{19}$). As reported by
Ben Taarit et al.($\underline{1}$), the reaction of $C_2H_4$ with $O^-$ gives rise to
an EPR spectrum which has been attributed to the complex

$$H_2C=\overset{\bullet}{C}^-\text{---}HO^-$$

where the $HO^-$ is a surface hydroxide ion. (The organic anion radical is bonded to a surface proton.) This species is believed to
be formed by the reactions

$$H_2C=CH_2 + O^- \rightarrow H_2C=\overset{\bullet}{C}H + OH^- \qquad (12)$$
and
$$H_2C=\overset{\bullet}{C}H + O^{2-} \rightarrow H_2C=\overset{\bullet}{C}^-\text{---}HO^-. \qquad (13)$$

Here again, the initial reaction is the abstraction of a hydrogen
atom by $O^-$.

The thermal desorption curve following the reaction of ethylene or propylene with $O^-$ shows no substantial products in the gas
up to 450°C, and at that temperature $CH_4$ is the primary product
($\underline{19}$). These results suggest the following reactions, using ethylene as an example:

$$H_2C=\overset{\bullet}{C}^-\text{---}HO^- + O^{2-} \rightarrow H_3C-C{\underset{O}{\overset{O^-}{\lessgtr}}} + 3e^- \qquad (14)$$

$$H_3C-C{\underset{O}{\overset{O^-}{\lessgtr}}} + OH^- \overset{\Delta}{\rightarrow} CH_4 + CO_3^{2-}. \qquad (15)$$

Separate experiments have shown that above 450°C acetate ions on
MgO decompose to form $CH_4$ and $CO_3^{2-}$ ($\underline{20}$).

The reaction of 1-butene with $O^-$, followed by the thermal decomposition of surface intermediates, leads to the formation of
butadiene as the main product. Thus, 1-butene appears to be more
similar to alkanes than to ethylene or propylene in its reaction
with $O^-$.

Reactions of $O_2^-$ with Alkanes and Alkenes. When compared with $O^-$,
the superoxide ion is a much less reactive species. Since the $O_2^-$
is thermally stable up to ~175°C, stoichiometric reactions must
be carried out at 175°C or lower temperatures ($\underline{12}$). Methane is
essentially unreactive with $O_2^-$ at 175°C, and the reaction with
ethylene and propane is slow. Essentially all of the $O_2^-$ ions re-

acted with propylene and butene over a period of 2 h at 175°C.

Following the reactions of $O_2^-$ with propylene, propane, or 1-butene at 175°C thermal decomposition at elevated temperatures (>300°C) gives rise to a variety of products including oxygen-containing molecules. In particular, the reaction of propylene with $O_2^-$ results in the formation of acetaldehyde and methanol as the partially oxygenated products. With propane as the reactant, acetone is detected at 175°C in addition to acetaldehyde and methanol, which are observed at higher temperatures. The reaction of 1-butene with $O_2^-$ ions results in the formation of 2-butanol together with methanol, acrolein, and acetaldehyde. The yeild of these products is generally 5-10%, based on the amount of hydrocarbon reacted.

Sufficiently high concentrations of $O_2^-$ may be achieved for detection of intermediates by infrared spectroscopy (12). It is evident from the infrared spectra that a number of species are formed following the reaction of $O_2^-$ with propylene at 175°C and subsequent heating to higher temperatures. Bands at 1606, 1387, and 1354 $cm^{-1}$ have been assigned to the 0-C-0 asymmetric and symmetric stretching modes of formate ions, which are eliminated by thermal treatment at 300°C. Bands at 1585 and 1422 $cm^{-1}$ are in good agreement with literature values for the 0-C-0 vibrations of acetate ions. Bands at 1648 and 1326 $cm^{-1}$ have been assigned to bidentate carbonate ions on MgO.

Although the origin of the oxygenated hydrocarbons is speculative, a reasonable mechanism has been proposed for the formation of $CH_4$ and $CO_2$ which results from the reaction of propylene with $O_2^-$. The initial reaction appears to be the abstraction of an allylic hydrogen

$$CH_2=CH-CH_3 + O_2^- \rightarrow CH_2 \cdots CH \cdots CH_2 + HO_2^- \qquad (16)$$

followed by the conversion of the allyl radicals to the acetate and formate ions, which were detected by IR:

$$CH_2 \cdots CH \cdots CH_2 + 5O^{2-} \rightarrow CH_3 - C\underset{O}{\overset{O^-}{\diagdown}} + HC\underset{O}{\overset{O^-}{\diagdown}} + OH^- + 7e^-. \qquad (17)$$

The acetate ion decomposes to $CH_4$ according to reaction 15, and the formate ion is transformed into $OH^-$ and $CO_3^{2-}$. The carbonate ions subsequently decompose to form $CO_2$. Epoxide ions formed by the reaction

$$CH_2 \cdots CH \cdots CH_2 + HO_2^- \rightarrow CH_2 \overset{O}{\diagup\diagdown} CH - CH_3 + 1/2O_2 + e^- \qquad (18)$$

may be responsible for the incorporation of oxygen into the hydrocarbons. In support of this concept the thermal desorption pattern of propylene oxide on MgO exhibited significant amounts of acetaldehyde and methanol.

Reactions of $O_3^-$ with Alkanes and Alkenes. Ozonide ions are intermediate in reactivity between $O^-$ and $O_2^-$ (20,21). On MgO they re-

act with $C_1$ and $C_4$ alkanes at 25°C with half-lives between 1.7 and
5.2 min (20). The reactions of $O_3^-$ with ethane and propane, followed by thermal decomposition of surface intermediates yields the
corresponding alkene, plus considerable amounts of $CH_4$ and $CO_2$ at
temperatures greater than 400°C. In the presence of gas-phase $O_2$,
the yield of hydrocarbons decreases, and the principal gas-phase
product is $CO_2$.

   Since the maximum concentration of $O_3^-$ is about an order of
magnitude greater than that of $O^-$, one can easily follow the formation of stable intermediates using IR spectroscopy. As shown
in Figure 3, after the reaction of $O_3^-$ with ethane, a band appeared
at 1095 $cm^{-1}$, which is attributed to a surface ethoxide ion.
Bands at 850, 980, 1318, and 1668, due to bidentate and unidentate
carbonate ions, were also present. When $O_2$ was introduced and
the sample heated to 150°C for 1 h, the absorption bands of the
ethoxide ions were replaced by those of acetate ions at 1430 and
1590 $cm^{-1}$.

   The formation of similar reaction products when alkanes react with either $O^-$ or $O_3^-$ suggests that the ozonide ion may first
dissociate according to the reverse of reaction 2, and the alkane
would then react with the $O^-$ ion. However, the lifetime for the
$O_3^-$ ion under vacuum is considerably longer than the lifetime for
the reaction of $O_3^-$ with an alkane. In addition, each alkane reacts with $O_3^-$ at a characteristic rate; therefore, it seems likely
that the alkane reacts directly with $O_3^-$, rather than indirectly
with $O^-$.

   It is believed that hydrogen atom abstraction is the first
step,

$$C_2H_6 + O_3^- \rightarrow \cdot C_2H_5 + OH^- + O_2 \qquad (19)$$

but the presence of $O_2$ results in side reactions such as

$$\cdot C_2H_5 + O_2 \rightleftarrows C_2H_5OO \cdot \qquad (20)$$

and

$$C_2H_5O^- + O_2 \rightarrow CH_3COO^- + H_2O \qquad (21)$$

The latter two reactions lead to the formation of $CO_2$, and the
acetate ion is responsible for the $CH_4$ which is produced according to reaction 15.

   Reactions between alkenes and $O_3^-$ on MgO also lead to nonselective oxidation (21). One would hope to gain insight into the
possible role of this ion in epoxidation catalysis, but rapid
surface reactions, for example between ethylene oxide and MgO,
make it difficult to obtain such information. The principal reaction products, $CH_4$ and $CO_2$, are believed to be formed in a manner analogous to reactions 12-15. The initial hydrogen abstraction again is effected by the $O_2^-$ ion.

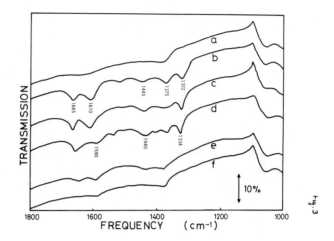

Figure 3.  Infrared spectra following the reaction of ethane
with $O_3^-$:  (a) MgO background;  (b) after the reaction at
25 °C;  (c) after thermal treatment under vacuum at 150 °C;
(d) 200 °C;  (e) 400 °C;  (f) 500 °C(20).

Comparison of Reactions Involving Oxygen Ions on MgO.  One of the
most significant observations in this series of studies was the
large difference in the reactivities of the three forms of oxygen
ions on MgO.  Taking ethylene as an example, $O^-$ ions reacted readily at -60 °C and $O_3^-$ ions reacted at 25 °C with a half-life of 5
min., whereas about two-thirds of the $O_2^-$ ions remained unreacted
after contact with $C_2H_4$ at 175 °C for 2 h.  These results, which
are qualitatively the same for other simple hydrocarbons, indicate
that the order of reactivity is $O^- \gg O_3^- \gg O_2^-$.
    The proposed surface intermediates are summarized in Table
II.  The intermediates in each reaction of alkanes includes alkoxide ions, regardless of the type of active oxygen species.  Although carboxylate ions are believed to be the intermediate in the
reactions of $C_2$ and $C_3$ alkenes, the type of carboxylate formed
with $O^-$ as a reactant is different from the type of carboxylate
ion formed when $O_2^-$ or $O_3^-$ was a reactant.  With $O^-$ ions the carbon
number of the carboxylate ions is the same as that of the hydrocarbon reactant, but with $O_2^-$ or $O_3^-$ the carboxylate ions have carbon numbers smaller than the parent hydrocarbon.  The reaction
schemes of $C_4$ alkenes are somewhat complicated, yet it appears
that they react in a manner more similar to $C_2-C_4$ alkanes than to
$C_2$ or $C_3$ alkenes.

Table II. Stoichiometric Reactions of Oxygen Ions
         with Hydrocarbons on MgO

| Ion | Reactant | Intermediate | Major Products |
|---|---|---|---|
| $0^-$ | $C_2$–$C_4$ alkanes<br>$C_2$, $C_3$ alkenes<br>$C_4$ alkenes | alkoxide ions[a]<br>carboxylate ions[a]<br>alkoxide ions[a]<br>carboxylate ions[a] | alkenes, $CO_2$<br>$CH_4$, $C_2H_4$, $CO_2$<br>butadiene, $CH_4$, $CO_2$ |
| $0_3^-$ | $C_2$–$C_4$ alkanes<br><br>$C_2$, $C_3$ alkenes<br>$C_4$ alkenes | alkoxide ions[a]<br>peroxy radicals[a]<br>carboxylate ions[b]<br>alkoxide ions[a]<br>peroxy radicals[a] | alkenes, $CH_4$, $CO_2$<br><br>$CH_4$, $CO_2$<br>butadiene, $CH_4$, $CO_2$ |
| | $C_3$ alkane<br><br>$C_2$ alkene<br>$C_3$ alkene<br><br>$C_4$ alkenes | alkoxide ions[a]<br><br>carboxylate ions[b]<br>carboxylate ions[b]<br>(epoxide)[a]<br>carboxylate ions[b]<br>(epoxide)[a] | alkenes, $CH_4$, acetone,<br>acetaldehyde, $CO_2$<br>$CH_4$, $CO_2$<br>$CH_4$, acetaldehyde,<br>methanol, $CO_2$<br>$CH_4$, 2-butanol, $CO_2$ butadiene, acetaldehyde |

[a] The carbon number is the same as that of the reactant.
[b] The carbon number is smaller than that of the reactant because of scission reactions.

## The Role of $0^-$ in Alkane Oxidation

Oxidative Dehydrogenation of Ethane. The dehydrogenation of alkanes also occurs, but in a catalytic manner, over molybdenum supported on silica (22,23). In addition to the stoichiometric reactions, the role of the $0^-$ ion in this catalytic reaction is further suggested by the observation that $N_2O$ is an effective oxidant at temperatures as low as 280°C, but no reaction is observed at these temperatures with $O_2$ as the oxidant (22). It should be noted that at moderate temperatues $N_2O$ gives rise to $0^-$, whereas $O_2$ yields $0_2^-$ over $Mo/SiO_2$. Under steady-state conditions the rates of formation of $C_2H_4$ were in the ratio of 7:1 at 375°C and 3.7:1 at 450°C when $N_2O$ and $O_2$ were used as the oxidants, respectively (23).

    With $N_2O$ as the oxidant the mechanism may be described as:
This mechanism involves a nonselective (22) and a selective (23)
sequence of reactions.  The nonselective reaction 22 is required
to produce $Mo^V$.  Once formed, the $Mo^V$ will react with $N_2O$ to yield
$Mo^{VI}O^-$ via a one-electron transfer.  The resulting cycle (23) may
be repeated many times until a two-electron transfer occurs and
$Mo^{VI}=O$ is formed.

$$(22)$$

$$(23)$$

## Partial Oxidation of Methane.

As described previously, $O^-$ ions on
$Mo/SiO_2$ react with $CH_4$, forming methyl radicals, which in turn
give rise to methoxide ions.  The methoxide ions further react to
methanol.  These reactions form part of a catalytic cycle which
leads to the partial oxidation of methane (8,18).
The complete cycle is the following:

$$8 Mo^{VI} + 4O^{2-} + CH_4 \longrightarrow 8 Mo^{V} + CO_2 + 2H_2O \qquad (24)$$

$$2 Mo^{V} + N_2O \longrightarrow 2 Mo^{VI} + O^{2-} + N_2 \qquad (25)$$

where the nonselective reactions 24 and 25 determine the kinetics of the process. The selectivity is determined, however, by the number of times the cycle is completed before an undesirable two-electron transfer occurs. Because of the elevated temperatures required for reaction 24, the methanol may undergo secondary reactions to form formaldehyde and oxides of carbon. In addition, it is believed that in the absence of water, the methoxide ion is an intermediate for the formation of formaldehyde.

Recent experimental results are summarized in Table III. The catalyst used here was molybdenum supported on Cab-O-Sil. We have experienced difficulty in repeating the previous results, which indicated that up to 50% selectivity to methanol could be achieved even at conversions of ~15% (18). In the more recent study it has been possible to obtain this selectivity only at conversions of 1%. Nevertheless, at sufficiently low conversions methanol is the principal product, provided steam is present in the system.

Table III.    Conversion and Selectivity during Methane Oxidation[a]

| T(°C) | Conv.(%) | Selectivity (%) | | | |
|---|---|---|---|---|---|
| | | $CH_3OH$ | HCHO | CO | $CO_2$ |
| 532 | 2.0 | 13 | 61 | 21 | 4 |
| 559 | 4.0 | 7 | 47 | 41 | 4 |
| 580 | 7.1 | 5 | 36 | 53 | 6 |
| 600 | 12.1 | 3 | 26 | 61 | 9 |

[a]Catalyst was 2 g of 1.7% by wt. Mo/Cab-O-Sil, calcined in air at 600°C and reduced in CO at 600°C.  Gas feed was 72 torr $CH_4$, 277 torr $N_2O$, 266 torr $H_2O$, and 145 torr He at a flow rate of 1.33 $mL \cdot s^{-1}$.

## Conclusions

Of the three types of paramagnetic ions which have been studied, it appears that the $O^-$ ion is the most important in oxidation catalysis because of its ability to activate alkanes in a selective manner.  Nitrous oxide, an uncommon oxidant, has been the source of $O^-$ ions in these studies; however, the spontaneous formation of $O^-$ ions may occur at certain defect centers, particularly at elevated temperatures.  We have recently observed, for example, the formation of a bulk V-type center (an $O^-$ ion adjacent to a magnesium ion vacancy) upon heating high-purity MgO in the presence of oxygen (24).  Kunz and co-workers (25) have provided theoretical evidence that the $O^-$ ion is a stable species when associated with cation vacancies on NiO and CoO.  Thus, in heterogeneous catalysis, $O^-$ ions may play a role in the abstraction of hydrogen from strong C-H bonds.

## Acknowledgments

The work reported here was mainly supported by the National Science Foundation under Grant No. CHE-8112893.

## Literature Cited

1.  Ben Taarit, Y.; Symons, M.C.R.; Tench, A.J.; J. Chem. Soc., Faraday Trans.1, 1977, 73, 1149.
2.  Aika, K.; Lunsford, J.H.; J. Phys. Chem. 1978, 82, 1794.
3.  Lipatkina, N.I.; Shvets, V.A.; Kazansky, V.B.; Kinet. Katal. 1978, 19, 979.
4.  Balistreri, S.; Howe, R.F., in "Magnetic Resonance in Colloid and Interface Science"; Fraissard, J.P.; Resing, H.A., Eds.; Reidel: Boston, 1980; pp. 489-494.
5.  Wong, N.-B.; Lunsford, J.H.; J. Chem. Phys. 1971, 55, 3007.

6.  Ben Taarit, Y.; Lunsford, J.H.; Chem. Phys. Letters 1973, 19.
    348: Shvets, V.A.; Kazansky, V.B., J. Catal. 1972, 25, 123:
    Kolosov, A.K.; Shvets, V.A.; Kazansky, V.B., Chem. Phys. Let-
    ters 1975, 34, 360.
7.  Ben Taarit, Y.; Lunsford, J.H.; J. Phys. Chem. 1973, 77, 780.
8.  Liu, R.-S.; Liu, H.; Liew, K.Y.; Johnson, R.E. Lunsford, J.H.;
    to be published.
9.  Lunsford, J.H.; Jayne, J.P.; J. Chem. Phys. 1966, 44, 1487.
10. Lunsford, J.H.; Catal. Rev. 1973, 8, 135.
11. Cordischi, D.; Indovina, V.; Occhiuzzi, M.J.; Chem. Soc.,
    Faraday Trans. 1, 1978, 74, 456.
12. Iwamoto, M.; Lunsford, J.H.; J. Phys. Chem. 1980, 84, 3079.
13. Tench, A.J.; Holroyd, P.J.; Chem. Commun.1968., 471.
14. Wong, N.-B.; Lunsford, J.H.; J. Chem. Phys. 1972, 56, 2664.
15. Iwamoto, M.; Lunsford, J.H.; Chem. Phys. Letters 1979, 66, 48.
16. Bohme, D.K.; Fehsenfeld, F.C.; Can. J. Chem. 1969, 47, 2717.
17. Aika, K.; Lunsford, J.H.; J. Phys. Chem. 1977, 81, 1393.
18. Liu, R.-S.; Iwamoto, M.; Lunsford, J.H.; J. Chem. Soc. Chem.
    Commun. 1982, 78.
19. Aika, K,; Lunsford, J.H.; J. Phys. Chem. 1978, 82, 1794.
20. Takita, Y.; Lunsford, J.H.; J. Phys. Chem. 1979, 83, 683.
21. Takita, Y.; Iwamoto, M.; Lunsford, J.H.; J. Phys. Chem. 1980,
    84, 1710.
22. Ward, M.B.; Lin, M.J.; Lunsford, J.H.; J. Catal. 1977, 50, 306.
23. Yang, T.-J.; Lunsford, J.H.; J. Catal. 1980, 63, 505.
24. Driscoll, D.J.; Martir, W.; Lunsford, J.H.; to be published.
25. Surratt, G.T.; Kunz, A.B., Phys. Rev. Letters 1978, 40, 347:
    Wepfer, G.G.; Surratt, G.T.; Weidman, R.S.; Kunz, A.B.; Phys.
    Rev. B 1980, 21, 2596.

RECEIVED January 9, 1984

# ZEOLITE CHARACTERIZATION

# The Future and Impact of Quantum Mechanical Calculations in the Description and Characterization of Zeolites

PAUL G. MEZEY

Department of Chemistry, University of Saskatchewan, Saskatoon, Canada S7N 0W0

Computational quantum chemistry gives a detailed description of the bonding, charge distribution, energy relations, structural preferences and geometry changes of molecules. With the increased potential of computational techniques one can expect a rapid increase in the number of studies where experimental and theoretical techniques are applied in combination, one complementing the other. One compelling economic reason: computer prices keep declining whereas the costs of experimental research keep increasing. For simple molecules it is already feasible to carry out preliminary investigations using computational quantum chemistry, in order to decide which experiment is likely to lead to a desired result.

Using the more advanced quantum chemical computational methods it is now possible to determine the fundamental electronic properties of zeolite structural units. The quantum chemical basis of Loewenstein's "aluminum avoidance" rule is explored, and the topological features of energy expectation value functionals within an abstract "nuclear charge space" model yield quick estimates for energy relations for zeolite structural units.

The exceptional catalytic properties and structural features of zeolites are a powerful stimulus for both experimental and theoretical research. With the advent of the computer age and with the spectacular development of advanced quantum chemical computational methods in the last decade, one may expect that molecular quantum theory will find more and more practical and even industrial applications. The most rapid progress is expected to occur along the borderline of traditional experimental and theoretical chemistry, where experimental and computational (theoretical) methods can be combined in an efficient manner to solve a variety

0097–6156/84/0248–0145$06.00/0
© 1984 American Chemical Society

of problems.  Whereas experiment will always remain the ultimate
test of any theory, theoretical computational quantum chemistry
can provide answers to problems for which no experimental technique is available at present.  E.g., geometry changes of reacting
molecules along various reaction paths, and properties of transition structures can be studied and analysed in fine detail, using
computational quantum chemistry (1), whereas experimental information on these problems is scarce at best.  In other problems,
e.g. in the assignments of vibrational spectra, and in particular,
in the evaluation of interaction force constants, the accuracy of
quantum chemical calculations often surpasses that of experimental methods (2).

Among the advances contributing to the recent progress of
computational quantum chemistry there are two factors of particular importance.  On the one hand, there is a very large volume of
accumulated experience with ab initio calculations on thousands
of molecules, that lends confidence in evaluating the reliability
of calculated results.  Although even "state-of-the-art" quantum
chemical calculations of certain molecular properties (e.g. bond
dissociation energies) are often less accurate than the experimental data, nevertheless, the expected errors can be estimated
based upon previous computational experience.  On the other hand,
efficient algorithms and computer programs have been developed
for the calculation of not only the energies of molecules but
also of analytic energy derivatives (2-4).  That is, energies and
also forces and force constants can be calculated rather
accurately for almost any geometry of the molecule, independently
whether it is the equilibrium geometry of the molecule, or some
highly distorted geometry, that may occur only for an instant in
the classical model of a reaction or conformational process.
Evidently, experimental information is seldom, if ever available
for such distorted molecular geometries, although the relative
feasibility of various reaction processes does critically depend
on the energy content of such distorted molecular arrangements
and on the internal forces acting upon them.  Quantum chemical
calculations can give a detailed account why and how a given
reaction or distortion of a molecular structure occurs using
relatively simple mathematical models (5).  In fact, a more
rigorous global topological analysis of the energy function, which
function is defined for all possible molecular geometries (6),
leads to a description of all possible molecules composed from a
given set of N nuclei and a fixed number of k electrons, as well
as to the network of all possible reactions between these molecules.  Such a global quantum chemical approach is expected to
find applications in computer-aided quantum-chemical synthesis
design (7-9).

The rapidly changing economy of experimental vs. theoretical
research clearly shows the changing direction of future research
efforts:  whereas the prices of most chemicals and laboratory
equipment keep increasing, the cost of computation is on a

consistent and even accelerating decline.  It is indeed fortunate
that innovative, advanced quantum chemical computational methods
and computer programs have become available at the same time,
when the speed, accessibility and affordability of computers have
shown such spectacular improvements.  Computational quantum
chemistry is bound to have a significant impact on a wide range
of chemical fields due to these developments.

Although even the smaller structural units of zeolites are
large enough to tax the most advanced quantum chemical computational methods to their limits, nevertheless, it is now possible
to determine the fundamental electronic properties of zeolite
structural units.  In addition to their unique geometrical (in
fact, topological) properties, the electronic structure and
charge distribution of zeolites are of fundamental importance in
explaining their catalytic and other chemical properties.

In the following we shall briefly review some of the recent
applications of computational quantum chemistry to zeolites, in
particular, some studies on the quantum chemical origin of
Loewenstein's aluminum avoidance rule, and on the role of counter
ions in stabilizing various structural units in zeolite lattices.
These calculations are often extremely time consuming, nevertheless, the scope of their application is continuously expanding.
By contrast, only "back-of-an-envelope" calculations are needed
to apply some recently proven quantum chemical theorems (10-12)
which give a variety of energy relations between zeolite
structural units.

Quantum Chemical Calculations on the Electronic Structure of
Zeolite Clusters

The relative merits of ab initio and semiempirical methods

Most ab initio quantum chemical molecular orbital calculations
involve, in some form, the solution of the Hartree-Fock equations.
Following Roothaan (13,14) these equations are usually given in
a matrix form that for a closed shell molecule takes the deceiv-
ingly simple form:

$$\underline{F}\,\underline{C} = \underline{S}\,\underline{C}\,\underline{E} \tag{1}$$

These equations, derived from the Schrödinger equation of Quantum
Mechanics, can be solved iteratively for matrices $\underline{C}$ and $\underline{E}$, containing as elements the appropriately normalized molecular
orbital (MO) coefficients and orbital energy eigenvalues of eq.
(1), respectively.  Whereas the overlap matrix $\underline{S}$ is relatively
easy to determine, the evaluation of the Fock matrix $\underline{F}$ may
involve several million integrations even for a small zeolite
structural unit containing about a dozen atoms.  Furthermore, the
evaluation of $\underline{F}$ and the solution of the eigenvalue equation (1)
must be repeated iteratively, since the Fock matrix $\underline{F}$ depends on

matrix $\underline{C}$ as well.  If the iterative sequence converges, that is,
the same $\underline{C}$ matrix is obtained in solving the eigenvalue problem
(1) that has been used to build the $\underline{F}$ matrix in the previous
iterative step, then the MO's built from elements of $\underline{C}$ are called
the self consistent molecular orbitals.  The accuracy of molecular energies and other molecular properties calculated from this
solution is often sufficient for a stereochemical analysis or for
the analysis of the main features of charge densities.  However,
if more accurate results are needed, then these MO's can be used
to evaluate the effects of <u>electron correlation</u>, e.g. by the
methods of configuration interaction (C.I.).

In a quantum chemical analysis of a reaction process or a
conformational change these calculations can be repeated for a
sequence of molecular geometries, thus energy profiles and charge
density changes can be calculated pointwise along various reaction paths.  Analytic or numerical first and second energy derivatives, obtained from equation (1), can also be calculated (the
analytic calculation involves a large number of additional inte-
grals), that gives at each molecular geometry the slopes and
curvatures of energy profiles, or in higher dimensions, on energy
surfaces and hypersurfaces.

The problems of the quantum chemical analysis of zeolite
structural units are somewhat different from the more common
problems of applied computational quantum chemistry.  On the one
hand, a typical zeolite structural unit of chemical interest is
rather large, rendering the dimensions of the matrices in eq. (1)
very high and the integrations and the iterative solution
extremely time consuming.  Furthermore, the effects of van der
Waals interactions and electrostatic contributions from the
surrounding lattice sites are difficult to evaluate.  On the
other hand, the rigidity of zeolite lattices can be exploited in
the calculations, since it is often sufficient to consider position and geometry changes only for the ions and small molecules
in the cavities, and not for the lattice itself.  Thus quantum
chemical calculations can determine the most stable location of
counter ions, energy barriers to their site-change processes,
optimum location and orientation of small molecules within the
cavities, etc., without assuming major lattice rearrangements.

In the earliest quantum chemical studies on zeolites <u>semi-
empirical</u> methods have been used (<u>15-25</u>).  These methods
(Extended Hückel Method, CNDO, INDO, MINDO, etc.) are less time
consuming than a full <u>ab initio</u> calculation, since they involve
various and often drastic simplifications in approximating and
solving eq. (1).  Consequently, the semiempirical techniques are
in general less reliable than the <u>ab initio</u> method, although by
a suitable "calibration" of certain parameters, they can mimic
the exact solution, and often lead to useful results.  Whereas
their limited reliability is a potentially serious disadvantage,
these methods are applicable to much larger clusters than the
more accurate <u>ab initio</u> methods, (<u>27-32</u>) without requiring
extremely large amounts of computer time.

Ab initio MO studies on the ordering of Si and Al atoms in zeolite
frameworks and on the stabilizing effects of counter ions

It is well known that the ordering of Si and Al tetrahedral
centers in alumosilicates is not entirely random. Following a
suggestion by Pauling (33) Loewenstein proposed the "aluminum
avoidance rule", stating that in tetrahedral alumosilicate frameworks two Al atoms do not bound to the same oxygen atom when Si
atom is available (34). Whereas on electrostatic grounds such a
restriction on the Al and Si atom distribution is not unexpected,
some exceptions to this rule have been pointed out and evidence
for the existence of Al-O-Al moieties has been reported (35-38).
    Using 3G and 4-31G type basis sets (39-41), ab initio
quantum chemical calculations have been carried out for several
small structural units of zeolites, with a variety of observed
and hypothetical Si-Al distributions (29-32). The results of
these studies can be summarized in a series of hypothetical
Si → Al exchange reactions within these structural units. The
calculated internal energy changes for the reactions involving
two neighbouring tetrahedra, are as follows:

$$(HO)_3Si-O-Si(OH)_3 + [AlH_4]^- \rightarrow [(HO)_3Si-O-Al(OH)_3]^- + SiH_4,$$

$$\Delta E = -60 \text{ kcal mol}^{-1} \qquad\qquad (2)$$

$$[(HO)_3Si-O-Al(OH)_3]^- + [AlH_4]^- \rightarrow [(HO)_3Al-O-Al(OH)_3]^{2-} + SiH_4,$$

$$\Delta E = +56 \text{ kcal mol}^{-1} \qquad\qquad (3)$$

For reactions involving four tetrahedra, arranged in a four
membered ring, the following results have been obtained:

$$[H_8Si_3AlO_4]^- + [H_8Si_3AlO_4]^- \rightleftarrows H_8Si_4O_4 + [H_8SiAlSiAlO_4]^{2-},$$

$$\Delta E = +88 \text{ kcal mol}^{-1} \qquad\qquad (4)$$

$$[H_8Si_3AlO_4]^- + [H_8Si_3AlO_4]^- \rightleftarrows H_8Si_4O_4 + [H_8Si_2Al_2O_4]^{2-},$$

$$\Delta E = +120 \text{ kcal mol}^{-1} \qquad\qquad (5)$$

$$[H_8Si_2Al_2O_4]^{2-} \rightleftarrows [H_8SiAlSiAlO_4]^{2-},$$

$$\Delta E = -32 \text{ kcal mol}^{-1} \qquad\qquad (6)$$

Evidently, arrangements where two Al atoms share a common O atom
are much less stable than those cluster models where Loewenstein's
rule is followed.
    The quantum chemical calculations strongly favour the

Lowenstein-rule-allowed arrangements on the basis of the energetic
stability of electronic structures. However, these calculations
also give information on the stereochemical stability of various
clusters. The internal energy variation of three possible double
tetrahedral systems with Si-O-Si, Si-O-Al and Al-O-Al linkages,
respectively, has been calculated as the function of bond angle
$T_1$-O-$T_2$ at the common oxygen atom, linking the two tetrahedral
centers. Whereas for the Si-O-Si and Al-O-Si systems the free
clusters favour a bent $T_1$-O-$T_2$ arrangement with a bond angle close
to the experimental value (29), the free cluster containing an
Al-O-Al linkage favours on open, linear arrangement, that cannot
be accommodated without a drastic distortion of the zeolite
lattice. Evidently, Al-O-Al linkages in zeolites are strongly
disfavoured on the grounds of both electronic and stereochemical
stability.

Such Al-O-Al linkages in zeolites, however, are not alto-
gether impossible, if suitable counter ions are present, as
indicated by the calculations on the following hypothetical
reactions, analogous to reactions (2) ... (6).

$$(HO)_3Si-O-Si(OH)_3 + Li^+ + [AlH_4]^- \rightarrow [(HO)_3Si-O-Al(OH)_3]^- +$$

$$Li^+ + SiH_4, \quad \Delta E = -183 \text{ kcal mol}^{-1} \tag{7}$$

$$[(HO)_3Si-O-Al(OH)_3]^- + Li^+ + [AlH_4]^- \rightarrow [(HO)_3Al-O-Al(OH)_3]^{2-} +$$

$$Li^+ + SiH_4, \quad \Delta E = -78 \text{ kcal mol}^{-1} \tag{8}$$

$$[H_8Si_3AlO_4]^- + [H_8Si_3AlO_4]^- Li^+ \rightleftarrows H_8Si_4O_4 + [H_8SiAlSiAlO_4]^{2-} Li^+,$$

$$\Delta E = -40 \text{ kcal mol}^{-1} \tag{9}$$

$$[H_8Si_3AlO_4]^- + [H_8Si_3AlO_4]^- Li^+ \rightleftarrows H_8Si_4O_4 + [H_8Si_2Al_2O_4]^{2-} Li^+,$$

$$\Delta E = -18 \text{ kcal mol}^{-1} \tag{10}$$

$$[H_8Si_2Al_2O_4]^{2-} Li^+ \rightleftarrows [H_8SiAlSiAlO_4]^{2-} Li^+,$$

$$\Delta E = -22 \text{ kcal mol}^{-1} \tag{11}$$

$$[H_8Si_2Al_2O_4]^{2-} Be^{2+} \rightleftarrows [H_8SiAlSiAlO_4]^{2-} Be^{2+},$$

$$\Delta E = -14 \text{ kcal mol}^{-1} \tag{12}$$

If counter ions are explicitly included in the calculations (for
simplicity, ions $Li^+$ and $Be^{++}$ have been used, in a location near
the bridging O atom, (29,30)), then the calculated energy values
for clusters containing Al-O-Al linkages indicate a significant
stabilization that increases with increasing cation charge.

## Simple quantum chemical upper and lower bounds for energies of cluster models

Not all quantum chemical calculations on zeolite clusters involve necessarily millions of integrations, and in the case of iso-electronic chemical systems fulfilling certain geometrical criteria, almost trivial back-of-an-envelope type calculations can yield rigorous upper and lower energy bounds. Fortunately, some zeolite structural units fulfill these geometric criteria.

These energy bounds are based on two theorems (10-12) which utilize the fact that in the quantum chemical energy expectation value functional the nuclear charges can be regarded as continuous variables. A series of energy relations can be derived for iso-electronic molecules which contain different nuclei, or the same nuclei in different positions. A corollary of the first theorem states (eq.32 in (10)) that

$$(1 - \alpha_n) E_e(M_1) + \alpha_n E_e(M_n) \leq E_e(M) \tag{13}$$

where $E_e$ stands for electronic energy, $M$, $M_n$ and $M_1$ are iso-electronic molecules (in the simplest case $M_1$ is a single atom) and number $\alpha_n$ fulfills the following constraints:

$$0 < \alpha_n < 1 \tag{14}$$

and

$$Z_i(M) - \alpha_n Z_i(M_n) \geq 0 \tag{15}$$

for every nuclear charge $Z_i(M)$ and $Z_i(M_n)$ of molecule $M$ and $M_n$, respectively. This is a rigorous result and is valid for the electronic ground states and for any excited electronic state that is the lowest lying state of the given type (e.g. lowest singlet, lowest triplet state, etc.). Since the nuclear repulsion energy component $E_n$ of the total energy $E_t$ can be calculated easily, using Coulomb's Law, upper and lower bounds for the $E_t = E_e + E_n$ total energy can be obtained just as easily (12). We shall illustrate the application of the above simple relation with the example of the following three systems:

$$asa = \begin{matrix} H & H & H \\ | & | & | \\ H-Al^{\ominus}-O-Si-O-Al^{\ominus}-H \\ | & | & | \\ H & H & H \end{matrix} + Be^{2+} \tag{16}$$

$$aas = \begin{matrix} H & H & H \\ | & | & | \\ H-Al\overset{\ominus}{-}O-Al\overset{\ominus}{-}O-Si-H \\ | & | & | \\ H & H & H \end{matrix} + Be^{2+} \tag{17}$$

$$M_1 = Er \tag{18}$$

For the actual nuclear charges involved in asa = M and aas = $M_n$ the largest possible value for $\alpha_n$, satisfying conditions (14) and (15), is

$$\alpha_n = 13/14 \tag{19}$$

Note that the nuclear charge of the erbium atom ($M_1$ = Er) does not affect the choice of $\alpha_n$. One obtains the following energy constraints for the two cluster models:

$$\frac{1}{14} E_e(Er) + \frac{13}{14} E_e(aas) \leq E_e(asa) \tag{20}$$

If, however, the roles of the two cluster models are interchanged, one obtains

$$\frac{1}{14} E_e(Er) + \frac{13}{14} E_e(asa) \leq E_e(aas) \tag{21}$$

since the same $\alpha_n$ value of 13/14 obviously fulfills conditions (14) and (15) for the new assignment of roles as well. The combination of the two inequalities (20) and (21) results in the following relations:

$$\frac{1}{14} E_e(Er) + \frac{13}{14} E_e(aas) \leq E_e(asa) \leq \frac{14}{13} E_e(aas) - \frac{1}{13} E_e(Er) \tag{22}$$

and equivalently

$$\frac{1}{14} E_e(Er) + \frac{13}{14} E_e(asa) \leq E_e(aas) \leq \frac{14}{13} E_e(asa) - \frac{1}{13} E_e(Er) \tag{23}$$

If energy values are available for the single atom Er and for one of the cluster models, e.g. for system asa of the alternating Al, Si, Al arrangements then result (23) gives both upper and lower energy bounds for system aas, that system is not favoured according to Loewenstein's aluminum avoidance rule.

In the above inequalities the electronic energy of the erbium atom is much lower than that of cluster models (aas) and (asa). Although the coefficients 1/14 and 1/13 of terms involving $E_e(Er)$ are small, the large negative value of $E_e(Er)$ renders the calculated energy bounds rather loose and of little practical value.

However, more useful energy relations can be obtained if the single atom is replaced by one or several molecules, satisfying the constraints of the same general relation (eq. 32 in (10)). If, for example, $M_1$ is a molecule, isoelectronic and isoprotonic with $M_n$ and M, furthermore $\alpha_n$ fulfills the following constraints (in addition to (14)):

$$(1 - \alpha_n)\, Z_i(M_1) + \alpha_n Z_i(M_n) = Z_i(M) \qquad (24)$$

for every triplet of nuclear charges, then inequality (13) is valid. Since in this case the chemical structures compared are more similar than in the examples involving single atoms, the resulting energy bounds are much tighter.

As an illustration of relation (13), subject to constraint (24), we shall consider the following zeolite structural units:

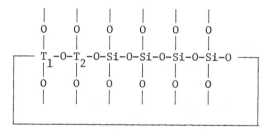

i.e., isoelectronic ring structures containing six tetrahedral centres. We shall use the following short notations

(ssssss)    if $T_1 = T_2 = Si$                           (25)

(asssss)    if $T_1 = Al \quad T_2 = Si$                  (26)

(aasss)     if $T_1 = T_2 = Al$                           (27)

(msssss)    if $T_1 = Mg, \ T_2 = Si$                     (28)

(smssss)    if $T_1 = Si, \ T_2 = Mg$                     (29)

The isoelectronic condition implies that a neutral (ssssss) can be compared only to the mononegative anion (assssss) and to doubly negative anions of the remaining three structures. Note that the last two structures, containing Mg impurities, are equivalent, and we formally distinguished them only for convenience in labelling.

Taking $\alpha_n = 0.5$, $M_1 = $ (msssss), $M_n = $ (smssss) and M = (aassss) conditions (14) and (24) are satisfied and one obtains

$$E_e(msssss) = 0.5\, E_e(msssss) + 0.5\, E_e(smssss) \leq E_e(aassss) \qquad (30)$$

that is the expected result.  An energy relation of more chemical interest can be obtained with a different assignment of roles. Taking $\alpha_n = 0.5$, $M_1 = (mssss)$, $M_n = (sssss)$ and $M = (assss)$, conditions (14) and (24) are fulfilled, and one obtains

$$0.5 \, E_e(mssss) + 0.5 \, E_e(sssss) \leq E_e(assss) \qquad (31)$$

For zeolite structural units of the above size detailed ab initio calculations are prohibitively expensive even with the currently available most advanced computer programs.  Convexity relation (13), and the resulting energy bounds, on the other hand, are easily applicable to a variety of similar problems, and require only few elementary algebraic operations.

Future prospects

The currently available quantum chemical computational methods and computer programs have not been utilized to their potential in elucidating the electronic origin of zeolite properties.  As more and more physico-chemical methods are used successfully for the description and characterization of zeolites, (e.g. (42-45)), more questions will also arise where computational quantum chemistry may have a useful contribution towards the answer, e.g. in connection with combined approaches where zeolites and metal-metal bonded systems (e.g. (46,47)) are used in combination.  The spectacular recent and projected future improvements in computer technology are bound to enlarge the scope of quantum chemical studies on zeolites.  Detailed studies on optimum intercavity locations for a variety of molecules, and calculations on conformation analysis and reaction mechanism in zeolite cavities are among the promises what an extrapolation of current developments in computational quantum chemistry and computer technology holds out for zeolite chemistry.

Literature Cited

1.  Mezey, P.G., The Topological Model of Non-rigid Molecules and Reaction Mechanisms, in Symmetries and Properties of Non-rigid Molecules:  A Comprehensive Survey, Elsevier Sci. Publ. Co., Amsterdam, 1983.
2.  Pulay, P., Direct Use of Gradients for Investigating Molecular Energy Surfaces, in Applications of Electronic Structure Theory, Schaefer, H.F., Ed.; Plenum Press, New York, 1977.
3.  Binkley, J.S.; Whiteside, R.A.; Krishnan, R.; Seeger, R.; DeFrees, D.J.; Schlegel, H.B.; Topiol, S.; Kahn, L.R.; Pople, J.A., Program Gaussian 80, Dept. Chemistry, Carnegie-Mellon University, Pittsburgh, PA  15213.
4.  Saxe, P.; Yamaguchi, Y.; Schaefer, H.F. J. Chem. Phys. 1982, 77, 5647.
5.  Bruno, A.E.; Steer, R.P.; Mezey, P.G. J. Comput. Chem. 1983, 4, 104.

6. Mezey, P.G. Theor. Chim. Acta 1982, 60, 409; 1982, 62, 133.
7. Mezey, P.G. J. Chem. Phys. 1983, 78, 6182.
8. Mezey, P.G. Can. J. Chem. 1983, 61, 956 (Volume dedicated to Prof. H. Gunning).
9. Mezey, P.G., Reaction Topology: Manifold Theory of Potential Surfaces and Quantum Chemical Synthesis Design, in Chemical Applications of Topology and Graph Theory, Elsevier Sci. Publ. Co., Amsterdam, 1983.
10. Mezey, P.G. Theor. Chim. Acta 1981, 59, 321.
11. Mezey, P.G. Int. J. Quantum Chem. 1982, 22, 101.
12. Mezey, P.G. Mol. Phys. 1982, 47, 121.
13. Roothaan, C.C.J. Rev. Mod. Phys. 1951, 23, 69.
14. Roothaan, C.C.J. Rev. Mod. Phys. 1960, 32, 179.
15. Gibbs, G.W.; Louisnathan, S.J.; Ribbe, P.H.; Philips, M.W. in MacKenzie, W.S.; Zussman, J., Eds., The Feldspars, Manchester Univ. Press, Manchester, 1974, pp. 49-67.
16. Gibbs, G.W.; Meagher, E.P.; Smith, J.V.; Pluth, J.J., in Katzer, J.R., Ed., Molecular Sieves II, American Chemical Society, Symposium Ser. 40, Washington, 1977, pp. 19-29.
17. Cohen, J.P.; Ross, F.K.; Gibbs, G.V. Am. Mineral, 1977, 62, 67.
18. Tossel, J.A.; Gibbs, G.V. Phys. Chem. Minerals, 1977, 2, 21.
19. Tossel, J.A.; Gibbs, G.V. Acta Crystallogr., 1978, Sect. A34, 463.
20. Seregina, V.A.; Lygin, V.I.; Gryaznova, Z.V. Dokl. Akad. Nauk. SSR, 1976, 226, 640.
21. Mortier, W.J.; Geerlings, P.; Van Alsenoy, C.; Figeys, H.G. J. Phys. Chem. 1979, 83, 855.
22. Beran, S.; Dubský, J. J. Phys. Chem. 1979, 83, 2538.
23. Dubský, J.; Beran, S.; Bosácek, V. J. Mol. Catal. 1979, 6, 321.
24. Beran, S. Chem. Phys. Letters 1982, 91, 86.
25. Beran, S. J. Phys. Chem. 1983, 87, 55.
26. Klier, K.; Hutta, P.J.; Kellerman, R.; Katzer, J.R., Ed., Molecular Sieves II, American Chemical Society, Symposium Ser. 40, Washington 1977, pp. 108-119.
27. Sauer, J.; Hobza, P.; Zahradnik, R. J. Phys. Chem. 1980, 84, 3318.
28. Sauer, J.; Fielder, K.; Schirmer, W.; Zahradnik, R.; Rees, L.V.C., Ed., Proc. Fifth Internat. Conference on Zeolites, Heyden, London, 1981, pp. 501-509.
29. Hass, E.C.; Mezey, P.G.; Plath, P.J. J. Mol. Structure 1981, 76, 389.
30. Hass, E.C.; Mezey, P.G.; Plath, P.J. J. Mol. Structure 1982, 87, 261.
31. Hass, E.C.; Plath, P.J.; Mezey, P.G., in Computational Theor. Org. Chem., Reidel, New York, 1981, pp. 403-408.
32. Mezey, P.G.; Flakus, H., to be published.
33. Pauling, L. J. Am. Chem. Soc. 1929, 51, 1010.
34. Loewenstein, W. Am. Mineral. 1954, 39, 92.

35.  Smith, J.V. Feldspar Minerals, Springer-Verlag, New York, 1974, pp. 78,79.
36.  Smith, J.V. Adv. Chem. Ser. 1971, 101, 171.
37.  Johansson, G. Acta Chem. Scand. 1966, 20, 505.
38.  Haladjian, J; Roziere, J. J. Inorg. Nucl. Chem. 1973, 35, 3821.
39.  Hehre, W.J.; Stewart, R.F.; Pople, J.A. J. Chem. Phys. 1969, 51, 2657.
40.  Ditchfield, R.; Hehre, W.J.; Pople, J.A. J. Chem. Phys. 1971, 54, 724.
41.  Hehre, W.J.; Lathan, W.A. J. Chem. Phys. 1972, 56, 5255.
42.  Derouane, E.G., New aspects of molecular shape-selectivity: catalysis by zeolite ZSM-5; Imelik, B.; Naccache, C.; Taarit, Y. Ben; Vedrine, J.C.; Coudrier, G.; Praliaud, H., Eds., Catalysis by Zeolites, Studies in Surface Science and Catalysis 5, Elsevier Sci. Publ. Co. Amsterdam, 1980, pp. 5-18.
43.  Fejes, P.; Kiricsi, I.; Hannus, I.; Tihanyi, T.; Kiss, A., Poisoning of acidic centres in zeolites with sodium azide, Imelik, B.; Naccache, C.; Taarit, Y. Ben; Vedrine, J.C.; Coudrier, G; Praliaud, H., Eds., Catalysis by Zeolites, Studies in Surface Science and Catalysis 5, Elsevier Sci. Publ. Co. Amsterdam, 1980, pp. 135-140.
44.  Rabo, J.A., Zeolite Chemistry and Catalysis, American Chem. Soc., Washington, 1976.
45.  Breck, D.W., Zeolite Sieves - Structure, Chemistry and Use, Wiley-Interscience, New York, 1974.
46.  Ozin, G.A., Farad. Soc. Symp. Chem. Soc. 1980, 14, 1.
47.  King, R.B. Acc. Chem. Res. 1980, 13, 243.

RECEIVED September 20, 1983

# The Preparation and Characterization of Aluminum-Deficient Zeolites

JULIUS SCHERZER[1]

Filtrol Corporation, 3250 E. Washington Boulevard, Los Angeles, CA 90023

The preparation methods of aluminum-deficient zeolites are reviewed. These methods are divided in three categories: (a) thermal or hydrothermal dealumination; (b) chemical dealumination; and (c) combination of thermal and chemical dealumination. The preparation of aluminum-deficient Y and mordenite zeolites is discussed. The structure and physico-chemical characteristics of aluminum-deficient zeolites are reviewed. Results obtained with some of the more modern methods of investigation are presented. The structure, stability, sorption properties, infrared spectra, acid strength distribution and catalytic properties of these zeolites are discussed.

The discovery of the new class of high-silica zeolites "pentasil" during the last decade has attracted considerable interest due to the important applications of these zeolites in catalysis. The best known member of this family of zeolites is ZSM-5, developed in the Mobil laboratories. The unusual properties of pentasil zeolites have rekindled the interest in other high-silica zeolites, prepared by dealumination of low-silica zeolites. In this paper we shall review the preparation methods of aluminum-deficient zeolites, and shall discuss the properties of these materials, with emphasis on recent advances in their characterization.

[1]Current address: Union Oil of California Company, Science and Technology Division, Brea, CA 92621.

General Preparation Methods of Aluminum-Deficient
Zeolites

Zeolite dealumination was first reported by Barrer and Makki
(1), who progressively removed aluminum from clinoptilolite by
treating the zeolite with hydrochloric acid of different
strengths.  Subsequent dealumination studies were carried out
primarily on mordenite (2-5) and Y zeolites.
    High-silica, aluminum-deficient zeolites have been
prepared by the following methods (Table I):
    A)  Thermal or hydrothermal treatment of zeolites.  This
results in partial framework dealumination, but the aluminum
remains in the zeolite cages or channels.
    B)  Chemical treatment of zeolites.  Such a treatment can
be carried out with a variety of reagents and results in the
removal of aluminum from the zeolite in a soluble or volatile
form.
    C)  Combination of thermal and chemical treatments.
    We shall examine more closely each of these methods.

Thermal dealumination.  The method involves calcination of
the ammonium (or hydrogen) form of the zeolite at relatively
high temperatures (usually over 500°C) in the presence of steam.  This results in the expulsion of tetrahedral aluminum
from the framework into non-framework positions, but does not
remove the aluminum from the zeolite.  The process consists
essentially in a high-temperature hydrolysis of Si-O-Al bonds
and leads to the formation of neutral and cationic aluminum
species (Figure 1A).
    An example of such thermal dealumination is the formation
of ultra-stable Y zeolites (USY zeolites).  McDaniel and Maher
(6) reported the preparation of two types of ultrastable Y
zeolites:  (a)  one type prepared by the hydrothermal
treatment of an $NH_4$, Na-Y zeolite (USY-A) and (b)  another
type involving an additional ammonium exchange and a second
high-temperature treatment (USY-B).
    Kerr (7-9) has shown the critical role of the calcination
environment and bed geometry in the formation of USY zeolites
("deep bed" vs."shallow bed"calcination).  Ward (10) prepared
USY zeolites by calcining ammonium Y zeolites in flowing
steam. The work done by Kerr and Maher et al. (11) has clearly
demonstrated that USY zeolites are formed as a result of
aluminum expulsion from the framework at high  temperatures in
the presence of steam. The nature of the non-framework aluminum species has not been completely clarified.  Obviously,
their composition will be strongly affected by the preparation
procedure of the USY zeolite.  Table II shows different
oxi-aluminum species assumed to be  formed during thermal
dealumination of the zeolite framework.

TABLE I

PREPARATION METHODS OF ALUMINUM-DEFICIENT ZEOLITES

I.   Dealuminated Y Zeolites

    A.   Hydrothermal treatment of $NH_4$ Y zeolites (6)
        (formation of ultrastable Y zeolites)

    B.   Chemical treatment

        1.   Reaction with chelating agents (8)
        2.   Reaction with $CrCl_3$ in solution (19)
        3.   Reaction with $(NH_4)_2SiF_6$ in solution (107)
        4.   Reaction with $SiCl_4$ vapors (27) or other halides (108)
        5.   Reaction with $F_2$ gas (102)

    C.   Hydrothermal and chemical treatment

        Reaction of ultrastable Y zeolites with:

        1.   Acids (e.g. HCl, $HNO_3$) (28)
        2.   Bases (e.g. NaOH) (17)
        3.   Salts (e.g. KF) (30)
        4.   Chelating agents (e.g. EDTA) (32)

II.  Dealuminated Mordenite

    A.   Chemical treatment

        1.   Reaction with acids (2)
        2.   Reaction with $SiCl_4$ vapors (69)

    B.   Hydrothermal and chemical treatment

        1.   Steaming and acid leaching (4)
        2.   Repeated steaming and acid leaching (5)

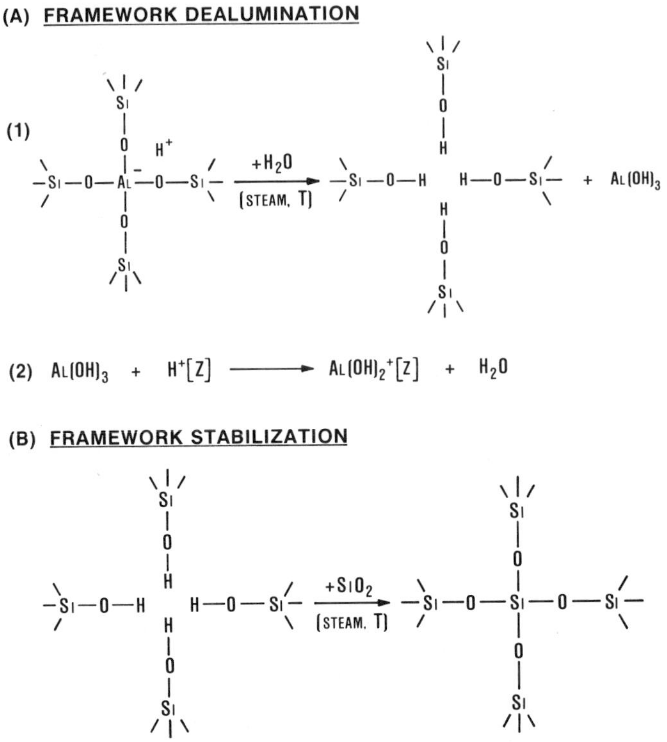
(A) FRAMEWORK DEALUMINATION
(1)
+H2O
(STEAM, T)
+ AL(OH)3
(2) AL(OH)3 + H+[Z] → AL(OH)2+[Z] + H2O
(B) FRAMEWORK STABILIZATION
+SIO2
(STEAM, T)

Figure 1. Reaction mechanism for hydrothermal dealumination and stabilization of Y zeolites.

**Table II. FRAMEWORK and NON-FRAMEWORK ALUMINUM SPECIES in USY ZEOLITES**

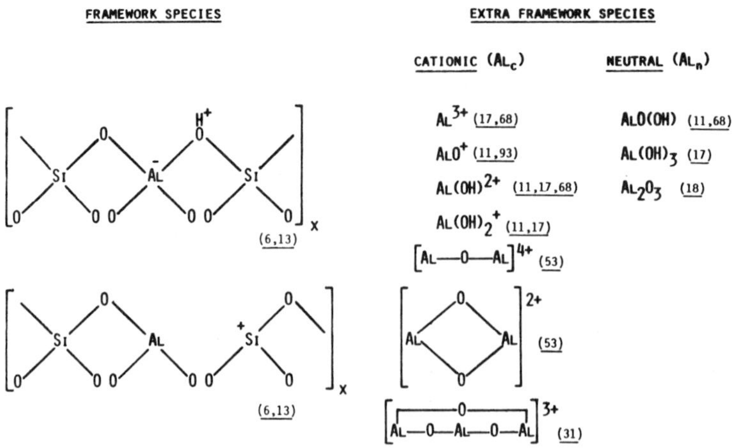
FRAMEWORK SPECIES
EXTRA FRAMEWORK SPECIES
CATIONIC (ALc)
NEUTRAL (ALn)
(6,13)
(6,13)
AL3+ (17,68)
ALO+ (11,93)
AL(OH)2+ (11,17,68)
AL(OH)2+ (11,17)
[AL—O—AL]4+ (53)
2+ (53)
3+ (31)
ALO(OH) (11,68)
AL(OH)3 (17)
AL2O3 (18)

It was also shown that thermal treatment of an ammonium
zeolite under steam causes not only framework dealumination,
but also a structural rearrangement in the zeolite framework.
The defect  sites left by dealumination are filled to a large
extent by silica, which leads to a very stable, highly
silicious framework (11,12) (Figure 1B).  Defect sites not
filled by silica are occupied by "hydroxyl nests" (13).

Chemical dealumination.  In this case dealumination is
achieved (1) by reacting the zeolite with a suitable reagent
in solution  (aqueous or non-aqueous); or (2)  by reacting the
zeolite with a reagent in vapor phase at high temperature.
Dealumination in solution was accomplished by reacting the
zeolite with solutions of acids, salts or chelating agents.

Reactions with acids.  Hydrochloric acid was used in the
dealumination of clinoptilolite (1), erionite (14) and mordenite (2,3,15,92). In the case of Y zeolite, dealumination
with mineral acids was successful only after conversion of the
zeolite into the ultrastable form (vide infra).  Barrer and
Makki (1)  were the first to propose a mechanism for the
removal of aluminum from mordenite by mineral acids.  It
involves the extraction of aluminum in a soluble form and its
replacement by a nest of four hydroxyl groups as follows:

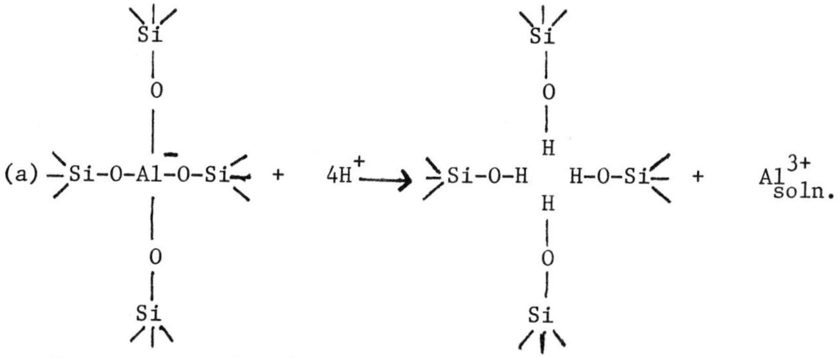

Reactions with salts.  This procedure is more limited and
is illustrated by the use of chromium chloride solutions under
reflux for partial dealumination of Y and X zeolites (19), as
well as of erionite (20).  It is assumed that in this case a
partial substitution of chromium for aluminum takes place,
leading to the formation of Si-O-Cr bonds in the framework
(19).  Up to 40 percent of aluminum was removed by this
method.  Zeolites can also be dealuminated with solutions of
ammonium fluorosilicate (107).

Reaction with chelating agents. Such reactions have been
used primarily for partial dealumination of Y zeolites. In
1968, Kerr (8,21) reported the preparation of aluminum-
deficient Y zeolites by extraction of aluminum from the
framework with EDTA. Using this method, up to about 50
percent of the aluminum atoms was removed from the zeolite in
the form of a water soluble chelate, without any appreciable
loss in zeolite crystallinity. Later work (22) has shown that
about 80 percent of framework aluminum can be removed with
EDTA, while the zeolite maintains about 60 to 70 percent of
its initial crystallinity. Beaumont and Barthomeuf (23-25)
used acetylacetone and several amino-acid-derived chelating
agents for the extraction of aluminum from Y zeolites.
Dealumination of Y zeolites with tartaric acid has also been
reported (26). A mechanism for the removal of framework
aluminum by EDTA has been proposed by Kerr (8). It involves
the hydrolysis of Si-O-Al bonds, similar to the scheme in
Figure 1A, followed by formation of a soluble chelate between
cationic, non-framework aluminum and EDTA.

High-temperature reactions with volatile compounds. By
reacting Y zeolites with silicon tetrachloride vapors at high
temperatures, Beyer and Belenykaia (27) were able to prepare
highly dealuminated Y zeolites. Framework substitution of
silicon from $SiCl_4$ for aluminum takes place, while the
resulting $AlCl_3$ is volatilized. More recently, the same
method was applied successfully in the preparation of aluminum-
deficient mordenite, with a degree of dealumination of 26
percent (69). However, mordenite proved much more difficult
to dealuminate with $SiCl_4$ than Y zeolites, possibly due to
lesser site accessibility to $SiCl_4$ molecules. Dealumination
has also been achieved by reacting zeolites with other
volatile halides (108).
       More recently, dealumination was achieved by fluorination
of zeolites at ambient temperature with a dilute fluorine-
in-air stream, followed by high-temperature calcination (102).
The suggested reaction mechanism involves the formation of
different aluminum-fluorine compounds along with zeolites
containing hydroxyl and fluorine nests. During the high-
temperature calcination, it is assumed that silica insertion
occurs, similar to the scheme in Figure 1B.

Combination of thermal and chemical dealumination. This
is a two-step method which was applied in the preparation of
aluminum-deficient mordenite (4,5) and Y zeolites (28,29). In
some instances the two-step treatment was repeated on the same
material, in order to obtain a higher degree of dealumination
(5,28).
       The reaction mechanism during the thermal treatment step
is similar to the one already described for thermal dealumination. High temperatures and steam will enhance the expulsion

of aluminum from the framework.  The chemical treatment in the
two-step process involves the solubilization primarily of
non-framework aluminum generated during the thermal treatment,
although some framework aluminum can also be removed.  The
non-framework aluminum can be in the form of cationic and
neutral species, the amount and composition of which depends
upon the conditions of the preceding thermal treatment.
Examples of aluminum-deficient Y zeolites prepared by this
method are shown in Table III (18).

In the case of mordenite, aluminum was solubilized in the
two-step process primarily with mineral acids (5), while in
the case of Y zeolites it involved acids (28,29), bases (17)
or salts (30,31).  The use of chelating agents has also been
reported (32).

The solubilization of cationic and neutral aluminum
species by acids can be illustrated by the following reactions
(Z = zeolite):

(b)  $Al(OH)_2^+[Z] + 3H^+ \longrightarrow H[Z] + Al^{3+}_{soln.} + 2H_2O$

(c)  $AlO(OH) + 3H^+ \longrightarrow Al^{3+}_{soln.} + 2H_2O$

In reaction (b) an ionic exchange and solubilization of the
aluminum species takes place.  In reaction (c) only the
solubilization of neutral aluminum species takes place.  If
the chemical treatment with acid in the two-step process also
involves the solubilization of framework aluminum, reaction
(a) takes place.

If a base (e.g. NaOH) is used in the chemical treatment,
soluble aluminates are the reaction products.  When salts like
KF or NaCl are used, the dealumination is essentially an ion
exchange process, in which $K^+$ and $Na^+$ ions are substituted
for cationic aluminum species.

When the two-step process is repeated on the same material, the thermal treatment following the chemical dealumination results in further expulsion of aluminum from the
framework into zeolite cages or channels.  The solubilization
of non-framework aluminum during the first chemical treatment
appears to facilitate further framework dealumination during
the subsequent thermal treatment due to the altered steric and
electrostatic parameters in the zeolite channels.  The newly
formed non-framework aluminum species can be readily solubilized by acid treatment.  This cyclic method has allowed the
almost total removal of aluminum from mordenite (5).

TABLE III

ALUMINUM DEFICIENT Y ZEOLITES
PREPARED BY STEAM/ACID TREATMENT (18)[a]

| Zeolite Type | $SiO_2/$ $Al_2O_3$ | $Al_2O_3$ Wt.% | $Al/$ U.C.[b] | % Al Removed | Unit Cell,Å |
|---|---|---|---|---|---|
| USY-A | 4.85 | 25.80 | 56 | (~ 50)[c] | 24.52 |
| USY-B | 4.85 | 25.79 | 56 | (~ 90)[c] | 24.35 |
| DAY | 6.7 | 20.24 | 41.5 | 26 | 24.35 |
| DAY | 11 | 13.30 | 24.9 | 55.5 | 24.34 |
| DAY | 17 | 9.06 | 16.1 | 71.25 | 24.33 |
| DAY | 28 | 5.74 | 9.7 | 82.7 | 24.26 |
| DAY | 45 | 3.61 | 6.1 | 89.1 | 24.26 |
| DAY | 80 | 2.07 | 3.4 | 94 | 24.25 |
| DAY | 142 | 1.18 | 1.9 | 96.6 | 24.25 |
| DAY | 180 | 0.93 | 1.5 | 97.3 | 24.25 |

[a] Starting Zeolite NaY:$Na_{56}(AlO_2)_{56}(SiO_2)_{136} \cdot 250H_2O$.
[b] Based on 136 Si per unit cell.
[c] Al in non-framework positions.

A modification of the above cyclic method has proved more effective in the dealumination of Y zeolites. An almost aluminum-free, Y-type structure was obtained by using a process involving the following steps: a)  calcination, under steam, of a low-soda (about 3 wt.% $Na_2O$), ammonium exchanged Y zeolite; b) further ammonium exchange of the calcined zeolite; c) high-temperature calcination of the zeolite, under steam; d) acid treatment of the zeolite.  Steps a) and c) lead to the formation of ultrastable zeolites USY-A and USY-B, respectively.  Acid treatment of the USY-B zeolite can yield a series of aluminum-deficient Y zeolites with different degrees of dealumination, whose composition depends upon the conditions of the acid treatment.  Under severe reaction conditions (5N HCl, 90°C) an almost aluminum-free Y-type structure can be obtained ("silica-faujasite") (28,29).

## Structural and Physico-Chemical Characteristics of Aluminum-Deficient Zeolites

Aluminum-deficient Y zeolites.  The properties of aluminum-deficient Y zeolites, including ultrastable zeolites, have been reviewed in several papers (9,33-35).  During the last several years, new techniques have been applied to study these materials.  This led to a better understanding of their structural characteristics and of the correlations between structure and properties.  We shall discuss the structure and properties of aluminum-deficient Y zeolites, with the emphasis on more recently published results.

Structural characteristics.  The structure of Y zeolites consists of a negatively charged, three dimensional framework of $SiO_4$ and $AlO_4$ tetrahedra, joined to form an array of truncated octahedra.  These truncated octahedra (β-cages or sodalite cages) are joined at the octahedral faces by hexagonal prisms resulting in tetrahedral stacking.  This type of stacking creates large cavities (α-cages or supercages) with a diameter of ~13Å.  The supercages can be entered through any of four tetrahedrally distributed openings (12-membered rings), each having a diameter of ~8Å.  The supercages, connected through 12-membered rings, form the large-pore system of the zeolite (Figure 2).  The structure comprises also a small-pore system, made up of sodalite cages and the connecting hexagonal prisms.  The 6-membered rings of the sodalite cages have a diameter of 2.2Å.  A typical unit

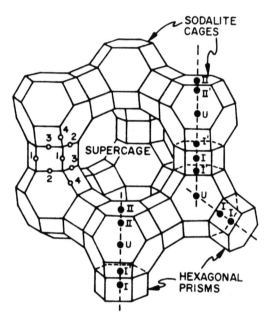

Figure 2.   Y zeolite framework showing oxygen type (0) and
            non-framework (●) locations (101).

cell content of NaY zeolite is $Na_{56}[(AlO_2)_{56}(SiO_2)_{136}]\cdot 250\ H_2O$ (66).

It has already been mentioned that the formation of ultrastable Y zeolites has been related to the expulsion of Al from the framework into the zeolite cages in the presence of steam (8,9), and the filling of framework vacancies by silicon atoms (11,12). This results in a smaller unit cell size and lower ion- exchange capacity (6). It also results in a shift of X-ray diffraction peaks to higher 2θ values. Ultrastable Y zeolites prepared with two calcination steps (USY-B) have a more silicious framework than those prepared with a single calcination step (USY-A). Furthermore, since fewer aluminum atoms are left in the USY-B framework, its unit cell size and ion-exchange capacity are also lower and most of the non-framework aluminum is in neutral form (18).

Several reaction mechanisms have been suggested to explain the framework stabilization of zeolites upon dealumination:

a) Formation of new Si-O-Si bonds at the hydroxyl nests by elimination of water. This mechanism was first suggested by Barrer and can be represented schematically by the following reaction:

b) Silica transport and insertion into vacancies left by dealumination. This mechanism was first suggested by Maher et al. (11) and is represented schematically in Figure 1B. According to this mechanism, the silica required to fill the framework vacancies originates in those parts of the zeolite crystal which collapse during the hydrothermal treatment. The silica freed from the collapsed framework migrates under steam towards the vacancies of the remaining framework and by filling them, increases framework stability.

Recent sorption studies tend to support this interpretation. Based on sorption studies on USY-zeolites, Lohse et al.
(36) concluded that whole sodalite units are destroyed during
the dealumination and rebuilding of the framework in USY-
zeolites. Such sodalite units provide the silica necessary to
fill the large number of framework vacancies left by dealumination. The consumption of whole sodalite cages leads to the
formation of "secondary" pores (vide infra).

c) The so-called T-jump reaction has been suggested by von
Ballmoos (103) to explain framework stabilization upon
dealumination under steam or in acid solution (T stands for
framework Si or Al). The T-jump reaction mechanism assumes
that the vacancies created by framework dealumination gradually migrate from the interior of the zeolite crystal to its
surface, by exchanging places with neighboring T atoms. Thus
the integrity of the bulk crystal is restored (Scheme I).

Lippmaa et al. (37,38) and Thomas et al. (39-41) have
recently investigated the structure of aluminum-deficient Y
zeolites, using magic angle spin $^{29}$Si-NMR and $^{27}$Al-NMR
spectroscopy. These relatively new techniques have proven
useful in structural studies of zeolites and have provided
interesting details concerning the structure of aluminum-
deficient Y zeolites. $^{29}$Si-NMR spectra can provide information
about the composition and Si-Al distribution in the zeolite
framework. $^{27}$Al-NMR spectra provide information about the
distribution of both framework and non-framework Al species in
ultrastable and aluminum-deficient zeolites.

The $^{29}$Si-NMR spectra of zeolites can exhibit up to five
distinct signals, which can be assigned to the five possible
types of $SiO_4$ tetrahedra with different numbers of $AlO_4$
tetrahedra connected to them (37,38). The signals marked
Si(nAl), (n=0-4), correspond to $SiO_4$ tetrahedra connected to
n $AlO_4$ tetrahedra. From the chemical shifts and corresponding
signal intensities, the presence and quantitative distribution
of the five Si(n Al) groupings as well as the $SiO_2/Al_2O_3$
ratio in the zeolite framework can be determined. By using
the $^1$H-NMR cross-polarization (CP) technique, it is also
possible to obtain information with regard to Si-OH groups in
defect centers of the aluminum-deficient framework.

The $^{29}$Si-NMR spectra obtained for NaY, USY-A, USY-B and
DAY-16 (acid-dealuminated USY-B with $SiO_2/Al_2O_3$ = 16)
show the effect of progressive dealumination on the short
range environment of Si atoms in the zeolite framework (106)
(Figure 3). The spectrum of the parent NaY has two strong
signals, corresponding to Si(2 Al) and Si(1 Al), and two weak
signals, corresponding to Si(3 Al) and Si(0 Al) groupings.
The number of Si(4 Al) in this zeolite is negligible. In the

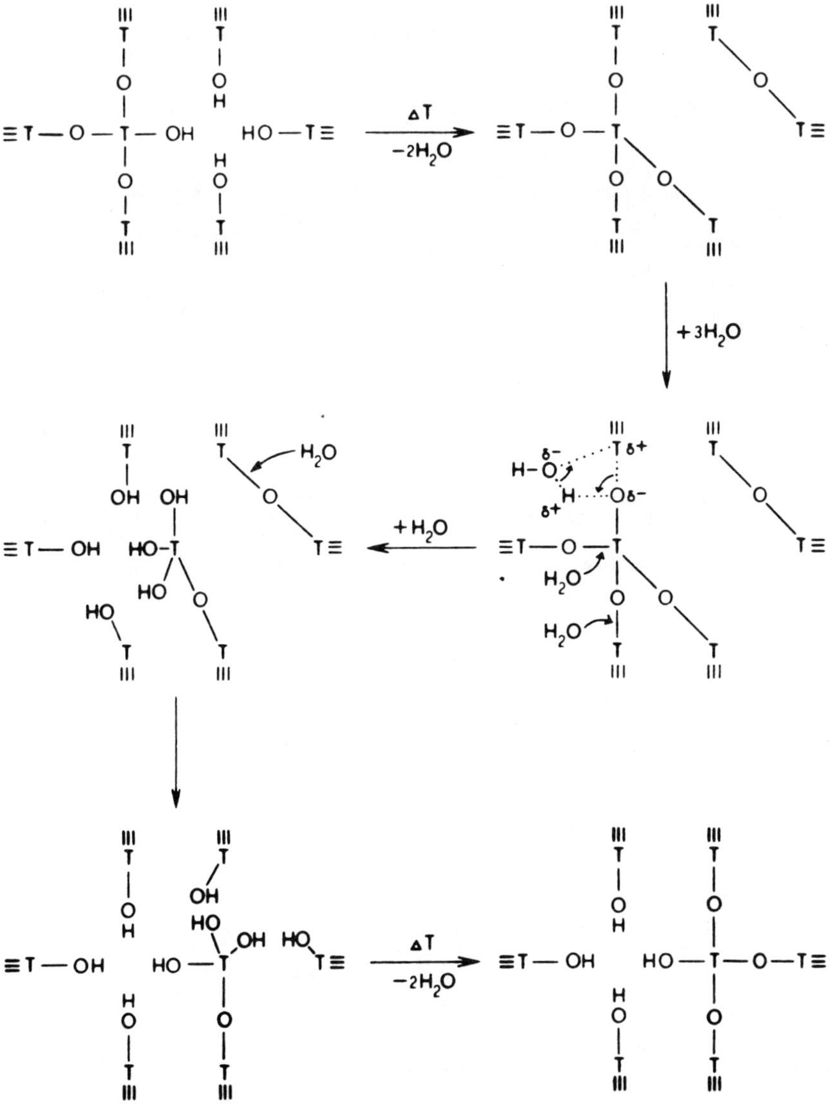

Scheme I. T-jump mechanism (high-temperature steaming) ($\underline{103}$).

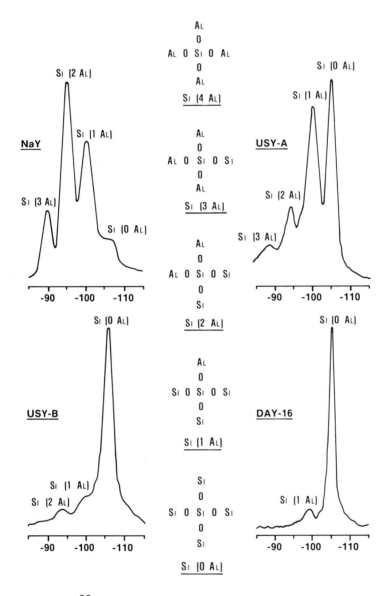

Figure 3.    $^{29}$Si–MASNMR spectra of Y zeolites with different
Al–content in the framework.

spectrum of USY-A, the predominant groupings are Si(0 Al) and
Si(1 Al), characteristic of a framework with high $SiO_2/Al_2O_3$
ratio.   In the spectrum of USY-B, the predominant grouping is
Si(0 Al), while other signals are very weak.   This is indicative of very advanced framework dealumination.   The spectrum
of DAY-16 has a very strong signal corresponding to Si(0 Al)
and a very weak signal for Si(1 Al), characteristic of a
framework with very few Al atoms.

Using $^{29}Si$-NMR spectra to calculate the $SiO_2/Al_2O_3$
ratio in the framework and chemical analysis to determine the
overall $SiO_2/Al_2O_3$ ratio, Lippmaa et al. (38) concluded
that in their sample of USY-A zeolite, 33 Al/u.c. are in the
framework and 24 Al/u.c. are in non-framework positions (42
percent framework dealumination).   In the USY-B zeolite
prepared from the same parent NaY, Lippmaa et al. found 4
Al/u.c. in the framework and 53 Al/u.c. in non-framework
positions (93% framework dealumination).

Lippmaa et al. (38) have also shown that the spectrum of
acid-leached USY-A zeolite has a stronger Si(0 Al) signal and
a weaker Si(2 Al) signal than the corresponding USY-A spectrum.
This was explained by the extraction of some framework Al
atoms in addition to non-framework Al.   In the vacancies
created in the framework by Al extraction, $(SiO)_3$ Si-OH
groups are formed, as was found from the CP spectrum.

Thomas et al. (39,41) recorded the $^{29}Si$-NMR spectrum
of an aluminum-deficient Y zeolite prepared by reacting NaY
zeolite with $SiCl_4$ vapors.   The spectrum showed a single
sharp peak, characteristic of Si(0 Al) groupings, and indicative of an essentially aluminum-free faujasite structure.

Magic angle spin $^{27}Al$-NMR spectroscopy applied to
ultrastable and aluminum-deficient Y zeolites has provided
further information with regard to the location and distribution of Al atoms in these zeolites.   The spectra show two
distinct peaks for tetrahedral and octahedral Al, respectively
(40,41) (Figure 4).   Using this technique, Bosaček et al. (42)
found that the amount of aluminum in non-framework positions
increases from 5 to 50 percent going from 300°C shallow-bed
pretreatment of a 84 percent $NH_4$ exchanged NaY zeolite to
500°C deep-bed treatment.   Zeolites activated under deep-bed
conditions at 300° have the non-framework Al located mainly in
the large cavities.   Those deep-bed activated at 500°C have
the non-framework Al located preferentially in the small
cavities.

In recent years several surface analytical techniques
have been developed and applied to the study of surface

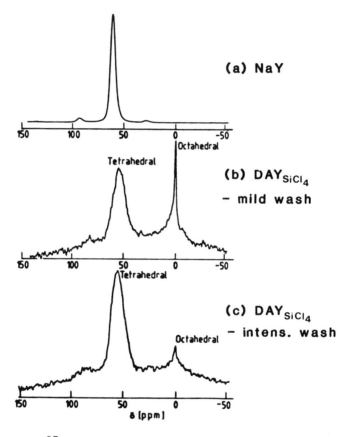

Figure 4.    $^{27}$Al-MASNMR spectra of (a) NaY zeolite; (b)
partially dealuminated NaY with SiCl$_4$, and
moderate washing; (c) same sample as in (b) but
after extensive washing (40).

compositions of zeolites. Dwyer et al. used secondary ion mass spectrometry (SIMS) (43) and fast atom bombardment mass spectroscopy (FABMS) (44) to study the surface composition and depth profiles of different zeolites, in order to asses the compositional uniformity of these materials. In these studies, the surface Si/Al ratios (obtained from SIMS or FABMS data) are compared to bulk Si/Al ratios (obtained from X-ray fluorescence analysis). The depth profile of USY zeolite steamed at 600°C showed dislodging of Al species from the framework and an enrichment of the surface in aluminum (Figure 5). A similar observation was made for steamed H-ZSM-5 and H-mordenite derived from its ammonium form.

Aluminum-deficient Y zeolites prepared by reacting Y zeolites with $SiCl_4$ vapors at 500°C also showed an enrichment of the surface in aluminum (44). The X-ray data show a shift of diffraction peaks to higher $2\theta$ values, consistent with a more silicious framework (27). However, the X-ray pattern also indicates some structural differences between this material and the one prepared by the steam/acid treatment. This can be related to the fact that the Si atoms substituting Al in the framework during the $SiCl_4$ treatment originate outside the zeolite (i.e. from $SiCl_4$), while in the steam/ acid treatment the corresponding silicon atoms originate in other parts of the zeolite crystals. This can also explain the absence of "secondary" pores in the material prepared with $SiCl_4$, as shown by sorption isotherms for different hydrocarbons (27).

Dwyer et al. (43) have also reported that dealumination of Y zeolites by a steam/acid leaching process produces a more uniform composition than dealumination by EDTA. The later method caused a depletion of Al in the outermost surface layer, producing a compositional gradient in the zeolite crystals. The conclusions reached by J. Dwyer in his studies of aluminum-deficient zeolites using the FABMS method are summarized in Table IV.

X-ray studies carried out by Gallezot et al. (46) on a 53 percent EDTA-dealuminated Y zeolite, have shown that the aluminum extraction does not leave any vacancies in the framework after calcination at 400°C in flowing, dry oxygen and nitrogen (46). It was suggested that a local re-crystallization of the framework occurs even in the absence of steam. The silica necessary for the process presumably originates in the destroyed surface layers of the crystallite and diffuses into its interior.

Stability. Ultrastable Y zeolites, prepared by the hydrothermal treatment of ammonium Y zeolites, have considerable thermal and hydrothermal stability (6). The high

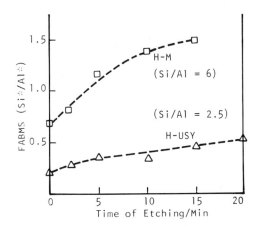

Figure 5.    Depth profiles of zeolites heated under steam: 600°C, $P_{H_2O}$ = 1 atm, 14 hours (44).


TABLE IV

SURFACE COMPOSITION AND DEPTH PROFILE

OF DEALUMINATED ZEOLITES (43,44)

| TREATMENT | DEPTH PROFILE |
| --- | --- |
| Steaming (H–M,H–Y,H–ZSM–5) | Al–Rich Surface |
| Acid (H–M,H–ZSM–5) | Al–Deficient Surface |
| Steaming + Acid (H–M,H–ZSM–5) | Al–Deficient Surface |
| EDTA (HY) | Al–Deficient Surface |
| SiCl$_4$ (HY) | Al–Rich Surface |
| Steaming + Acid (H–Y) | Near–Uniform Al Distribution |

stability is due to the expulsion of Al from the framework
into zeolite cages and the filling of framework vacancies with
Si atoms. USY-B zeolites are more stable than USY-A zeolites,
due to the more silicious framework of USY-B.

Aluminum-deficient Y zeolites prepared by partial removal
of aluminum with a chelating agent (e.g. EDTA) also show
improved thermal and hydrothermal stability compared to the
parent zeolite. The optimum stability was found in the range
of 25 to 50 percent of framework Al extraction (8). However,
the maximum degree of dealumination is also affected by the
$SiO_2/Al_2O_3$ ratio in the parent zeolite: a higher ratio
appears to allow more advanced dealumination without loss of
crystallinity (8,25,45). Above 50 or 60 percent dealumination, significant loss of crystallinity was observed. Calcination
of the aluminum-deficient zeolite resulted in a material with
a smaller unit cell size and lower ion-exchange capacity
compared to the parent zeolite.

Aluminum-deficient Y zeolites prepared by acid-leaching
of USY-B zeolites show very high thermal, hydrothermal and
chemical stability (28). For example, a material prepared by
this method, with a $SiO_2/Al_2O_3$ ratio of 192, maintained
very good crystallinity even after calcination at 1150°C. The
thermal stability of such materials surpasses that of USY
zeolites. Their stability towards acids is also high: even
after boiling in hydrochloric acid for several hours these
materials maintain good crystallinity. Aluminum- deficient Y
zeolites prepared by acid-leaching of USY-A zeolites are
generally less stable than those prepared from USY-B zeolites.

The unusually high stability of DAY zeolites prepared
from USY-B and having $SiO_2/Al_2O_3$ ratios over 100 indicates
that the non-framework aluminum species present in USY-B play
no role in enhancing the stability of this zeolite. It is the
highly silicious framework, in which most of the aluminum has
been replaced by silicon atoms, that is responsible for the
high stability of USY-B zeolites and of corresponding DAY
zeolites. In zeolites with a lesser degree of framework
dealumination (i.e. in USY-A), the non-framework aluminum
species appear to play a role in the stabilization of the
zeolites, since their removal results in materials of lesser
stability (28).

DAY zeolites prepared from Y zeolites and $SiCl_4$ also
show high stability, a smaller unit cell and resistance to
mineral acids (27).

Sorption. Lohse et al. (36,47,48) have studied the
sorption properties of acid-dealuminated Y zeolites that
contain about 99% $SiO_2$, and compared them with those of the

parent NaY and USY-B zeolites. The nitrogen adsorption
isotherms, measured at -195°C, show that while adsorption on
NaY reaches rapidly saturation at low pressure ($p/p_o \simeq$
0.05), the adsorption on USY-B and DAY increases with increasing $p/p_o$ values. For $p/p_o < 0.5$ the adsorption on DAY is
lower and for $p/p_o > 0.5$ it is higher than on NaY zeolite.
NaY yields a completely reversible type I isotherm, characteristic of micropore filling common in many zeolites. However,
USY-B and DAY yield an isotherm close to type IV. Similar
differences in adsorption isotherms were observed for n-hexane,
cyclohexane, n-pentane and benzene. Furthermore, many of the
isotherms measured on DAY zeolites showed hysteresis loops
(Figure 6).

The differences in the shape of isotherms have been
attributed to the formation of secondary pores during framework dealumination. The hysteresis loops observed were
attributed to capillary condensation in the secondary pores.
The adsorption capacity of DAY is greater than that of USY-B
zeolites, due to the removal of Al species from the zeolite
pores.

Pore size distribution data obtained from adsorption
isotherms and from mercury porosimetric measurements show that
in addition to the micropores characteristic of the parent
zeolite, the DAY zeolites contain secondary pores with radii
of 1.5nm (supermicropores) and 10nm (mesopores) (36,47). The
secondary pores in USY-B have a radius of 5nm. It was also
shown that the micropore volume of DAY amounts to about 75
percent of that of NaY (36). Due to dealumination and the
formation of secondary pores, the total pore volume of DAY is
considerably larger than that of NaY zeolite (0.56 vs. 0.29
cc/g).

The micropore volumes of DAY and USY-B are roughly the
same (about 0.21 cc/g), while the secondary pore volume of DAY
is considerably higher (0.34 vs. 0.12cc/g). This led Lohse et
al. to conclude that the non-framework Al species are located
in the secondary pores. The partial blocking of the mesopores
by Al species in USY-B is considered the reason for their
smaller radius (5nm), while the absence of supermicropores in
USY-B is explained by the total blocking of such pores.

Yoshida et al. (45) have investigated the changes in pore
geometry in EDTA-dealuminated Y zeolites. From nitrogen
adsorption studies they concluded that the pore sizes are
distributed over a broader range with the progress of dealumination. In more advanced dealuminated zeolites, pores of over
100Å were indentified. From differential heats of immersion
measurements and pore size distribution data, Yoshida et al.
concluded that the larger pores are formed by the partial
destruction of α cages in the framework. However, in a nearly

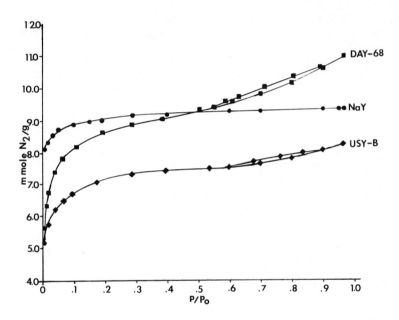

Figure 6.    Adsorption isotherms for nitrogen on NaY, USY-B and DAY-68 zeolites at 195.6°C (18).

amorphous sample with 75 percent dealumination, over 66 percent of the α cages were still present.

Beyer and Belenykaia (27) have investigated the sorption
properties of DAY zeolites prepared from Y zeolite and SiCl$_4$
vapors. They reported a very low adsorption capacity for
water and ammonia, similar to that of the almost aluminum-free
silicalite (49). The low adsorption capacity for water is
indicative of a hydrophobic zeolite surface. The adsorption
isotherms for n-butane, benzene and n-hexane obtained on the
aluminum-deficient zeolite have a shape similar to those
obtained on NaY zeolite and are characteristic for micropore
structures. They show the absence of secondary pores in this
DAY zeolite.

DAY zeolites obtained by fluorination of Y zeolites are
also hydrophobic (102). This is due to the low concentration
of OH groups in these materials as shown by their I.R.
spectra.

Infrared spectra. The infrared spectra of ultrastable Y
zeolites have been investigated by Ward (10), Jacobs and
Uytterhoeven (50,53), Scherzer and Bass (51) and Peri (52).
It was reported that hydrothermal treatment of ammonium Y
zeolites generates strong absorption bands around 3700 cm$^{-1}$
and 3600 cm$^{-1}$ in the OH stretching region of the infrared
spectrum, while the bands present in the original, vacuum-
calcined ammonium Y zeolite (at about 3640 and 3540 cm$^{-1}$)
are greatly reduced in intensity or eliminated. The intensity
of the band of 3750 cm$^{-1}$, attributed to silanol groups or
amorphous silica impurities, is increased by dealumination.
Since the new bands did not disappear after solubilization of
non-framework Al with NaOH, it was assumed that they are due
to OH groups attached to the framework and generated during
the hydrothermal dealumination of the framework (50,51). The
OH groups responsible for the bands at 3700 and 3600cm$^{-1}$
are non-acidic towards ammonia (50,53) and pyridine (50,51),
but the 3600 cm$^{-1}$ band shows weak acidity towards NaOH
solutions (53). The OH stretching region of the infrared
spectrum of USY-B zeolites indicates a high degree of
dehydroxylation, except for silanol groups (51).

The mid-infrared spectra of USY-zeolites show a shift to
higher frequencies of bands associated with the framework
tetrahedra, due to the decrease in aluminum content of the
framework (51,52). A sharpening of the bands in the spectra
of USY-B type zeolites has been observed and attributed to an
increase in the degree of order within the framework (51).

The infrared spectra of EDTA-dealuminated Y zeolites show
bands in the OH stretching region similar to those encountered
in HY zeolites: at about 3750, 3640 and 3540 cm$^{-1}$ (50,54).
However, the OH groups responsible for the 3640 and 3540
cm$^{-1}$ bands in the spectra of the aluminum-deficient zeolites

interact with pyridine, while in the case of HY zeolite only
the band at 3640cm$^{-1}$ is affected by pyridine (54,55). This
suggests that dealumination makes the OH groups responsible
for the 3540 cm$^{-1}$ band more accessible to pyridiue molecules.
Furthermore, Beaumont et al. (54) reported that after
adsorption and evacuation of pyridine at temperatures up to
350°C, a new band was observed at 3600cm$^{-1}$ and a weak band
at 3670 cm$^{-1}$ was sharpened. The mid-infrared spectra of
EDTA-dealuminated Y zeolites showed increases in band frequencies similar to those observed in the spectra of USY
zeolites (55-57).

Bosáček et al. (29) have investigated the infrared
spectra of advanced dealuminated Y zeolites prepared by
acid-leaching USY zeolites. In the OH stretching region of
the spectra they found the same bands that are known to exist
in HY zeolites, except for different band intensities: a
strong band at 3735 cm$^{-1}$ and weak bands at 3630 and 3540
cm$^{-1}$ (Figure 7). The last two bands, which are characteristic of faujasite type zeolites, are likely to be generated by
OH groups located in the vicinity of the remaining Al atoms in
the framework. There is also a broad band in the 3000 to
3700 cm$^{-1}$ region, characteristic of OH groups interacting
through hydrogen bonds. These are most likely groups located
in the (OH)$_4$ "nests" generated by acid-leaching of Al from
the framework. An increase in outgasing temperature of the
DAY zeolite results in a gradual disappearance of the broad
band, indicating a relatively poor stability of the "nests" of
OH groups. $^1$H-NMR measurements support this interpretation.

The mid-infrared spectra in the skeletal region of
different acid-dealuminated USY zeolites have also been
investigated (18,29) (Figure 8). The sharp absorption bands
in the spectra of acid-dealuminated zeolites are indicative of
a highly crystalline structure. Scherzer and Humphries (18)
have shown that the T-O (T=Si,Al) asymmetric stretching
frequency in the 1000-1100 cm$^{-1}$ region increases in the
following order: NaY< USY-A <USY-B ≃ DAY-17<DAY-180. The
increase in frequency reflects the corresponding decrease in
Al content in the framework. USY-B and DAY-17 have absorption
peaks essentially at the same frequency, suggesting that mild
acid-leaching of USY-B removes primarily non-framework Al. In
highly advanced dealuminated zeolites (i.e. DAY-180), Al is
removed from both non-framework and framework positions.

The mid-infrared spectra of aluminum-deficient Y zeolites
prepared by the reaction of Y zeolites with SiCl$_4$ are similar
to those prepared by steam/acid treatment (27). As in the

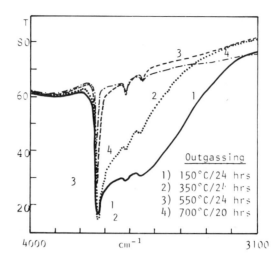

Figure 7.    Infrared spectra of DAY zeolites in the OH
             stretching region after treatment in vacuum at (1)
             150°C for 24 hrs; (2) 350°C for 24 hrs; (3) 550°C
             for 6 hrs; and (4) 700°C for 20 hrs (29).

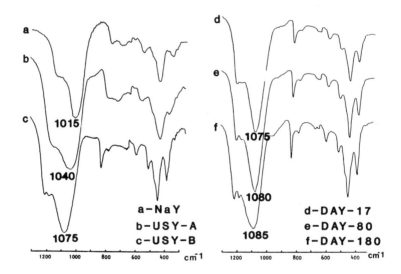

Figure 8.    Infrared spectra in the framework region for
             zeolites: (a) NaY; (b) USY-A; (c) USY-B; (d)
             DAY-17; (e) DAY-80 (f) DAY-180 (18).

previous case, the absorption bands are considerably sharper
than those of NaY zeolite, indicating a higher "degree of
ordering" in the framework.

Acid properties. The acid properties of zeolites,
including those of aluminum-deficient zeolites, have been
described in several reviews (e.g. 33-35). The methods used
to study the acidity of aluminum-deficient Y zeolites include
infrared spectroscopy (primarily pyridine and ammonia sorption
studies), n-butylamine titrations in the presence of Hammett
or arylmethanol indicators, and to a lesser extent
potentiometric titrations and calorimetric measurements.

Pyridine sorption studies have shown the presence of both
Bronsted and Lewis acid sites in USY zeolites, although to a
lesser extent than in the corresponding HY zeolite (51,53).
Acidity is maintained even after strong dehydroxylation of
USY-B at 820°C. Rehydration of the calcined material did not
regenerate significantly Bronsted acid sites, due to irreversible changes in the zeolite framework (51).

Using the n-buthylamine titration method, Scherzer and
Humphries (18) have shown that USY-B zeolites have considerably less acidity than USY-A zeolites. This is due to the
more advanced thermal dealumination of USY-B, which reduces
both Bronsted and Lewis type acidity.

Pyridine sorption studies on EDTA-dealuminated Y zeolites
at various temperatures (54,58), as well as measurements of
differential heats of adsorption of ammonia on aluminum-
deficient Y zeolites (57,59) have led to the conclusion that
aluminum-deficient Y zeolites have stronger acid sites than
the parent zeolite.

Beaumont and Barthomeuf (23-25) have extensively investigated the acidity and acid strength distribution in EDTA-
dealuminated Y zeolites and in USY zeolites. Based on data
obtained from titrations of dealuminated NaHY zeolites with
n-butylamine in the presence of Hammett and arylmethanol
indicators, they suggested the presence of two types of Al
atoms, associated with weak and strong acid sites, respectively. Starting from zeolites with 56 Al/u.c., progressive
dealumination initially removed Al atoms associated with weak
acid sites ($3 \times 10^{-4} H_2SO_4$). Aluminum-deficient zeolites
with about 35 Al/u.c. or less had primarily strong acid sites
(>88% $H_2SO_4$). The number of these sites decreased with
further dealumination. The number of strongly bound Al atoms,
35, associated with strong acid sites, is in reasonable
agreement with the results obtained by Kerr (7), Jacobs and
Uytterhoeven (50) and Maher et al. (11), who showed that
between 30 and 40 Al/u.c. remain in framework positions in
ultrastable Y zeolites prepared with a single thermal

treatment. Beaumont and Barthomeuf (24) suggested that each
framework aluminum atom in USY zeolites is associated with a
strong acid site. Dempsey (60) provided an electrostatic
interpretation and Kerr (61) a chemical interpretation of
Beaumont and Barthomeuf's findings.

Scherzer and Humphries (18) have investigated the acid
properties of advanced dealuminated Y zeolites prepared by
steam/acid leaching. The acidity was measured on a series of
DAY zeolites with $SiO_2/Al_2O_3$ ratios between 5 and 180,
using the n-butylamine titration method in non-aqueous solution. It was concluded that total acidity ($H_o \lesssim 3.3$) goes
through a maximum when about 25-30 percent of aluminum is
removed, while strong acidity ($H_o \lesssim -5.6$) reaches a maximum
when 80-90 percent of aluminum is removed. Mild acid-dealumination is essentially an ionic exchange of cationic Al for
hydronium ions, leading to an increase of both total and
strong acidity. Stronger dealumination removes neutral Al as
well as some of the framework Al, thus reducing the total
acidity of the zeolite. The stronger acidity observed in more
advanced dealuminated zeolites is explained by the lower
density of acid sites and the easier accessibility of such
sites to the organic base following removal of extreneous
materials from the zeolite pores. The almost aluminum-free
material ("silica-faujasite") has practically no acidity.
Potentiometric titration in aqueous solution showed the same
trend in the change of total acidity with progressive dealumination. The initial increase and subsequent decrease in
total acidity with progressive dealumination is accompanied by
a corresponding increase and subsequent decrease in ion
exchange capacity for silver ions (18).

It was also shown than DAY zeolites prepared by different
methods have different acid strength distributions (18). This
was related to structural differences between these zeolites.

Catalytic properties. Relatively few studies have been
published with regard to catalytic applications of aluminum-
deficient Y zeolites. Several studies of catalytic cracking
over such zeolites have been reported. Topchieva and T'huoang
(58) reported that progressive dealumination of Y zeolites
with EDTA results first in an increase and then a decrease in
catalytic activity. The highest activity for cumene cracking
was observed for dealuminated samples with a $SiO_2/Al_2O_3$ ratio
of 8, corresponding to about 33 Al/u.c. Tsutsumi et al (57,59)
reported a similar maximum in activity with progressive
dealumination.

Barthomeuf and Beaumont (25) have found that the activity
for iso-octane cracking remains unchanged for NaHY zeolites in
which up to 33 percent of Al has been extracted with chelating
agents. This corresponded to a dealumination from 56 to 37
Al/u.c. Further dealumination resulted in a decrease in

activity.  This was in agreement with the data obtained by the
same authors from acidity measurements, which showed that the
first 30 percent of Al atoms extracted are associated with
weak acid sites.  The authors concluded that dealumination
beyond 33 percent results in the destruction of strong acid
sites.  This, in turn, caused a decline in catalytic activity,
since no loss of crystallinity was observed.

Several studies have been published describing the
catalytic properties of USY and acid-dealuminated Y zeolites.
McDaniel and Maher (6) have shown that USY zeolites have
catalytic cracking activity, using the cracking of tetralin as
an example.  Jacobs et al. (62) have investigated the cumene
cracking and toluene disproportionation over ultrastable and
EDTA-dealuminated Y zeolites.  They concluded that only a
fraction of the acidic hydroxyl groups are catalytically
active and that the sites on which cumene cracking occurs are
different from those on which toluene disproportionation
occurs.  The latter reaction appears to require stronger acid
sites.  The catalytic activity also appeared to be affected by
the formation of Lewis acid sites, the degree of dealumination
and the non-framework Al species.

Scherzer and Ritter (63) have compared the catalytic
cracking of gas oil over catalysts containing rare earth
exchanged Y zeolites (REY), ultrastable Y zeolites (USY) and
rare earth exchanged USY zeolites  (RE-USY).  It was shown
that the activity decreased in the order REY > RE-USY > USY.
Although the USY zeolite was less active, it showed good
gasoline selectivity, gave lower coke yields and higher $C_3$
and $C_4$ olefine yields, as compared to the rare earth exchanged
zeolites.  Furthermore, the gasoline fraction obtained with
the USY zeolite had a higher content in aromatic and cyclic
allylic hydrocarbons (Table V).  Rare earth exchange of the
USY zeolite increased its cracking activity, while its selectivity was shifted in the direction of REY's selectivity.

The catalytic behaviour of steam/acid leached Y zeolites
appears to be similar to that of USY zeolites.  Bremer et al.
(64) have reported that at equal conversions, cracking of gas
oil over a DAY zeolite results in lower coke yields as compared
to those obtained over REY zeolites.

The differences in selectively between REY and USY
zeolites have been related to the lower total acidity (lower
site density) in USY zeolites (63,65).  A similar interpretation can be applied to explain the selectivity characteristics
observed for DAY zeolites (64).

A high selectivity for n-hexane isomerisation over DAY
zeolites containing 0.5 percent Pt has also been reported
(64).

TABLE V

GAS OIL CRACKING OVER CATALYSTS

WITH USY AND RE,HY ZEOLITES (63)

(PILOT PLANT DATA)

Test Conditions:   40 WHSV,493°C,4 Cat/Oil,West Texas Gas Oil
Catalyst:   Zeolite in Semisynthetic Matrix
Catalyst Deactivation:   825°C,12 Hours,20% Steam, Atm. Pres.

| Zeolite Type | USY | RE,HY |
|---|---|---|
| Zeolite, Wt.% | 25 | 15 |
| Conversion, Vol.% | 70 | 75 |
| $H_2$, Wt.% | 0.04 | 0.05 |
| $C_1 + C_2$ Wt.% | 1.4 | 2.1 |
| $C_3$ Olefin, Vol.% | 6.8 | 6.5 |
| $C_4$ Olefin, Vol.% | 4.9 | 4.8 |
| $I-C_4$, Vol.% | 4.2 | 6.4 |
| $C_5^+$ Gasoline, Vol.% | 61.5 | 62.0 |
| Gravity, °API | 56.0 | 55.3 |
| Aniline Point, °F | 75 | 83 |
| Bromine No. | 81 | 46 |
| Coke, Wt.% | 3.8 | 6.2 |

The bi-functional conversion of 2,2,4-trimethylpentane over Pt/DAY has been recently reported by Jacobs et al. (104). It was compared to the corresponding conversion over Pt/H-ZSM-5 and Pt/H-ZSM-11. All three zeolites had the same chemical composition. The authors found that 2,2,4-trimethylpentane underwent β-scission over Pt/DAY, while the formation of feed isomers was favored over the other two catalysts. The differences in reaction products were related to differences in the pore geometry of the zeolites. A similar study was carried out with n-decane.

A fairly large number of patents has been issued describing the application of aluminum-deficient Y zeolites in different areas of catalysis. Ultrastable Y zeolites have been used in the preparation of catalysts applied in hydrocarbon cracking, e.g. (94,95); hydrocracking, e.g. (96,97); hydrotreating, e.g. (98) and disproportionation, e.g. (99).

Catalysts containing aluminum-deficient Y zeolites prepared by Al extraction with chelating agents have been used for cracking and hydrocracking hydrocarbons, e.g. (100). Similar applications have been described for catalysts containing aluminum-deficient Y zeolites prepared by the steam/EDTA-leaching method, e.g. (32).

Correlations between preparation method and properties. A review of the physico-chemical characteristics of aluminum-deficient Y zeolites has shown that certain characteristics are common to all DAY zeolites, regardless of preparation method, while other characteristics are strongly affected by the preparation method used.

Characteristics common to most DAY zeolites, regardless of preparation method, are as follows: a) Structural Si(0 Al) groups are predominant. b) A composition gradient is present in most DAY zeolite crystals. c) Unit cell sizes are smaller compared to the parent zeolite. d) The X-ray diffraction peaks are shifted to higher 2θ values. e) Thermal and hydrothermal stability increase (provided the zeolite maintains good crystallinity during dealumination). f) The frequency of some I.R. lattice vibrations is shifted to higher values. g) The intensity of the acid OH-brands in the I.R. spectra decreases. h) The total acidity of the zeolite decreases. i) The acid strength increases. j) The catalytic site density decreases.

Other characteristics of DAY zeolites are strongly affected by the preparation method used. The effect of the preparation method on certain characteristics of DAY zeolites is shown in Table VI.

TABLE VI

CORRELATION BETWEEN PREPARATION METHOD

AND PROPERTIES OF DAY

| Properties | | Preparation Method | | |
| --- | --- | --- | --- | --- |
| | | EDTA | Steam/ Acid | SiCl$_4$ |
| Crystallinity | | | | |
| Dealuminated | 60% | Good | Good | Good |
| Dealuminated | 60% | Poor | Good | Good |
| Composition Gradient | | Al- Deficient Surface | Al- Deficient Surface | Al- Rich Surface |
| Type of Residual Al* | | Al (T) | Al(T)+Al(O) | Al(T)+Al(O) |
| Mineral Acid Resistance | | No | Yes | Yes |
| Pore System | | Micro + Secondary Pores | Micro + Secondary Pores | Micropores |
| Adsorption Properties | | Type IV Isotherm | Type IV Isotherm | Type I Isotherm |
| H$_2$O Affinity | | Hydro- philic | Hydro- philic | Hydro- phobic |

*Al(T) = Tetrahedral Al; Al(O) = Octahedral Al.

## Aluminum-deficient mordenite

Structural characteristics. Both natural and synthetic
mordenite have an orthorhombic structure that consists of
parallel, 12-membered ring channels in the c-direction, having
an eliptical cross-section with dimensions of 6.7 x 7.0 Å
(Figure 9). Smaller 8-membered ring channels with dimensions
of 2.9 x 5.7 Å, which are perpendicular to the main channels,
are too small to allow the movement of molecules from one main
channel to another. Mordenite has been synthesized in a "large
-port" and "small-port" form that have different sorption
properties. A typical unit cell content is $Na_8[(AlO_2)_8(SiO_2)_{40}]$.
$24\ H_2O$.

Using the X-ray diffraction method, Olsson and Rollmann
(67) have measured the unit cell dimensions of acid-dealuminated $NH_4$-mordenite zeolites with $SiO_2/Al_2O_3$ ratios varying
from $10^4$ to 100. They found that lattice contractions which
occur on aluminum removal are strongly anysotropic and non
linear, regardless of the preparation method used. Thus, the
removal of the initial 25 percent of aluminum (down to 6
Al/u.c.) leaves the lattice parameter "a" essentially unaffected, but strongly reduces the "b" parameter and to a lesser
extent the "c" parameter. Intermediate levels of dealumination
change all three lattice parameters about equally. This was
interpreted in terms of an Al atom distribution in three basic
types of siting in the mordenite structure. Al atoms in
different sites show different susceptibility to removal by
acid. The authors also assume that the structural defects
left by Al removal are replaced by new Si-O-Si bonds, as
suggested by Barrer and Klinovski (68). This explains why
most of the Al atoms can be removed from the mordenite framework with only minor changes in the X-ray diffraction pattern.

More recently, Thomas et al. (69) investigated the
structural characteristics of aluminum-deficient mordenite,
prepared by reacting large-port Na-mordenite with $SiCl_4$
vapors at 700°C. The authors used high-resolution magic angle
spin $^{29}$Si-NMR and $^{27}$Al-NMR techniques to prove that part
of the Al atoms in the framework are replaced by Si atoms
during the preparation process. The $^{29}$Si-NMR spectrum of
the aluminum-deficient mordenite was less well-resolved than
the corresponding spectrum of aluminum-deficient Y zeolites.
This was explained by the presence of four kinds of crystallo-
graphically nonequivalent sites in mordenite. The $^{29}$Si-NMR
spectrum of the parent Na-mordenite shows two relatively
strong signals, corresponding to Si(0 Al) and Si(1 Al)
groupings, and a shoulder attributed to Si(2 Al) groupings.

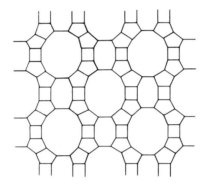

Figure 9.    Structure of mordenite framework as viewed along
             c-axis (66).

The spectrum of the aluminum-deficient mordenite shows no
signal for Si(2 Al) and a considerable decrease in intensity
of the Si(1 Al) signal, due to framework dealumination.


   Stability.  Being a fairly high-silicious zeolite,
mordenite generally has high thermal stability.  It was
reported (77) that progressive acid dealumination results in
an increase in thermal stability, followed by a decrease.
Maximum stability was reached for a $SiO_2/Al_2O_3$ ratio of
about 19.


   Sorption.  The sorption properties of aluminum-deficient
mordenite are strongly affected by the dealumination procedure
used and by the degree of dealumination.  Materials prepared
by procedures that do not involve high temperature treatments
show a relatively high sorption capacity for water (15,70),
due to the presence of silanol groups, which are hydrophilic
centers.  However, aluminum-deficient mordenite zeolites
prepared by methods requiring heat treatment show a lower
sorption capacity for water due to fewer silanol groups.  This
was shown by Chen (71), who studied the sorption properties of
aluminum-deficient mordenite prepared by the two-step method.
He found that these zeolites were hydrophobic and attributed
this to the absence of silanol groups and the formation of
Si-O-Si bonds in the vacancies generated by dealumination.
The formation of such bonds during the heat treatment of
dealuminated mordenite has also been suggested by Rubinshtein
et al. (72-74), in some instances without the intermediate
formation of SiOH groups.  The hydrophobic nature of the
zeolite also increases with progressive dealumination.  Chen
(71) has shown that aluminum-deficient mordenite zeolites with
$SiO_2/Al_2O_3$ ratios over 80 absorb little or no water at
low pressure.  These highly silicious zeolites are truly
hydrophobic and in this respect are similar to highly silicious
zeolites prepared by direct synthesis (e.g. ZSM-5) (75).
   Chen (71) has also reported that aluminum-deficient
mordenite adsorbs cyclohexane.  With progressive dealumination
the sorption capacity first increases, goes through a maximum
at $SiO_2/Al_2O_3$ ratios between 10 and 20, and then decreases
to a constant value above a $SiO_2/Al_2O_3$ ratio of 40.  The
initial increase is attributed to the removal of extraneous
materials from the mordenite channels -as was initially
assumed by Eberly et al. (76) -while the decrease at higher
$SiO_2/Al_2O_3$ ratios is due to unit cell shrinking. A similar
maximum in adsorption capacity for cumene was found by
Piguzova et al. (77) in aluminum-deficient mordenite with
$SiO_2/Al_2O_3$ ratio of 19.  Sand et al. (4) reported that
with progressive dealumination, cumene adsorption increases at
low pressures, but decreases at high pressures.

The variation in adsorption capacity for mordenite zeolites with different $SiO_2/Al_2O_3$ ratios has been related by Eberly et al. to changes in diffusivity (76,78). Moderate acid-dealumination decreases the resistance to diffusivity either by increasing the effective micropore diameter of the zeolite or by removing amorphous material from the channels. This explains the diffusion of large molecules like n-decane and decaline into the zeolite micropores (Figure 10) and the adsorption of 1,3,5-triethylbenzene with a critical molecular diameter of 8.6 Å. However, the shrinking of the unit cell with progressive dealumination can have the opposite effect. This is especially true for mordenite zeolites with high $SiO_2/Al_2O_3$ ratios and those exposed to high temperature treatment.

Infrared spectra. The infrared spectra of aluminum-deficient mordenite have been investigated by several authors (72-74, 78,79). Eberly et al. (78) have found that the spectrum of aluminum-deficient H-mordenite shows only a band at 3740 cm$^{-1}$, while regular H-mordenite has an infrared spectrum with bands at about 3740 and 3590 cm$^{-1}$. In the spectrum of regular H-mordenite, Karge (80) observed OH bands at 3610 cm$^{-1}$ generated by acidic OH groups, and at 3650 cm$^{-1}$ due to non-acidic OH groups. Upon dealumination, Kiovsky et al. (79) reported that the band at 3650 cm$^{-1}$ increased in intensity and attributed this to the formation of additional silanol groups (hydroxyl nests). The findings of other authors (78), who did not observe such an increase in intensity of the absorption band, was explained by a higher pretreatment temperature of the sample. In the framework region of the infrared spectrum (400 to 1200 cm$^{-1}$) Eberly et al. (78) have found that dealumination causes a shift of absorption bands to higher wave numbers, reflecting the higher silica content of the framework (Figure 11). A similar shift in band frequencies was observed in the spectra of calcined $NH_4$-mordenite, suggesting framework dealumination by a "deep bed" effect (81).

Acid properties. It was reported that the acidity of aluminum-deficient mordenite, measured by ammonia chemisorption, decreases linearly with aluminum content (78,82,83). Kühl (84) reported a decrease in the number of acid sites in calcined or steamed H-mordenite, due to thermal dealumination of the framework. Barthomeuf et al. (85), using data obtained from ammonia and pyridine sorption, concluded that the total acidity and the number and strength of Lewis sites decreases upon acid-dealumination of H-mordenite to a $SiO_2/Al_2O_3$ ratio of 18, while the Bronsted acidity remained unchanged.

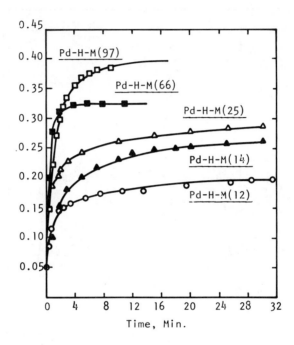

Figure 10.    Adsorption of decalin at 93°C and 0.12 mm pressure (78).

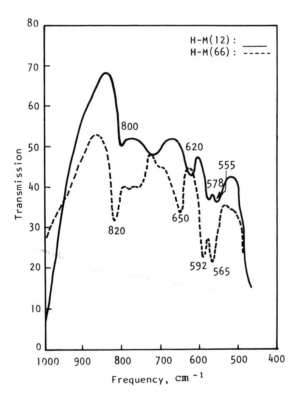

Figure 11.   Mid-infrared spectra of HM (12) and HM (66),  (78).

Dry air pretreatment of the aluminum-deficient zeolite does
not change significantly the acidity, while wet air pretreatment decreases the total acidity by a factor of 2-2.5.
The decrease in acidity is due to the selective elimination of
Bronsted acid sited characterized by the band at 3600 $cm^{-1}$
in the infrared spectrum.  The authors claim that at the same
time new, strongly acidic sites are formed under these conditions.

Using calorimetric and n-butylamine titrations, Ghosh
and Curthoys (105) have reported that aluminum-deficient
mordenite exhibits strong acid sites.  However, the location
of these sites is uncertain.

Catalytic properties.  Dealumination of mordenite causes
changes in the catalytic properties of the zeolite.  Topchieva
et al. (86) investigated cumene cracking over acid-leached
H-mordenite with $SiO_2/Al_2O_3$ ratios of 10, 13 and 18, and
concluded that catalytic activity increases with increasing
$SiO_2/Al_2O_3$ ratio.  Using hexane cracking to characterize
the catalytic cracking activity of aluminum-deficient mordenite
with $SiO_2/Al_2O_3$ ratios between 15 and 18, Weller and
Brauer (87) also found an increase in cracking activity for
the aluminum-deficient zeolites.

Using the steam/acid leaching method, Sand et al. (4)
prepared a series of aluminum-deficient mordenite zeolites
with $SiO_2/Al_2O_3$ ratios varying from 12 to 1200, and
investigated their adsorption, cracking, hydrocracking and
isomerization properties.  Although the X-ray pattern suggested
no major disruption of the crystal lattice, adsorptive and
catalytic properties were significantly altered, especially
for materials with high $SiO_2/Al_2O_3$ ratios.  Both cracking
and hydrocracking activity for cumene decreased with decreasing
Al content at high $SiO_2/Al_2O_3$ ratios.  Initial dealumination
of H-mordenite resulted in increased activity for 1-butene
isomerization, with a maximum at $SiO_2/Al_2O_3$ ratio of
about 30, followed by a decline in activity upon further
dealumination.  The authors assume that the disappearance of
acid sites with advanced dealumination accounts for loss of
activity in both cracking and isomerization.  On the other
hand, Eberly et al. (3,89) reported that the cracking activity
for cumene and the hydrocracking activity for large molecules
such as decalin and n-decane increases with acid-dealumination
when the $SiO_2/Al_2O_3$ ratio is increased from 12 to 64.
This increase in activity was attributed to a lower diffusion
resistance of the aluminum-deficient zeolites, which resulted
from the removal of amorphous material from the zeolite
channels.  However, the hydroisomerization of n-pentane

decreased when the $SiO_2/Al_2O_3$ ratio of the zeolite was
increased from 12 to 96 (78).

Weiss et al. (88) also found a decrease in catalytic
activity for cumene cracking over aluminum-deficient mordenite
zeolites with very high $SiO_2/Al_2O_3$ ratios. However, the
rate of deactivation was significantly reduced over the highly
dealuminated zeolites (with 0.1 percent $Al_2O_3$). It was
assumed that due to the lower density of active sites on
highly aluminum-deficient mordenite, the condensation of
high-molecular weight species (capable of blocking micropores)
will be catalysed at a much lower rate, thus reducing the
fouling rate of the catalyst. Furthermore, the open pore
structure of aluminum-deficient mordenite will facilitate the
removal of condensation products once formed.

Karge and Ladebeck (90) studied the alkylation of benzene
with olefins over aluminum-deficient, beryllium exchanged
mordenite and found a considerable extension of the lifetime
of the catalyst, as compared to H-mordenite. The authors were
able to carry out quite efficiently the alkylation reaction as
well as the transalkylation of ethylbenzene at relatively low
temperatures.

Gnep et al. (91) investigated toluene disproportionation
and coke formation over chemically dealuminated mordenite.
They found that aluminum-deficient mordenite ($SiO_2/Al_2O_3$
ratio of 18), pre-calcined in a flow of dry air, is twice as
active for toluene disproportionation and coke formation as
compared to regular mordenite. The aluminum-deficient form
also has a higher rate of deactivation. It is assumed that
the formation of new, strong acid sites during the process of
moderate dealumination is responsible for the observed change
in activity.

### Correlations Between Physico-chemical Characteristics and Catalytic Properties - Concluding Remarks

Published data by different authors suggest that the catalytic
activity of aluminum-deficient mordenite zeolites depends
primarily upon the following factors:

Preparation method. Mild acid-dealumination will
generally result in a more active material than the parent
zeolite due to: (a) removal of amorphous materials from the
zeolite channels, thus lowering the diffusion resistance for
the reacting molecules; and (b) generation of stronger acid
sites during the dealumination process, which enhances the
catalytic activity of the zeolite for acid-catalyzed reactions.
However, thermal dealumination will generally result in less

active materials due to (a) the elimination of acid sites; (b) the generation of extraneous materials in the zeolite channels (reduced diffusivity rate); and (c) stronger reduction in the unit cell size (narrower pore openings).

Degree of dealumination.  Moderate dealumination generally increases the catalytic activity or leaves it unchanged, while advanced dealumination leads to a decrease in activity.  Such a decrease is due to a loss of active sites with advanced framework dealumination.

Acidity.  Total acidity, type of acid sites (Bronsted or Lewis), and acid strength distribution affect the catalytic behavior of the zeolite in acid-catalyzed reactions.  Progressive dealumination will tend to decrease the total acidity of the zeolite, while at the same time modify the acid strength distribution.  Mild acid-dealumination appears to shift the acid strength distribution towards higher catalytic activity, possibly due to the formation of stronger acid sides.  At the same time, the lower total acidity (lower density of acid sites) will reduce the coking rate in some cracking reactions, thus reducing the deactivation rate of the catalyst.  Advanced dealuminated zeolites show a decline in catalytic activity, due in part to the drastic reduction in total acidity of the zeolite.

Diffusivity and effective pore diameter.  Acid-dealumination will alter the pore geometry and increase the effective pore diameter, thus reducing the resistance to diffusivity.  On the other hand, thermal dealumination will tend to decrease the effective pore diameter, primarily by generating amorphous materials in the zeolite channels and thus increase the resistance to diffusivity.  Since in many instances catalytic activity is enhanced by a lower resistance to diffusivity, mildly acid-dealuminated mordenite will allow the diffusion of larger molecules into its pores and show a higher activity in their catalytic conversion.  Larger pores will also facilitate the removal of larger condensation products and reduce the deactivation rate of the catalyst. However, advanced dealumination will result in shrinking of the unit cell and a narrowing of the pores, thus reducing the catalytic activity of the zeolite.

These are some of the more important factors that should be considered in order to explain the catalytic activity of aluminum-deficient mordenite zeolites. In general, the same factors will affect the catalytic properties of aluminum-deficient Y zeolites, although fewer data are available with regard to the catalytic properties of these materials.

Literature Cited

(1)   R.M. Barrer and M.B. Makki, Can. J. Chem., 42, 1481
      (1964).
(2)   I.M. Belen'kaya, M.M. Dubinin and I.I. Krishtofori,
      Isv. Akad. Nauk SSSR, Ser. Khim, 2164 (1967).
(3)   P.E. Eberly, Jr. and C.N. Kimberlin, Jr., Ind. Eng.
      Chem. Prod. Res. Dev., 9, 335 (1970).
(4)   W.L. Kranich, Y.H. Ma, L.B. Sand, A.H. Weiss and I.
      Zwiebel, Adv. Chem. Ser., 101, 502 (1970).
(5)   N.Y. Chen and F.A. Smith, Inorg. Chem. 15, 295 (1976).
(6)   C.V. McDaniel and P.K. Maher, Conf. Mol. Sieves, 1967,
      Soc. of Chem. Ind., London, Monogr. 186 (1968); US
      Patent #3,293,192 (Dec. 20, 1966)
(7)   G.T. Kerr, J. Catal. 15, 200 (1969).
(8)   G.T. Kerr, J. Phys. Chem. 72, 2594 (1968).
(9)   G.T. Kerr, Adv. Chem. Ser., 121, 219 (1973).
(10)  J.W. Ward, J. Catal. 18, 348 (1970).
(11)  P.K. Maher, F.D. Hunter and J. Scherzer, Adv. Chem.
      Ser., 101, 266 (1971).
(12)  J.B. Peri, Proc. 5th Intern. Congr. on Catal., 329
      (1972).
(13)  J.B. Uytterhoeven, L.G. Christer and W.K. Hall, J.
      Phys. Chem. 69, 2117 (1967).
(14)  S.P. Zhdanov and B.G. Novikov, Dokl. Akad. Nauk SSSR,
      16, 1107 (1966).
(15)  M.M. Dubinin, G.M. Fedorova, V.M. Plavnik, L.I.
      Piguzova and E.N. Prokofeva, Izv. Akad. Nauk SSSR, Ser.
      Khim., 2429 (1968).
(16)  R.M. Barrer and D.L. Peterson, Proc. R. Soc. London,
      Ser. A, 280, 466 (1964).
(17)  G.T. Kerr, J. Phys. Chem., 71, 4155 (1967).
(18)  J. Scherzer and A. Humphries, Symposium on Advances in
      Zeolite Chemistry, Reprints, A.C.S. Meeting, Las Vegas,
      1982, p. 520.
(19)  W.E. Garwood, S.J. Lucki, N.Y. Chen, and J.C. Bailar,
      Jr., Inorg. Chem., 17, 610 (1978).
(20)  W.E. Garwood, N.Y. Chen, and J.C. Bailar, Jr. Inorg.
      Chem. 15, 1044 (1976).
(21)  G.T. Kerr, J. Phys. Chem. 73, 2780 (1969).
(22)  G.T. Kerr, A.W. Chester and D.H. Olson, Acta Phys.
      Chem. 24, 169 (1978).
(23)  R. Beaumont and D. Barthomeuf, J. Catal. 26, 218
      (1972).
(24)  R. Beaumont and D. Barthomeuf, J. Catal. 27, 45 (1973).

(25) R. Beaumont and D. Barthomeuf, J. Catal. 30, 288 (1973).

(26) A. Wiecznikowski and B. Rzepa, Rocz. Chem., 51, 1955 (1977).

(27) H.K. Beyer and I. Belenykaya, Catalysis by Zeolites, ed. B. Imelik et al. Elsevier, Amsterdam, 1980, p. 203.

(28) J. Scherzer, J. Catal. 54, 285 (1978).

(29) V. Bosaček, V. Patzelova, Z. Tvaruzkova, D. Freude, U. Lohse, W. Schirmer, H. Stach and H. Thamm, J. Catal. 61, 435 (1980).

(30) D.W. Breck and G.W. Skeels, Proc. 6th Intern. Congr. on Catalysis, London, 1976; v.2, p. 645.

(31) D.W. Breck and G.W. Skeels, Molecular Sieves, II; A.C.S. Symposium Series 40, 271 (1977).

(32) G.T. Kerr, J.N. Miale and R.J. Mikovsky, U.S. Patent #3,493,519 (Feb. 3, 1970).

(33) D. Barthomeuf, Molecular Sieves,II, A.C.S. Symposium Series 40, 453 (1977).

(34) H.A. Benesi and B.H.C. Winquist, Adv. in Catal., 27, 97 (1978).

(35) H.W. Haynes, Jr., Catal. Rev.Sci. Eng. 17(2), 273 (1978).

(36) U. Lohse, H. Stach, H. Thamm, W. Schirmer, A.A. Isirikjan, N.I. Regent, and M.M. Dubinin, Z. anorg. allg. Chem. 460, 179 (1980).

(37) E. Lippmaa, M. Magi, A. Samoson, M. Tarmak and G. Engelhardt, J. Am. Chem. Soc. 103, 4992 (1981).

(38) G. Engelhardt, U. Lohse, A. Samoson, M. Magi, M. Tarmak and E. Lippmaa, Zeolites, 2, 59 (1982).

(39) J. Klinowski, J.H. Thomas, M. Audier and S. Vasudevan, J.C.S. Chem. Comm. 570 (1981).

(40) J.M. Thomas, J. Gonzalez-Calbet, S. Ramdas, J. Klinowski, M. Audier, C.A. Fyfe and G.C. Gobbi, A.C.S. Meeting, Div. of Petr. Chem., Las Vegas, Preprints, v. 27(2) 531 (1982).

(41) J. Klinowski, J.M. Thomas, C.A. Fyfe, G.C. Gobbi and J.S. Hartman, Inorg. Chem. 22, 63 (1983).

(42) V. Bosaček, D. Freude, T. Frohlich, H. Pfeifer and H. Schmiedel, J. Coll. Interf. Sci. 85, No. 2, 502 (1982).

(43) J. Dwyer, F.R. Fitch, F. Machado, G. Oin, S.M. Smyth, and J.C. Vickerman, J.C.S. Chem. Comm. 422 (1981).

(44) J. Dwyer, F.R. Fitch, G. Qin, and J.C. Vickerman, J. Phys. Chem. 86, 4574 (1982).

(45) A. Yoshida, H. Nakamoto, K. Okanishi, T. Tsuru and H. Takahashi, Bull. Chem. Soc. Jpn. 55, 581 (1982).

(46) P. Gallezot, R. Beaumont and D. Barthomeuf, J. Phys. Chem. 78, 1150 (1974).

(47) U. Lohse and M. Mildebrath, Z. anorg. allg. Chem. 476, 126 (1981).

(48)     W. Schirmer, H. Thamm, H. Stach, and U. Lohse, Proc.
        Conf. on Properties and Applications of Zeolites,
        London, 1979, No. 33, p. 204.
(49)     E.M. Flanigen, J.M. Bennett, R.W. Grose, J.P. Cohen,
        R.L. Patton and R.M. Kirchner, Nature, 271, 512 (1978).
(50)     P.A. Jacobs and J.B. Uytterhoeven, J. Catal. 22, 193
        (1971).
(51)     J. Scherzer and J.L. Bass, J. Catal. 28, 101 (1973).
(52)     J.B. Peri, Catalysis, v.1 (ed. J.W. Hightower) Amer.
        Elsevier, New York, 1973, p. 329.
(53)     P.A. Jacobs and J.B. Uytterhoeven, J. Chem. Soc.,
        Faraday Trans. 69, 373 (1973).
(54)     R. Beaumont, P. Pichat, D. Barthomeuf, and Y.
        Trambouze, Catalysis, v.1 (ed. J.W. Hightower), Amer.
        Elsevier, New York, 1973, p. 343.
(55)     A. Bielanski, J.M. Berak. E. Czerwinska, J. Datka, and
        A. Drelinkiewicz, Bull. Acad. Pol. Sci. 23, 445 (1975).
(56)     P. Pichat, R. Beaumont, and D. Barthomeuf, J. Chem.
        Soc., Faraday Trans. 1, (70), 1402 (1974).
(57)     K. Tsutsumi, H. Kajiwara, and H. Takahashi, Bull. Chem.
        Soc. Jpn., 47, 801 (1974).
(58)     K.V. Topchieva and H.S. T'huoang, Kinet. Catal. 11, 406
        (1970).
(59)     K. Tsutsumi, H. Kijiwara, H.K. Koh, and H. Takahashi,
        Proc. 3rd Int. Conf. Mol. Sieves (ed. J.B.
        Uytterhoeven), Leuven Univ. Press., 1973, p. 358.
(60)     E. Dempsey, J. Catal. 33, 497 (1974).
(61)     G.T. Kerr, J. Catal. 37, 186 (1975).
(62)     P.A. Jacobs, H.E. Leeman, and J.B. Uytterhoeven, J.
        Catal. 33, 31 (1974).
(63)     J. Scherzer and R.E. Ritter, Ind. & Eng. Chem., Prod.
        Res. Dev., 17, 219 (1978).
(64)     H. Bremer, U. Lohse, W. Reschetilowski and K.P.
        Wendland, Z. anorg. allg. Chem. 482, 235 (1981).
(65)     C.L. Thomas and D.S. Barmby, J. Catal. 12, 341 (1968).
(66)     D.W. Breck, Zeolite Molecular Sieves, John Wiley &
        Sons, New York, 1974.
(67)     R.W. Olsson and L.D. Rollmann, Inorg. Chem. 16, No. 3,
        651 (1977).
(68)     R.M. Barrer and J. Klinowski, J. Chem. Soc. Faraday
        Trans. 1, 71, 690 (1975).
(69)     J. Klinowski, J.M. Thomas, M.W. Anderson, C.A. Fyfe and
        G.C. Gobbi, Zeolites, 3, 5 (1983).
(70)     R.M. Barrer and E.V.T. Murphy, J. Chem. Soc., A, 2506
        (1970).
(71)     N.Y. Chen, J. Phys. Chem. 80, 60 (1976).
(72)     B.I. Mikunov, V.I. Yakerson, L.I. Lafer and A.M.
        Rubinshtein, Izv. Akad. Nauk SSSR, Ser. Khim, 449
        (1973).

(73)   I.V. Mishin, A.L. Klyachko-Gurvich and A.M.
       Rubinshtein, Izv. Akad. Nauk SSSR, Ser. Khim., 445
       (1973).
(74)   I.V. Mishin, G.A. Piloyan, A.L. Klyachko-Gurvich and
       A.M. Rubinshtein, Izv. Akad. Nauk SSSR, Ser. Khim.,
       1343 (1973).
(75)   H. Nakamoto and H. Takahashi, Zeolites, 2, 67 (1982).
(76)   P.E. Eberly, Jr. and C.N. Kimberlin, Jr., Ind. Eng.
       Chem., Prod. Res. Dev. 9, 335 (1970).
(77)   L.I. Piguzova, E.N. Prokof'eva, M.M. Dubinin, N.R.
       Bursian, and Yu. A. Shavandin, Kinet. Catal. 10, 252
       (1969).
(78)   P.E. Eberly, Jr., C.N. Kimberlin, Jr., J. Voorhies,
       Jr., J. Catal. 22, 419 (1971).
(79)   J.R. Kiovsky, W.J. Goyette, and T.M. Noterman, J.
       Catal. 52, 25 (1978).
(80)   H. Karge, Z. Phys. Chem. Neue Folge, 76, 133 (1971).
(81)   T.J. Weeks, Jr., H.F. Hillery, and A.P. Bolton, J.
       Chem. Soc., Faraday Trans., 1, 71, 2051 (1975).
(82)   D.K. Thakur and S.W. Weller, J. Catal. 24, 543 (1972).
(83)   D.K. Thakur and S.W. Weller, Adv. Chem. Ser. 121, 596
       (1973).
(84)   G.H. Kühl, Adv. Chem. Ser. 40, 96 (1977).
(85)   C. Mirodatos, B.H. Ha, K. Otsuka and D. Barthomeuf,
       Proc. 5th Intern. Conf. on Zeolites, Naples, 1980 p.
       382.
(86)   K.V. Topchieva, B.V. Romanovsky, L.I. Piguzova, Ho Si
       Thoang, and Y.W. Bizreh, Proc. 4th Intern. Congr.
       Catalysis, Moscow, USSR, 1968.
(87)   S.W. Weller and J.M. Brauer, AIChE Annual Meeting,
       62nd, Washington, D.C., Nov. 16-20, 1969.
(88)   H.S. Bierenbaum, S. Chiramongkol and A.H. Weiss, J.
       Catal. 23, 61 (1971).
(89)   R. Beecher, A. Voorhies and P. Eberly, Jr., Ind. Eng.
       Chem., Prod. Res. Dev., 7, 203 (1968).
(90)   H.G. Karge and J. Ladebeck, Catalysis by Zeolites (ed.
       Imelic et al.) Elsevier Sci. Publ. Co., Amsterdam,
       1980, p. 151.
(91)   N.S. Gnep, M.L. Martin de Armando, C. Marcilly, B.H. Ha
       and M. Guisnet, Catalyst Deactivation (ed. Delmon and
       Froment), Elsevier Sci. Publ. Co., Amsterdam, 1980, p.
       79.
(92)   V.J. Frilette and M.K. Rubin, J. Catal. 4, 310 (1965).
(93)   G.H. Kühl, Proc. 3rd Intern. Conf. Mol. Sieves, (ed.
       J.B. Uytterhoeven), Leuven Univ. Press, 1973. p. 227.
(94)   E.M. Gladrow, W.E. Winter and W.L. Schuette, U.S.
       Patent #4,259,212 (March 31, 1981).

(95)    J.W. Ward, U.S. Patent #4,036,739 (July 19, 1977).
(96)    R. J. Bertolacini and L.C. Gutberlet, U.S. Patent
        #3,622,501 (Nov. 23, 1971).
(97)    R.C. Hansford, U.S. Patent #3,836,454 (Sept. 17, 1974).
(98)    J.R. Kittrell, U.S. Patent #3,536,605 (Oct. 27, 1970).
(99)    L.C. Gutberlet and R.J. Bertolacini, U.S. Patent
        #3,548,020 (Dec. 15, 1970).
(100)   G.T. Kerr, C.J. Plank and E.J. Rosinski, U.S. Patent
        #3,442,795 (May 6, 1969).
(101)   J. Scherzer, J.L. Bass, and F.G. Hunter, J. Phys.
        Chem., $\underline{79}$, 1194 (1975).
(102)   B.M. Lok, F.P. Gortsema, C.A. Messian, H. Rastelli, and
        T.P.J. Izod, A.C.S. Symposium Series No. 218, 1983, p.41.
(103)   R. von Ballmoos, The $^{18}$O-exchange method in zeolite
        chemistry, Salle & Sauerlander, Frankfurt am Main,
        1981; p. 185.
(104)   P.A. Jacobs, J.A. Martens, J. Weitkamp, and H.K. Beyer,
        Faraday Disc. of the Chem. Soc., "Selectivity in
        Heterogeneous Catalysis", No. 72, 353 (1981).
(105)   A.K. Ghosh and G. Curthoys, J. Chem. Soc., Faraday
        Trans. 1, $\underline{79}$, 147 (1983).
(106)   J. Scherzer and A. Humphries, Symposium on Advances in
        Catalytic Chemistry, II, Salt Lake City, Utah, 1982.
(107)   G.W. Skeels and D. W. Breck, Sixth Intern. Zeolite
        Conf., Reno, Nevada, 1983; Abstracts, B-8.
(108)   P. Fejes, I. Kiricsi, I. Hannus and G. Schobel,
        Magy. Kem. Foly., $\underline{89}$ (6), 264 (1983).

RECEIVED December 5, 1983

# Aluminum Distributions in Zeolites

ALAN W. PETERS

W. R. Grace & Company, Davison Chemical Division, Columbia, MD 21044

Previous work has described a statistical approach
to the structure of faujasite. Distributions of
local structures were obtained that permit
comparisons with the results of $^{29}$Si NMR
experiments.
The results of these calculations for higher ratio
materials, Si/Al ~2.0 to 3.0, are not entirely
consistent with the results of NMR experiments. In
current work we have used the $^{29}$Si NMR data to
determine a distribution of prisms that give a best
fit to the NMR data for faujasite at a variety of
silicon/aluminum ratios. These results show
clearly that structures with four membered rings
containing two diagonal aluminum atoms or six
membered rings containing meta aluminum atoms are
avoided.
The calculation permits a transformation of
the $^{29}$Si NMR intensities to give distributions
of aluminum atoms in faujasite with implications
for the numbers of strong and weak acid sites
available.

There have been several recent suggestions that synthetic
faujasite may have a short range ordering of Si and Al atoms at
Si/Al ratios between 1.4 and 3.0. The results are based on the
possibility of matching experimental $^{29}$Si MASNMR (magic angle
spinning NMR) spectra with a spectra calculated from a selection
of proposed specific local structures ([1-2]). The suggestion that
the Si and Al may be ordered in faujasite derives from early work
by Dempsey, Kuhl and Olson ([3]) and by Breck and Flanigen ([4]).
These studies showed that discontinuities occur in the
correlation of Si/Al ratio with unit cell size. This work was
subsequently confirmed ([5]).

0097-6156/84/0248-0201$06.00/0
© 1984 American Chemical Society

In the present work we have developed a procedure for using $^{29}$Si MASNMR intensity data to provide structural information. An exhaustive list of all 19 structurally distinct hexagonal prism building blocks was prepared, Figure 1. Of the 19, only 16 give distinct silicon/aluminum neighbor relationships. Given a particular combination of structures, it is not difficult to compute the expected $^{29}$Si MASNMR intensities. What is desired, however, is to be able to go the other way, to estimate the relative number of each of the various structural units from the intensity data. A non linear optimization procedure, the Box complex algorithm (6), can be used to select from the 16 possible structures the combination that best fits the intensity data. Structural information acquired in this way is empirical in the sense that it is implied by the fitting procedure plus the data. There is some ambiguity in the result to the extent that different combinations of structures can give similar intensity results. Several ambiguities of this nature are discussed in the present case.

Distributions of structures obtained by fitting the intensity data can be compared to a most probable distribution of the sixteen structures assumming equal a priori probabilities subject to the constraint that the correct Si/Al ratio must be given. A method for calculating the most probable distribution of these structures has been previously reported (7).

Comparisons between the most probable and the best fit distributions of structural units show that there appear to be systematic preferences for certain units. These preferences are especially pronounced at higher Si/Al ratios.


EXPERIMENTAL

Four samples of faujasite were synthesized at Si/Al ratios of 2.61, 2.80, 2.97 and 3.03 using published methods from seeded slurries (8-9) and using proprietary methods. One additional sample of Si/Al ratio 2.58 was purchased from Union Carbide. The samples were characterized by X-ray powder diffraction, by surface area measurements, and by wet chemical analysis. The results of these measurements are contained in Table I. The $^{29}$Si MASNMR experiments were performed using a Bruker WH-400 spectrometer at the University of Guelph. The detailed experimental procedures and the methods of measuring peak

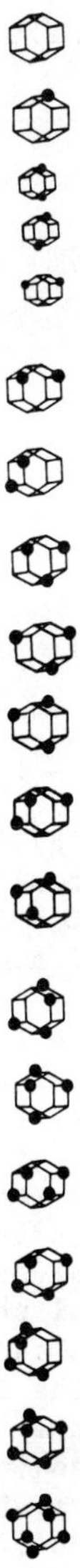

| Prism Type | | | Probability | $Si(OAl)_N$ | | | |
|---|---|---|---|---|---|---|---|
| | | | | N = 0 | 1 | 2 | 3 |
| $N_0^0$ | $M_0^0$ | | 1 | 12 | – | – | – |
| $N_1^0$ | $M_1^0$ | | 12 | 8 | 3 | – | – |
| $N_2^0$ | $M_2^0$ | | 24 | 4 | 6 | – | – |
| $N_2^0$ | $M_2^1$ | | 12 | 5 | 4 | 1 | – |
| $N_2^1$ | $M_2^0$ | | 12 | 6 | 2 | 2 | – |
| $N_3^0$ | $M_3^1$ | | 12 | 1 | 7 | 1 | – |
| $N_3^0$ | $M_3^3$ | | 4 | 3 | 3 | 3 | – |
| $N_3^1$ | $M_3^0$ | | 24 | 2 | 5 | 2 | – |
| $N_3^1$ | $M_3^1$ | | 24 | 3 | 3 | 3 | – |
| $N_3^2$ | $M_3^1$ | | 12 | 4 | 2 | 2 | 1 |
| $N_4^2$ | $M_4^0$ | | 6 | – | 4 | 4 | – |
| $N_4^2$ | $M_4^1$ | | 12 | – | 5 | 2 | 1 |
| $N_4^2$ | $M_4^2$ | | 6 | 2 | – | 6 | – |
| $N_4^2$ | $M_4^3$ | | 12 | 2 | 1 | 4 | 1 |
| $N_4^3$ | $M_4^2$ | | 12 | 2 | 2 | 2 | 2 |
| $N_5^4$ | $M_5^4$ | | 12 | 1 | – | 3 | 3 |
| $N_6^6$ | $M_6^6$ | | 2 | – | – | – | 6 |

Figure 1.  Arrangements of Al in Hexagonal Prisms

Table I.  Analytical Data Including $^{29}$Si MASNMR for
Union Carbide and for Synthesized Faujasites

| Source | Union Carbide SK-31 | Synthesized | | | |
|---|---|---|---|---|---|
| (Si/Al) Chemical | 2.71 | 2.61 | 2.97 | 2.91 | 3.04 |
| Surface Area, m$^2$/g | 900 | 944 | 961 | 900 | 903 |
| Unit Cell Size, A (XRD) | 24.66 | 24.67 | 24.58 | 24.62 | 24.57 |
| MASNMR Intensities[a] | | | | | |
| (Si/Al)$_{nmr}$ | 2.58 | 2.61 | 2.80 | 2.97 | 3.03 |
| Si(OAl)$_n$ | | | | | |
| n = 0 | 15.1 | 12.5 | 13.2 | 19.7 | 19.7 |
| 1 | 52.9 | 58.9 | 67.4 | 66.1 | 68.1 |
| 2 | 52.2 | 50.0 | 49.0 | 47.2 | 48.7 |
| 3 | 15.6 | 15.6 | 10.4 | 9.5 | 6.6 |
| 4 | 2.5 | 1.8 | 1.4 | 1.1 | 1.3 |

[a] Normalized to the number of silicon atoms per unit cell
calculated from the (Si/Al)$_{nmr}$ ratio.

intensities are described in reference (2). The results of
measurements performed on the four synthesized faujasites and on
the Union Carbide sample are also reported in Table I.  Intensity
measurements are normalized to the total number of silicon atoms
per unit cell.
     The optimization procedure used in fitting the intensity
measurements to combinations of structural units is a modified
version of the Box complex alogrithim described in reference
(6).  While a Fortran version of this alogrithim has been
published (10), the version used in this work was written in
APL.  In the application of the Box alogrithim to the faujasite
structure problem, one chooses an initial set of 20 to 25
distributions of the different structural prism units according
to the constraint that each distribution include a total of 16
prisms for one unit cell.  For each of the 16 prism types one
distribution is chosen containing only that type and no other.
This accounts for 16 distributions.  Also included in the initial
set is the most probable distribution computed according to
reference (7).  Examples of the most probable distribution are
listed in Table X.  The other three to eight distributions are
chosen at random using a random number generator to determine the
relative amount of each prism type.  This defines a starting
complex of distributions.  For each assummed distribution

intensities are calculated and compared with experimental
intensities by calculating the variance $\sigma^2$. The centroid
for all distributions is computed and the highest variance
distribution is reflected through the centroid to give a new
distribution point. The process is repeated until all
distributions have a similar variance. Typically 1000-2000
iterations are required. The result is typically independent of
the starting complex.

## RESULTS AND DISCUSSION

The random Al siting method of reference (*7*) was used to compute
$^{29}$Si NMR intensities for comparison with experimental results
reported in reference (*2*). The results in Table II show clearly
some discrepancy between the experimental and calculated results.
The variance $\sigma^2$ ranges from 35 to 329. The discrepancy is
greatest at higher Si/Al ratios where the experimental
distribution is much sharper than is expected of the maximum
probability distribution of silicon and aluminum atoms. These
results imply some ordering of the aluminum atoms in the lattice.

Table II.  Comparison of Calculated Maximum Probability
Distributions[a] of Silica Atoms in Faujasite
Having 0, 1, 2, 3, and 4 Nearest Neighbor OAl
Groups with Experimental Distributions from Reference 2

| (Si/Al)$_{nmr}$ | 1.14 | | 1.57 | | 1.98 | | 2.68 | |
|---|---|---|---|---|---|---|---|---|
| | | 2.76 | | 1.47 | | 0.896 | | 0.218 |
| Si(OAl)$_n$ | Exp | Calc | Exp | Calc | Exp | Calc | Exp | Calc |
| n = 0 | 1.9 | 0.7 | 4.8 | 5.3 | 5.7 | 10.2 | 8.5 | 18.8 |
| 1 | 1.4 | 5.2 | 11.7 | 15.7 | 25.8 | 24.1 | 43.0 | 34.2 |
| 2 | 6.2 | 4.5 | 27.5 | 22.2 | 36.5 | 29.3 | 39.5 | 29.3 |
| 3 | 26.5 | 23.1 | 36.1 | 33.0 | 24.5 | 25.9 | 9.0 | 14.6 |
| 4 | 64.0 | 66.4 | 19.9 | 23.8 | 7.5 | 10.5 | 0 | 3.1 |
| $\sigma^2$ | | 35 | | 69 | | 97 | | 329 |

[a] Normalized to 100.

In order to clarify the nature of the ordering, we
consider faujasite as consisting of connected hexagonal prisms
containing 12 atoms. It is possible to list prisms including all
possible silicon environments for a given aluminum atom content
consistent with Lowenstein's rule. This has been done in detail
in Figure 1 which provides an illustration of the configuration
of each of the 19 possible prism types. Each structure is
labeled with a symbol $N_s^r$ including a subscript s indicating
the number of aluminum atoms ranging from 0 to 6 and a
superscript r indicating the number of four membered rings in the
structure containing two aluminum atoms. A second symbol $M_s^t$

indicates the number s of aluminum atoms in the structure and
the number t of meta Al-O-Si-O-Al interactions across six
membered rings. Thus two symbols characterize the number and
type of nearest neighbor aluminum interaction for each type of
prism. The probability of each structure is the number of ways
of generating the structure assuming distinguishable lattice
sites and indistinguishable aluminum and silicon atoms. Three
of the structures have the same nearest neighbor aluminum
interactions and are characterized by the same symbol,
$N_2{}^0M_2{}^0$. In two other structures, $N_3{}^1M_3{}^1$ and $N_3{}^0M_3{}^3$, the
silicon atoms have the same nearest neighbor environment.

The fourth neighbor is generated when the prisms
are connected. Since each aluminum is connected to a silicon
according to Lowenstein's rule, there are $Q_S - Q_A$ silicon
atoms available for connecting to another silicon, where $Q_S$ =
number of silicon atoms and $Q_A$ = number of aluminum atoms.
Thus the probability of a Si-O-Si connecting bond is
$(Q_S-Q_A)/Q_S = 1-1/R$ and of a Si-O-Al bond is $Q_A/Q_S = 1/R$ where R
= $Q_S/Q_A$ = Si/Al atom ratio. This completely determines the
environment of each atom and therefore the relative intensities
of the five $^{29}Si$ NMR lines in a structure consisting of some
assumed distribution of building blocks.

There are some unsatisfactory aspects to this
procedure. In estimating maximum probability distributions,
there is a probability associated with the number of ways a
particular set of prisms can be hooked together, and this
probability will depend on the particular distribution of
prisms. Since this is not included in the calculation of
reference ($\underline{7}$), the maximum probability distribution may not be
exactly correct.

Another difficulty is that if three prisms are
connected in a particular way to form part of a single sodalite
cage, six nearest neighbors of the fourth prism are
determined. In accordance with Lowensteins rule this may
restrict the structure of the fourth prism. Lowensteins rule
means that there are constraints involved in the ways the
prisms can connect that are not considered here. However, this
will not restrict the possible distributions of a large enough
crystal and should not affect the results of this work.

A more serious difficulty is that in the process
of connecting hexagonal prisms together aluminum interactions
across six and four membered rings are generated. For example,
a structure generated entirely from prisms in which
interactions across four membered rings are avoided ($N^0$
type) will contain some four membered rings with diagonal
aluminum atoms generated in the process of randomly linking the
prisms. This will be reflected in the computed intensity
distribution and will result in an over estimate of the degree
of avoidance of specific structures.

Since an assumed distribution of structures per unit cell allows one to calculate the relative intensities of the five $^{29}$Si lines, one can choose the combination of structures which will best fit the NMR data. We have chosen to use the Box complex alogrithm to determine the distribution of structures that minimize the variance $\sigma^2$ between the experimental and computed relative $^{29}$Si intensities. Tables III, IV and V list results for sieves of various Si/Al ratios. The $^{29}$Si NMR data in Table III has appeared previously in the literature (2) and the data in Tables IV and V are new (Table I). At the lowest Si/Al ratio of 1.145, the best fit distribution is nearly the same as the maximum probability model with a low $\sigma^2 = 28$. At higher Si/Al ratios > 1.9 the best fit result continues to give a low $\sigma^2$ between 0 and 18. For the seven samples with Si/Al > 1.9, only four of the possible nine structures containing three and four aluminum atoms appear to consistently occurr. These are the structures $N_3{}^0 M_3{}^1$, $N_3{}^1 M_3{}^0$, $N_4{}^2 M_4{}^0$, and $N_4{}^2 M_4{}^1$ which minimize the number of nearest neighbor Al–O–Si–O–Al interactions across four or six membered rings.

In Table VI the occurrence of each prism type is summed over all the samples listed in Tables III, IV and V for those prisms containing two, three, or four aluminum atoms. The two preferred structures containing three aluminum atoms and the twocontaining four aluminum atoms are marked by asterisks. From these results it appears that a description of the $^{29}$Si NMR intensities of faujasite can be obtained using a relatively simple set of structures. At Si/Al < 2.0, prisms containing five and six aluminium atoms occur along with two of the five structures containing four aluminum atoms. At 2.0 < Si/Al < 3.0, two structures containing four Al occur along with two of the four possible structures containing three Al per prism. At Si/Al = 3.0, only two structures occur, $N_3{}^0 M_3{}^1$ and $N_3{}^1 M_3{}^0$. Tables VII, VIII and IX illustrate that using this limited selection of structures very good agreement with experiment is obtained. Table X gives results expected of the maximum probability distribution. It is clear from a comparison of these results that there are structural preferrences in faujasite.

Table III.  Correlations Between Published $^{29}$Si MASNMR Line Intensities[a] and Prism Types[b] in Faujasite

| (Si/Al) nmr | 1.15 | | 1.57 | | 1.98 | | 2.68 | |
|---|---|---|---|---|---|---|---|---|
| Si(OAl)n | Exp | Opt. | Exp | Opt. | Exp | Opt. | Exp | Opt. |
| n = 0 | 1.9 | 0.9 | 5.6 | 6.1 | 7.3 | 8.2 | 11.9 | 13.0 |
| 1 | 1.4 | 6.2 | 13.7 | 17.7 | 32.9 | 32.9 | 60.2 | 60.3 |
| 2 | 6.4 | 4.8 | 32.3 | 29.4 | 46.5 | 46.2 | 55.3 | 51.5 |
| 3 | 27.2 | 25.6 | 42.4 | 42.8 | 31.2 | 31.1 | 12.6 | 13.8 |
| 4 | 65.6 | 65.9 | 23.3 | 22.6 | 9.6 | 9.6 | 0 | 1.1 |
| $\sigma^2$ | — | 28 | — | 25 | — | 1 | — | 18 |
| **Prism Type** | | | | | | | | |
| N$_0$° | | 0 | | 0 | | 0 | | 0 |
| N$_1$° | | 0 | | 0 | | 0 | | 0 |
| N$_2$° M$_2$° | | 0 | | 0 | | 0 | | 0 |

| | | | |
|---|---|---|---|
| $N_2^0 M_2^1$ | 0 | .1 | .1 |
| $N_2^1 M_2^0$ | 0 | .1 | .4 | .0 |
| $N_3^0 M_3^1$ | 0 | .2 | 1.3 | 4.9 |
| $N_3^{0,1} M_3^{3,1}$ | 0 | .4 | .4 | .2 |
| $N_3^1 M_3^0$ | 0 | .2 | .1 | 6.1 |
| $N_3^2 M_3^1$ | 0 | .1 | .1 | .3 |
| $N_4^2 M_4^0$ | .1 | 1.6 | 2.8 | 2.0 |
| $N_4^2 M_4^1$ | .2 | 1.4 | 4.1 | 1.7 |
| $N_4^2 M_4^2$ | .2 | 1.1 | .1 | .2 |
| $N_4^2 M_4^3$ | .1 | .9 | 2.3 | .1 |
| $N_4^3 M_4^2$ | .2 | .3 | 1.6 | .2 |
| $N_5^4$ | 4.9 | 8.2 | 2.2 | .1 |
| $N_6^6$ | 10.0 | 1.3 | 0.5 | 0 |

a  Normalized to the number of silicon atoms per unit cell calculated from the $(Si/Al)_{nmr}$ ratio.

b  Figure 1

Table IV.   Correlation Between Si MASNMR Line Intensities[a] and Prism Types[b] in Medium Si/Al Ratio Faujasite

| (Si/Al) nmr Si(OAl)n | 2.58 | | 2.61 | |
|---|---|---|---|---|
| | Exp | Opt. | Exp | Opt. |
| n = 0 | 15.1 | 15.3 | 12.5 | 13.2 |
| 1 | 52.9 | 53.7 | 58.9 | 58.3 |
| 2 | 52.2 | 51.2 | 50.0 | 50.2 |
| 3 | 15.6 | 17.1 | 15.6 | 15.1 |
| 4 | 2.5 | 1.5 | 1.8 | 2.0 |
| $\sigma^2$ | – | 5 | – | 1.4 |

Prism Type

| | | | | |
|---|---|---|---|---|
| $N_0{}^0$ | | 0 | | 0 |
| $N_1{}^0$ | | 0 | | 0 |
| $N_2{}^0 \ M_2{}^0$ | | .1 | | .4 |
| $N_2{}^0 \ M_2{}^1$ | | .2 | | .2 |
| $N_2{}^1 \ M_2{}^0$ | | .2 | | .0 |
| $N_3{}^0 \ M_3{}^1$ | | 1.2 | | 2.2 |
| $N_3{}^{0,1} \ M_3{}^{3,1}$ | | .6 | | .1 |
| $N_3{}^1 \ M_3{}^0$ | | 8.2 | | 7.4 |
| $N_3{}^2 \ M_3{}^1$ | | .1 | | .1 |
| $N_4{}^2 \ M_4{}^0$ | | 2.4 | | 1.0 |
| $N_4{}^2 \ M_4{}^1$ | | 1.6 | | 4.2 |
| $N_4{}^2 \ M_4{}^2$ | | .4 | | 0 |
| $N_4{}^2 \ M_4{}^3$ | | .2 | | 0 |
| $N_4{}^3 \ M_4{}^2$ | | .2 | | .1 |
| $N_5{}^4$ | | .4 | | .1 |
| $N_6{}^6$ | | .1 | | .1 |

[a]  Normalized to the number of silicon atoms per unit cell calculated from the $(Si/Al)_{nmr}$ ratio.

[b]  Figure 1

Table V.   Correlation Between Si MASNMR Line Intensities[a] and Prism Types[b] in High Si/Al Ratio Faujasite

| (Si/Al) nmr | 2.80 | | 2.97 | | 3.03 | |
|---|---|---|---|---|---|---|
| Si(0Al)n | Exp | Opt. | Exp | Opt. | Exp | Opt. |
| n = 0 | 13.2 | 13.2 | 19.7 | 19.7 | 19.7 | 18.7 |
| 1 | 67.4 | 67.5 | 66.1 | 66.4 | 68.1 | 68.2 |
| 2 | 49.0 | 49.0 | 47.2 | 47.1 | 48.7 | 47.8 |
| 3 | 10.4 | 10.7 | 9.5 | 10.2 | 6.6 | 9.2 |
| 4 | 1.4 | 1.1 | 1.1 | 0.4 | 1.3 | 0 |
| $\sigma^2$ | – | .2 | – | 1.2 | – | 10 |

**Prism Type**

| | 2.80 | 2.97 | 3.03 |
|---|---|---|---|
| $N_0{}^0$ | 0 | 0 | 0 |
| $N_1{}^0$ | .1 | .1 | .1 |
| $N_2{}^0 \; M_2{}^0$ | .2 | .7 | .5 |
| $N_2{}^0 \; M_2{}^1$ | .2 | .4 | .2 |
| $N_2{}^1 \; M_2{}^0$ | .1 | .0 | .1 |
| $N_3{}^0 \; M_3{}^1$ | 8.7 | 3.5 | 5.4 |
| $N_3{}^{0,1} \; M_3{}^{3,1}$ | .4 | .1 | .3 |
| $N_3{}^1 \; M_3{}^0$ | 3.4 | 9.7 | 8.7 |
| $N_3{}^2 \; M_3{}^1$ | .0 | .0 | .0 |
| $N_4{}^2 \; M_4{}^0$ | .3 | .4 | .3 |
| $N_4{}^2 \; M_4{}^1$ | 2.1 | .8 | .0 |
| $N_4{}^2 \; M_4{}^2$ | .1 | .1 | .2 |
| $N_4{}^2 \; M_4{}^3$ | .0 | .0 | .1 |
| $N_4{}^3 \; M_4{}^2$ | .1 | .1 | .0 |
| $N_5{}^4$ | .2 | .0 | .0 |
| $N_6{}^6$ | .1 | .0 | .0 |

[a]   Normalized to the number of silicon atoms per unit cell calculated from the $(Si/Al)_{nmr}$ ratio.

[b]   Figure 1

Table VI. Total Numbers of Selected Prism Types Observed to Occur in Eight Samples of Faujasites with Si/Al Ratio 1.9 or Greater from Tables III and IV. Preferred Structures are Indicated by an Asterisk.

| Prism Type[a] | Probability[a] | Total Prism Observed |
|---|---|---|
| $N_2{}^0$ $M_2{}^0$ | 24 | 1.9 |
| $N_2{}^0$ $M_2{}^1$ | 12 | 1.5 |
| $N_2{}^1$ $M_2{}^0$ | 12 | 0.9 |
| *$N_3{}^0$ $M_3{}^1$ | 12 | 27.4 |
| $N_3{}^{0,1}$ $M_3{}^{3,1}$ | 28 | 2.5 |
| *$N_3{}^1$ $M_3{}^0$ | 24 | 43.8 |
| $N_3{}^2$ $M_3{}^1$ | 12 | 0.7 |
| *$N_4{}^2$ $M_4{}^0$ | 6 | 10.9 |
| *$N_4{}^2$ $M_4{}^1$ | 12 | 16.1 |
| $N_4{}^2$ $M_4{}^2$ | 6 | 2.4 |
| $N_4{}^2$ $M_4{}^3$ | 12 | 3.7 |
| $N_4{}^3$ $M_4{}^3$ | 12 | 2.8 |

[a] Normalized to the number of silicon atoms per unit cell calculated from the $(Si/Al)_{nmr}$ ratio.
[b] Figure 1

Table VII. Fitting $^{29}Si$ MASNMR Line Intensities[a] in Low Si/Al Ratio Faujasite with a Simplest Set of Prism Types[b]

| (Si/Al) nmr | 1.15 | | 1.57 | | 1.98 | |
|---|---|---|---|---|---|---|
| Si(0Al)n | Exp | Opt. | Exp | Opt. | Exp | Opt. |
| n = 0 | 1.9 | 0.6 | 5.6 | 3.7 | 7.3 | 7.4 |
| 1 | 1.4 | 4.6 | 13.7 | 14.9 | 32.9 | 32.7 |
| 2 | 6.4 | 6.7 | 32.3 | 31.8 | 46.5 | 46.7 |
| 3 | 27.2 | 25.7 | 42.4 | 42.9 | 31.2 | 31.0 |
| 4 | 65.6 | 65.4 | 23.4 | 23.3 | 9.6 | 9.7 |
| $\sigma^2$ | – | 14 | – | 6 | – | 2 |
| Prism Type | | | | | | |
| $N_3{}^0$ $M_3{}^1$ | | – | | – | | 2.5 |
| $N_3{}^1$ $M_3{}^0$ | | – | | – | | 3.2 |
| $N_4{}^2$ $M_4{}^0$ | | 1.2 | | 2.9 | | 3.5 |
| $N_4{}^2$ $M_4{}^1$ | | 0 | | 2.4 | | 0.7 |
| $N_5{}^4$ | | 4.5 | | 10.1 | | 6.2 |
| $N_6{}^6$ | | 10.2 | | 0.7 | | – |

[a] Normalized to the number of silicon atoms per unit cell calculated from the $(Si/Al)_{nmr}$ ratio.
[b] Figure 1

Table VIII.    Fitting $^{29}$ Si MASNMR Line Intensities[a] in
Faujasite with a Simplest Set of Four Prism Types[b]

| (Si/Al) nmr | 2.58 | | 2.61 | | 2.68 | | 2.80 | |
|---|---|---|---|---|---|---|---|---|
| Si(OAl)n | Exp | Opt. | Exp | Opt. | Exp | Opt. | Exp | Opt. |
| n = 0 | 15.1 | 13.1 | 12.5 | 12.5 | 11.9 | 11.5 | 13.2 | 13.1 |
| 1 | 52.9 | 55.4 | 58.9 | 58.6 | 60.2 | 60.3 | 67.4 | 67.4 |
| 2 | 52.2 | 53.2 | 50.0 | 50.6 | 55.3 | 54.3 | 49.0 | 49.7 |
| 3 | 15.6 | 16.1 | 15.6 | 15.0 | 12.6 | 13.4 | 10.4 | 10.6 |
| 4 | 2.5 | 0.8 | 1.8 | 2.0 | 0 | 0 | 1.4 | 6.9 |
| $\sigma^2$ | – | 14 | – | 0.9 | – | 2 | – | 0.7 |
| Prism Type | | | | | | | | |
| $N_3{}^0 \ M_3{}^1$ | | 0 | | 1.2 | | 4.8 | | 6.8 |
| $N_3{}^1 \ M_3{}^0$ | | 10.8 | | 9.5 | | 6.7 | | 6.8 |
| $N_4{}^2 \ M_4{}^0$ | | 3.1 | | 0 | | 0 | | 0 |
| $N_4{}^2 \ M_4{}^1$ | | 2.1 | | 5.2 | | 4.4 | | 2.4 |

[a] Normalized to the number of silicon atoms per unit cell calculated from the $(Si/Al)_{nmr}$ ratio.
[b] Figure 1

Table IX.    Fitting $^{29}$Si MASNMR Line Intensities[a] in
Faujasite with a Simplest Set of Two Prism Types[b]

| (Si/Al) nmr | 2.97 | | 3.03 | |
|---|---|---|---|---|
| Si(OAl)n | Exp | Opt. | Exp | Opt. |
| n = 0 | 19.7 | 19.6 | 19.7 | 18.7 |
| 1 | 66.1 | 66.2 | 68.1 | 68.3 |
| 2 | 47.2 | 48.2 | 48.7 | 47.8 |
| 3 | 9.5 | 10.0 | 6.6 | 9.2 |
| 4 | 1.1 | 0 | 1.3 | 0 |
| $\sigma^2$ | – | 2.6 | – | 10 |
| Prism Type | | | | |
| $N_3{}^0 \ M_3{}^1$ | | 2.4 | | 4.1 |
| $N_3{}^1 \ M_3{}^0$ | | 13.6 | | 11.9 |

[a] Normalized to the number of silicon atoms per unit cell calculated from the $(Si/Al)_{nmr}$ ratio.
[b] Figure 1

We have shown that four of the nine possible prisms containing three or four aluminum atoms are sufficient to describe the $^{29}$Si MASNMR data. Are all four necessary? It is difficult to answer that question since distributions derived from each of the four identified structures can be approximated by combinations of other stuctures. For example there is a combination of $N_3{}^0M_3{}^1$ and $N_3{}^0M_3{}^3$ that can give the same relative distribution of silicon environments as $N_3{}^1M_3{}^0$.

Table X.  Experimental $^{29}$Si MASNMR Intensities Compared to Intensities[a]
Calculated from the Most Probable Distributions
of Prism Types[b] Assuming Lowenstein's Rule
and Appropriate Constraints (Ref. 7)

| (Si/Al) nmr Si(OAl)n | 1.145 Exp[c] | Calc[d] | 1.57 Exp[c] | Calc[d] | 1.98 Exp[c] | Calc[d] | 2.97 Exp[e] | Calc[d] |
|---|---|---|---|---|---|---|---|---|
|  | — | 2.76 | — | 1.47 | — | .897 | — | 31.9 |
| n = 0 | 1.9 | 0.7 | 5.6 | 6.3 | 7.3 | 13.0 | 19.7 | 31.9 |
| 1 | 1.4 | 5.3 | 13.7 | 18.4 | 32.9 | 30.8 | 66.1 | 52.6 |
| 2 | 6.4 | 4.7 | 32.3 | 26.1 | 46.5 | 37.4 | 47.2 | 39.7 |
| 3 | 27.2 | 23.7 | 42.4 | 38.7 | 31.2 | 33.0 | 9.5 | 16.6 |
| 4 | 65.6 | 68.1 | 23.4 | 27.9 | 9.6 | 13.4 | 1.1 | 3 0 |
| $\sigma^2$ | — | 40 | — | 94 | — | 137 | — | 440 |

| Prism Type |  |  |  |  |  |  |  |  |
|---|---|---|---|---|---|---|---|---|
| $No^{o}$ |  | 0 |  | 0 |  | 0 |  | .08 |
| $N1^{o}$ |  | 0 |  | .01 |  | .16 |  | .95 |
| $N2^{o} M2^{o}$ |  | 0 |  | .13 |  | .49 |  | 1.92 |
| $N2^{o} M2^{1}$ |  | 0 |  | .06 |  | .25 |  | .96 |
| $N2^{1} M2^{o}$ |  | 0 |  | .06 |  | .25 |  | .96 |

| | | | | |
|---|---|---|---|---|
| $N_3^0 M_3^1$ | .02 | .28 | .61 | .96 |
| $N_2^0 M_3^3$ | .01 | .09 | .20 | .32 |
| $N_3^1 M_3^0$ | .03 | .56 | 1.21 | 1.93 |
| $N_3^1 M_3^1$ | .03 | .56 | 1.21 | 1.93 |
| $N_3^2 M_3^1$ | .01 | .28 | .61 | .96 |
| $N_4^2 M_4^0$ | .13 | .61 | .74 | .49 |
| $N_4^2 M_4^1$ | .26 | 1.21 | 1.49 | .97 |
| $N_4^2 M_4^2$ | .13 | .61 | .74 | .49 |
| $N_4^2 M_4^3$ | .26 | 1.21 | 1.49 | .97 |
| $N_4^3 M_4^2$ | .26 | 1.21 | 1.49 | .97 |
| $N_5^4$ | 4.09 | 5.28 | 3.64 | .98 |
| $N_6^6$ | 10.77 | 3.83 | 1.49 | .16 |

a  Normalized to the number of silicon atoms per unit cell calculated from the $(Si/Al)_{nmr}$ ratio.
b  Figure 1
c  Reference 2
d  Reference 7
e  Table 1

The results suggest only that aluminium atoms tend to
avoid each other, and are not inconsistent with a proposal
that avoidance across four membered rings is most important or
a counterproposal that avoidance across six membered rings is
more important. If for example at Si/Al = 3.0, diagonal
aluminiums across four membered rings are avoided, the
structure

$$N_3{}^0 = 3/4\ N_3{}^0M_3{}^1 + 1/4\ N_3{}^0M_3{}^3$$

would be expected, that is $N_3{}^0M_3{}^1$ and $N_3{}^0M_3{}^3$ would occur in a
3 to 1 ratio. On the other hand, if meta aluminum interactions
were avoided we would expect the structure $M_3{}^0 = N_3{}^1M_3{}^0$.
Table XI compares experimental results with predictions based
on $N_3{}^0$ and $M_3{}^0$ structures. The experimental results are
approximately an average of the $N_3{}^0$ and $M_3{}^0$ predictions. In
any case the estimates obtained here are not reliable enough to
distinguish between the two cases.

Table XI.     Experimental $^{29}Si$ MASMNR RelativeIntensities
             Obtained for 3.0 Si/AlRatio Faujasite Compared to
         Intensity Distributions Given by Two Simple Structures
                          $N_3{}^0$ and $M_3{}^0$

| $Si(OAl)_n$ | Experimental | $N_3{}^0$ (Diagonal Al Avoidance) | $M_3{}^0$ (Meta Al Avoidance) |
|---|---|---|---|
| n = 0 | 20 | 16 | 21 |
| 1 | 67 | 72 | 64 |
| 2 | 48 | 48 | 48 |
| 3 | 8 | 8 | 11 |
| 4 | 1 | 0 | 0 |

As mentioned above since there is no provision in the
placement of the fourth neighbors for the effect of aluminum
interactions, the results of this work will over-estimate the
tendency of aluminum atoms to avoid each other in the faujasite
framework.

Conclusions

An analysis of published $^{29}Si$ MASNMR data and new data
reported here indicate that there are preferred structural

units that occur in synthesized faujasite and other structures
that do not occur. Those that maximize the distance between
aluminum atoms are the more likely.

Recent work has shown that strong and weak sites exist
in faujasite (11) and it has been proposed that relatively
isolated sites are the strong acid sites and are responsible
for cracking activity (12). The results obtained in this work
imply that strong acid sites are preferred, and that at Si/Al
ratios beyond about 3.0 to 4.0 nearly all sites are isolated or
strong. Either an increase or decrease in the Si/Al ratio from
3.0 to 4.0 will result in some loss in strong acid sites.

## Acknowledgments

I would like to acknowledge the efforts of Grant C. Edwards who
prepared the faujasites reported in Table I and Colin Fyfe of
Guelph University who obtained the $^{29}Si$ MASNMR intensity
data also reported in Table I.

## Literature Cited

1. Melchior, M. T.; Vaughan, D. E. W.; Jacobson, A. J. J. Am.
   Chem. Soc. 1982, 104, 4859-4864.
2. Klinowski, J.; Ramdas, S.; Thomas, J. M. J. Chem. Soc.,
   Faraday Trans. 2, 1982, 78, 1025-1050.
3. Dempsey, E.; Kuhl, G. H.; Olson, D. H. J. Phys. Chem. 1969,
   73, 387-390.
4. Breck, D. W.; Flanigen, E. M. "Molecular Sieves";
   Barrer, R. M.; Ed.; Society of the Chemical Industry;
   London 1968; pp.47-61.
5. Kuhl, G. H. in "Molecular Sieve Zeolites-1"; Flanigen,
   E. M.; Sand. L. B.; Eds. American Chemical Society:
   Washington, DC, 1971; Adv. Chem. Ser. No. 101, p. 199.
6. Box, M. J. Comput. J. 1965, 8 (42).
7. Peters, A. W. J. Phys. Chem., 1982, 86, 3489.
8. Elliott, C. H.; McDaniel, C. V. U.S. Patent 3,636,099, 1979.
9. Vaughan, D. E. W.; Edwards; G. C.; Barrett, M. G. U.S.
   Patent 4,178,352.
10. Kaester, J. C.; Mize, J. E." Optimization Techniques with
    Fortran"; McGraw Hill, 1973, New York, p. 368.
11. Barthomeuf, D. and Beaumont, R. J. Catal. 1973, 30, 288.
12. Dempsey, E. J. Catal. 1974, 33, 497.

RECEIVED October 3, 1983

# Factors Affecting the Synthesis of Pentasil Zeolites

ZELIMIR GABELICA and ERIC G. DEROUANE[1]

Facultés Universitaires de Namur, Laboratoire de Catalyse, Rue de Bruxelles 61, B-5000 Namur, Belgium

NIELS BLOM

Haldor Topsøe Research Laboratories, Nymøllevej 55, DK-2800 Lyngby, Denmark

Several techniques were used in combination to investigate the role of different synthesis variables governing the crystallization of high silica zeolitic materials of the pentasil family. The use of Al-rich ingredients and polymeric silica generates a small number of nuclei which grow involving a liquid phase ion transportation process and yield large ZSM-5 single crystals (synthesis of type A). When high Si/Al ratios and monomeric Na silicate are used (synthesis of type B), numerous nuclei rapidly yield very samll ZSM-5 microcrystallites, which appear directly within the hydrogel and which are not detectable by X-ray diffraction. Their formation and growth are strongly dependent on competitive interactions between alkali and $Pr_4N^+$ cations with aluminosilicate polymeric anions. Some intrinsic properties of the alkali cations such as their size, their structure-forming (breaking) role towards water or their salting-out power, modify the specificity of such interactions, which greatly affect the morphology, size, chemical composition and homogeneity of the resulting crystallites. $Et_4N^+$, $Pr_4N^+$ and $Bu_4N^+$ cations specifically direct the structure of ZSM-8, ZSM-5 and ZSM-11 zeolites respectively. When tripropylamine or tributylamine is used, mixed ZSM-5/ZSM-11 phases are formed. Their nature seems to be determined more by the zeolitic channel filling than by the location of the organic molecules at the channel intersections.

[1]Current address: Mobil Research and Development Corporation, Central Research Division, Princeton, NJ 08540.

0097-6156/84/0248-0219$09.25/0
© 1984 American Chemical Society

Since the first attempts of Barrer (1) and Kerr (2) to synthesize
counterparts of some natural zeolite materials in presence of te-
tramethylammonium hydroxyde, a wide number of new zeolites have
been prepared from various organic cation-containing mixtures (3).
Soon, zeolites with high Si/Al ratio appeared attractive because
of their potential thermal, hydrothermal and acid stabilities. In
that respect, a variety of organics have proved successful in producing Si-rich zeolites (3).

Zeolites ZSM-5 and ZSM-11 are the most commercially important
end-members of a continuous series of intermediate structures belonging to the so-called pentasil family (4,5). The first preparation of ZSM-5 was described in 1972 (6) and since then, a number
of elaborate synthesis recipes have been reported in the patent
literature. Because of the unique and fascinating activity and
(shape) selectivity of this material for a variety of catalytic
reactions currently processed in chemical industries, increasing
attention has been devoted to a better understanding of the various mechanisms that govern the synthesis of ZSM-5 (7-33).

As numerous factors affect simultaneously, in various ways,
the nature and the rate of the nucleation and crystallization processes of highly siliceous zeolites (3,34), our preliminary efforts
were directed towards the investigation of two different patent
procedures describing the preparation of ZSM-5 from the classical
$Na_2O-(Pr_4N)_2O-Al_2O_3-SiO_2-H_2O$ reaction mixtures (6,35). In particular, clathrate and template effects played by the $Pr_4N^+$ cations
during synthesis, were questioned (11). For the first time, evidence was obtained that two extreme synthesis mechanisms can govern the formation and growth of ZSM-5 crystallites. Either a liquid phase ion transportation (mechanism A) or a solid hydrogel
reconstruction (mechanism B) can occur, depending essentially on
the source of silica and on the relative concentrations of the
reactants. Very recently, the existence of these two types of mechanisms was also ascertained to occur in the case of the nucleation and growth of the $(NH_4^+, Pr_4N^+)$ZSM-5 zeolite, in absence of
alkali cations (30).

High crystallization rates and the possibility to stabilize
X-ray amorphous phases, which exhibit ZSM-5 like properties, were
among the reasons why we decided to investigate the procedure B
in more detail. In order to optimize the particle size, homogeneity, morphology and composition, we have questioned more systematically the influence of secondary synthesis variables such as
the pH, solvent viscosity or the nature of the alkali cation, added as chloride.

As far as we know, the influence of the viscosity of the medium on the size and distribution of the ZSM-5 crystallites has
never been reported in the literature. The strong dependence of
the induction period required for a zeolite nucleus to form, on
alkalinity (pH), has been emphasized by Barrer (36). Yet, ZSM-5
zeolites appear to behave rather differently than other Al-rich
zeolites. The $OH^-/SiO_2$ ratio was recognized by Rollmann as the

dominant variable in influencing nucleation and crystallization rates of ZSM-5, as well as the final crystal morphologies, sizes and agglomeration (37,38). Other studies (14,24) have demonstrated that an optimum $OH^-$ concentration is needed to maintain enough dissolved hydroxy Si and Al species but not in such excess as to inhibit the subsequent nucleation and growth of ZSM-5 crystallites.

Alkali ions (salts) influence the formation of the precursor gel for most of the synthetic zeolites (3,34,39,40). $Na^+$ ions were shown to enhance in various ways the nucleation process (structure-directing role) (40-42), the subsequent precipitation and crystallization of the zeolite (salting-out effect) (1) and the final size and morphology of the crystallites (34,43). Informations on the various roles played by the inorganic (alkali) cations in synthesis of ZSM-5, such as reported in some recent publications (7,8,10,14,17,29,30,44,45) remain fragmentary, sometimes contradictory and essentially qualitative.

Except for the work of Erdem and Sand, who investigated the ($Na^+$, $K^+$, $Pr_4N^+$) system (7,8), and the recent study of Nastro and Sand (29) of the ternary cationic ($NH_4^+$, $M^I$ ($M^I$ = $Li^+$, $Na^+$, $K^+$), $Pr_4N^+$) systems, no other competitive interaction between $Na^+$, $Pr_4N^+$ and an alkali cation other than Na, has been reported.

The last part of our work consists in extending our investigations of variables influencing the synthesis of type B, by examining the potentiality of various organic compounds to direct the formation of pentasil-type structures.

$Pr_4N^+$ cations were recognized to form complexes with silicate (46) or aluminosilicate (23) species and subsequently to cause replication of the so formed framework structure via stereo-specific interactions (template effect). During this process, they are progressively incorporated and stabilized within the zeolite framework (11,47,48). Rollmann (37) has shown that initial $Pr_4N^+$/$SiO_2$ ratios above 0.05 are required to fill the whole zeolitic pore volume. These optimum $Pr_4N^+$ concentrations have been recently redefined (18). The presence of nearly 4 $Pr_4N^+$ entities per unit cell of crystalline ZSM-5 has been confirmed experimentally(11,23).

While ZSM-11 and ZSM-8 (its supposed structural analogue (49, 50)) need specifically the presence of $Bu_4N^+$ (or $Bu_4P^+$) (51,52) and $Et_4N^+$ (53) to be formed, ZSM-5 can be prepared by using $Pr_4N^+$ cations, or its precursors, or a wide variety of other organic compounds, or even from reaction mixtures free of organic cations, in the presence of ($Pr_4N^+$) ZSM-5 seeds (54). On the other hand, some organic compounds may yield less specific zeolitic structures. For example, a number of ZSM-5/11 intergrowths could be prepared using triethylpropylammonium cations (55) or various other $Alk_4N^+$ cations, either as admixture or with the addition of other organics (56).

Since the actual role of the various organics in directing specific zeolitic structures remains confusing or at least obscure, we have tried to explain the nature and the structure of the different pentasil zeolites synthesized under identical experimental

conditions, using various $Alk_4N^+$ cations and the corresponding
trialkylamines.

## Experimental

Materials and synthesis procedures  In synthesis A (essentially
based on example 1 of ref. 6), a solution of sodium aluminate,
prepared by dissolving adequate amounts of pure aluminium powder
(Al 99 % pure, Fluka) in concentrated NaOH (U.C.B. anal. grade),
is added to a solution of silica (Davison grade 950), preliminar-
ily dissolved in a solution of $Pr_4NOH$ (20 % in $H_2O$, Fluka, purum).
The so obtained gel (pH ≃ 11.5) is stirred for 2 h before heating.
     In synthesis B (essentially based on example 1 of ref. 35),
an acidic solution, prepared by mixing Al sulfate ($Al_2(SO_4)_3$.
18 $H_2O$, Merck, purum), sulfuric acid (98 % $H_2SO_4$, Merck, anal.
grade) and $Pr_4NBr$ (Fluka, purum) predissolved in $H_2O$, is added to
a basic solution of aqueous sodium silicate (Natronwasserglass
Merck, 28.5 wt % $SiO_2$, 8.8 wt % $Na_2O$ and 62.7 wt % $H_2O$). The so
formed gel (pH ≃ 2) is stirred for 1 h before its pH is adjusted
to about 11 by adding dropwise a concentrated NaOH solution. The
new gel is stirred for 3 h, left for 65 h at room temperature and
then stirred again for 12 h, prior to heating.
     Due to the very time consuming of the above procedure, the
latter was modified as follows (synthesis B) : equal volumes of
silicate and $Al-Pr_4N^+-H_2SO_4$ solutions were added at the same rate
to an aqueous solution of NaCl (Merck, anal. grade). The gel which
is formed (pH ≃ 3.5), is stirred for 2 h before its pH is adjusted
to about 9-9.5 by further adding of the remaining excess of Na silicate solution. The so obtained gel is stirred for 3 h before
heating.
     The molar compositions of the final mixtures for the synthe-
ses A, B and B' are the following :

| | | $SiO_2$ | : | $Na_2O$ | : | $Al_2O_3$ | : | $H_2SO_4$ | : | $Pr_4N^+$ | : | NaCl | : | $H_2O$ |
|---|---|---|---|---|---|---|---|---|---|---|---|---|---|---|
| A | : | 24.2 | | 1.03 | | 1 | | – | | 14.4 | | – | | 410 |
| B | : | 89.1 | | 32.0 | | 1 | | 25.9 | | 9.25 | | – | | 4020 |
| B' | : | 96.5 | | 28.8 | | 1 | | 16.6 | | 9.0 | | 47.1 | | 2440 |

     The total amounts of hydroxide ions, calculated by summing
moles of $OH^-$ added (as NaOH, $Pr_4NOH$, sodium aluminate or sodium
silicate) and by substracting moles of acid added (as $H_2SO_4$ or aluminium sulfate), are 16.45, 19.3 and 8.8 for mixtures A, B and B'
respectively.  Their final ingredient molar ratios are compared
in Table I.

Table I.  Comparison of the ingredient molar ratios for various
ZSM-5 synthesis procedures

| Ingredient molar ratios | Synthesis type | | | Suggested range (37) | |
|---|---|---|---|---|---|
| | A | B | B' | broad | preferred |
| $OH^-/SiO_2$ | 0.68 | 0.20 | 0.10 | $10^{-8}-1.0$ | 0.01-0.2 |
| $Na^+/SiO_2$ | 0.085 | 0.72 | 1.08 | 0.1-1.5 | 0.1-0.6 |
| $Pr_4N^+/SiO_2$ | 0.60 | 0.10 | 0.09 | 0.01-0.6 | 0.02-0.2 |
| $H_2O/SiO_2$ | 17.0 | 45.10 | 25.3 | 5-200 | 15-50 |
| Si/Al | 12.2 | 44.6 | 48.3 | $6.2-\infty$ | 10-1000 |
| Al/Na | 0.97 | 0.031 | 0.019 | 1.6-0 | 1-0.003 |
| $Na_2O/(Pr_4N)_2O$ | 0.14 | 6.98 | 11.63 | 0.17-150 | 5-30 |

Nucleation and crystallization of samples prepared according
to procedures A and B' were studied as a function of time, by dividing the gelatinous mixtures into several portions and sealing
them in identical 15 ml PYREX containers.  The latter were heated
at 120°C (procedures A and B) and at 130°C (procedure B') in an autoclave, under autogenous pressure, with occasional shaking, for
given periods of time, and progressively removed in order to obtain materials with increasing degrees of crystallinity.  After
cooling, each sample (gel+zeolite) was filtered, washed with cold
distilled water and dried at 120°C for 14 h, before characterization.  100 % crystalline ZSM-5 zeolites were obtained after 312 h
(type A), after 48 h (type B) and after 110 h (type B').  In each
case it was checked that longer crystallization times did not lead
to more crystalline ZSM-5 or to transformations into other phases.
Table II gives parameters and operating conditions for the three
procedures.

Modified syntheses   To compare the influence of pH on crystallization, a gel portion, as prepared according to the procedure B',
was divided into three portions, where the pH is either left at 9,
raised to 11 or lowered to 7, by adding small amounts of concentrated NaOH or $H_2SO_4$ respectively.  To check the effect of viscosity
of the medium, the gel, as obtained in procedure B', was filtered
from its mother liquor and an equivalent volume of glycerol was
added to the still wet gel.  The addition of other alkali cations
than $Na^+$ to the initial mixture was realized by mixing equal volumes of silicate and acid  solutions (procedure B') with a solution
of the corresponding alkali metal chloride (LiCl, $NH_4Cl$, KCl, RbCl,
CsCl, all Merck, anal. grade).  The effect of the nature of the organic compound was checked by substituting $Pr_4NBr$ in the procedure B',
by equivalent molar amounts of $Bu_4NBr$ (Aldrich), $Et_4NBr$ (Aldrich),
tripropylamine ($Pr_3N$) (Aldrich), tributylamine ($Bu_3N$) (Merck,
purum) and a $Pr_3N$ : n-propyl bromide (Aldrich) mixture.  All the
syntheses were stopped after the gelatinous phase had disappeared,
yielding heavy crystalline products and transparent mother liquors.
The time required for the crystalline zeolite to separate  was

Table II. Comparison of the parameters and operating conditions for various ZSM-5 synthesis procedures

| Synthesis type | A | B | B' |
|---|---|---|---|
| Source of Al | Na aluminate (Al+NaOH) (basic) | Al sulfate+$H_2SO_4$+$Pr_4NBr$ (acid) | |
| Source of Si | $SiO_2$ in $Pr_4NOH$ (basic) | Na silicate (basic) | |
| Other ingredients | – | NaOH | NaCl |
| Gel aging time (h) | 2(basic) | 1(ac.)+80(basic) | 2(ac.)+3(basic) |
| Synthesis conditions | | Hydrothermal, PIREX tubes | |
| Heating temp. (°C) | 120 | 120 | 130 |
| Time(h) required for 100 % crystallinity | 320 | 48 | 110 |

considered as a qualitative evaluation of the kinetics of crystallization. Hydrothermal heading was stopped usually two days after
the crystalline phases had appeared. All the organic-containing
zeolite precursors were filtered, washed and dried at 120°C for
14 h.

Characterization of the solid phases Si, Al and Na contents were
determined by proton-induced γ-ray emission (PIGE) (11,57) or by
high resolution solid state $^{29}$Si-NMR spectroscopy (49,50,58)
(bulk analysis), by energy-dispersive X-ray analysis (EDX) (26)
(outer rim analysis) and by XPS (59) (surface analysis), Si and Al
being detected respectively to a depth of 8-10 μm, 1-2 μm, and
5-10 nm. Si, Al, Na, K, Rb and Cs contents were also estimated
by probing by EDX both zeolite pellets of 1 cm diameter and 0.1 cm
thickness (for average analyses) and individual zeolite crystallites dispersed and directly glued onto the support (for individual
analyses). The Li and $NH_4^+$ contents were determined respectively
by PIGE (57) and TG (28). Morphologies and particle sizes were
determined by Scanning Electron Microscopy (SEM) (11). The nature of the zeolites, their degree of crystallinity and the percentage of ZSM-5 in the as-synthesized intermediate (gel + zeolite)
phases were evaluated from X-ray diffraction (XRD) (11) and thermal
analysis data (60). In the latter case, it was assumed that both
the area of the DTA exotherm corresponding to the oxydative decomposition between 300 and 600°C of the $Pr_4N^+$ species occluded in
the crystalline precursor, and the corresponding weight loss, as
measured by TG, were proportional to the actual amount of $Pr_4N^+$
and hence to the % of ZSM-5 present. The water and the organic
content of the new pentasil phases were measured by TG/DTA, as
previously reported (11,28,60). The nature of the organic guest
molecule occluded within the precursors was identified by CP-MAS
$^{13}$C NMR spectrometry (28,61).

## Results and Discussion

Composition of synthesis procedures A, B, and B' Table I gives
the initial reactant molar ratios used in the A, B and B' procedures. Although all of them do fall within the range which favours the formation of ZSM-5 from $SiO_2-Al_2O_3-Pr_4N^+-Na_2O-H_2O$ mixtures (37), those used in procedure A differ substantially from
the ratios chosen for B or B'. Rollmann has emphasized the influence of most of these variables on the mechanism that govern
the nucleation and growth rates of ZSM-5 (37) which, in turn, influence the size and/or the morphology of the crystallites.
For example, as the $Pr_4N^+/SiO_2$ ratio increases, the polycrystalline agglomerate intergrowths, whose formation is favoured by a high
$OH^-/SiO_2$ value, are progressively replaced by an increasing
amount of well dispersed single crystals.
The mechanisms describing the formation of ZSM-5 from procedures
A and B have been previously compared and discussed extensively (11). They can be summarized as follows :

In the case of synthesis A, the depolymerization of the active silica to yield ultimately monomeric silicate anions, was shown
to be the essential rate limiting step. As a result, a small number of negatively charged monomers are formed. They can either
condense aluminate species to form aluminosilicate complexes or
interact directly with $Pr_4N^+$ ions. The latter are known to order
around them preferably Si-richer (alumino) silicate tetrahedral
units to form stable nuclei. Yet, growing ZSM-5 crystallites are
known to accomodate silica in preference (9,21,30,62) and any incorporation of aluminium into their framework is a disruptive,
difficult process (12,37,38). Consequently, the crystallization
of the so formed Si-rich nuclei will occur at least more rapidly
than a further nucleation, so that the number of crystallites initially formed stays nearly constant during the synthesis course.
The particles will grow with a relatively fast rate at the expense
of the still present aluminosilicate gel which, at the limit, can
be occluded within the growing crystallite. As the silicate species available in solution are progressively exhausted, the gel
will continue to dissolve and bring progressively Al-rich soluble
species to the outer layer of the growing particle. The resulting
large and well defined ZSM-5 crystallites are therefore expected
to present an inhomogeneous radial Al distribution. An attempt
to confirm this hypothesis was proposed on the basis of a simple
mathematical treatment of the Si/Al ratios, as obtained by PIGE
analyses (11).

In type B syntheses, a hydrogel is rapidly formed from solutions already composed of monomeric (or low oligomeric) silica and
alumina species. Yet, its composition is not too different from
that expected from the reagents ratios (Table I), as the supply of
silicate ions is not limited by any depolymerization process.
Smaller amounts of $Pr_4N^+$ ions, as well as substantially higher
Si/Al and Na/Si ratios than for the A synthesis (Table I), favour
a rapid nucleation. Structure-directing $Pr_4N^+$ cations, still present all throughout the gel, can interact intimately with the numerous reactive aluminosilicate anions and a direct recrystalliza-
tion process involving the solid hydrogel phase transformation (or
surface nucleation) is expected. Indeed, a rapid growth, yielding
a large number of small crystallites, which present a homogeneous
Al radial distribution, was confirmed experimentally (11).

Recent results obtained by Sand and coworkers (21,30) have
confirmed the existence of these two different mechanisms : in
a gel obtained from $SiO_2-Pr_4NOH$ solutions, ZSM-5 was shown to grow
within a liquid slurry, while it appears directly within a wax-
like gel formed when Na silicate and Al sulfate are used as reagents.
Figures 1 and 2 summarize the main features of type A and B syntheses, while Figure 3 reflects the different morphologies of 100 %
crystalline ZSM-5 particles grown from both procedures.

The synthesis of type B offers several advantages among which
the most promising for catalytic applications is the possibility

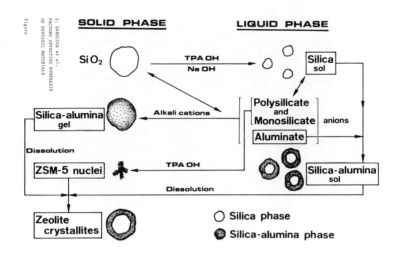

Figure 1.   Schematic representation of type A synthesis (Adapted from ref (11) and reproduced with permission, Elsevier Sci. Publ. Co.).

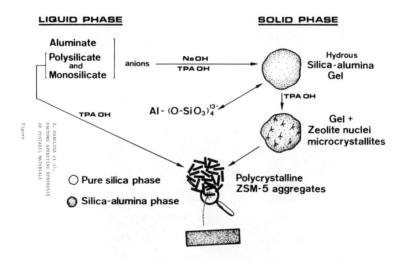

Figure 2.   Schematic representation of type B synthesis. (Adapted from ref (11) and reproduced with permission, Elsevier Sci. Publ. Co.).

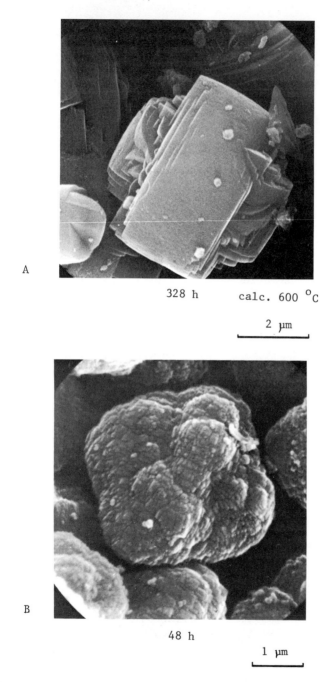

A

328 h        calc. 600 $^{o}$C

2 μm

B

48 h

1 μm

Figure 3.  Typical morphologies of 100 % crystalline A and B-type ZSM-5 (SEM micrographs) (Reproduced with permission, from ref (11), Elsevier Sci. Publ. Co.).

to prepare rapidly small 100 % crystalline ZSM-5 particles with
homogeneous Al distribution and also the possibility to stabilize
very small "X-ray amorphous" ZSM-5 crystallites, embedded within
amorphous aluminosilicate gel precursors (11).  In order to confirm and improve these properties, the operating conditions of
synthesis B were slightly modified.

Preparation of X-ray amorphous ZSM-5 crystallites according to
procedure B'  It is important that the gel formation takes place
as homogeneously as possible.  Because of the particular sensitivity of various silica and alumina species to the pH (63,64), the pH
range between 4.5 and 8.5 was avoided.  Nucleation was performed
at pH 3.5-4, where a low viscous gel containing essentially monomeric silica species is rapidly formed (65).The pH is then raised to
about 9, in order to form tetrahedral $Al(OH)_4^-$ entities and to
favour the further Al incorporation within the zeolitic framework.
NaCl was used to increase the (super)saturation of the gel, which
will flocculate into a macromolecular colloid (1) and initiate the
nucleation.  This procedure yields 100 % crystalline zeolite after
110 h hydrothermal heating at 130°C (Figure 4).  Longer heating
times (6 months) neither alter the purity nor the stability of the
crystallites (Figure 4).
     The crystallinity of the so formed intermediate phases was
checked by various physical methods.  No XRD crystallinity was detected after the first 45 hours of heating, while other techniques
such as Infrared (20,66,67), TG-DTA (11,32,67) or [13]C NMR (32,33),
which are sensitive to the presence of very small amounts of $Pr_4N^+$
species occluded in ZSM-5 crystallites, confirm that very small
size ZSM-5 particles are present in the early beginning of the synthesis process (Table III).  Further IR studies of the ZSM-5 ske-
leton vibration bands (20,66) or catalytic tests (6ɔ) have confirmed their presence in B-type procedures.  By contrast, and as expected, XRD and DTA techniques give identical crystallinity values
in the case of synthesis A (Table III).

Aluminium distribution in A and B' intermediate and final phases
Various compositional zonings have been reported in silica-rich
ZSM-5 zeolites prepared under particular conditions (11-13,19,26,
27,31,32,59,68) while other studies provided evidence for homogeneous Al distribution throughout the crystallites (69-71).  Obviously, the distribution of aluminium in ZSM-5 must depend on its
mechanism of crystallization.
In order to bring experimental confirmation of the Al radial distribution within the A-type crystallites and on homogeneous compositions of B-type phases (11), we have reexamined their composition
by using the PIGE, EDX and XPS techniques, which are selectively
able to probe the presence of Si and Al at various depths.  Table
III compares the Si/Al ratios for several intermediate phases
(gel + zeolite), isolated after different crystallization times,
from syntheses A and B'.

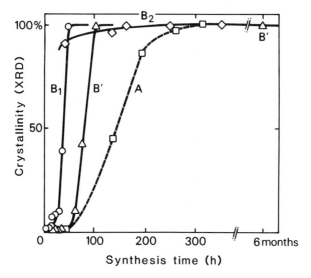

Figure 4.  Typical crystallization curves for A, B and B'-type syntheses.

Table III. Variations of composition and crystallinity in A and
B'-type intermediate and final phases.

| Synthesis | | Si/Al at. ratio | | | % crystallinity | | |
|---|---|---|---|---|---|---|---|
| Type | Cryst.time(h) | PIGE | EDX | XPS($\underline{73}$) | XRD | DTA | |
| | $0^*$ | $12.2^*$ | – | – | – | – | |
| | 48 | 1.8 | 3.5 | 5.0 | 0 | 0 | |
| $\underline{A}$ | 136 | 5.0 | 11.7 | 11.0 | 45 | 50 | |
| | 192 | 10.0 | 16.7 | 15.5 | 86 | 92 | |
| | 264 | 12.1 | – | – | 96 | – | |
| | 312 | 13.2 | 15.3 | 10.0 | 100 | 100 | |
| | $0^*$ | $45.0^*$ | – | – | – | – | |
| | 14 | 33.6 | 26.8 | 28.0 | 0 | 34 | |
| | 20 | 32.0 | 26.9 | – | 0 | 34 | |
| | 26 | 34.2 | 27.0 | – | 0 | 34 | XRD |
| $\underline{B'}$ | 32 | 34.1 | 27.3 | (34) | 0 | 38 | Amorphous |
| | 38 | 34.3 | 27.5 | – | 0 | 42 | Zeolites |
| | 42 | 34.3 | 28.6 | 20 | 0 | 43 | |
| | 49 | 34.3 | 28.0 | 24 | 12 | 49 | |
| | 64 | 32.0 | 26.7 | 24 | 38 | 63 | |
| | 110 | 36.1 | 27.0 | 23 | 100 | 100 | |
| | Average values : | 35 | 27.5 | | | | |

$^*$ Si/Al in aqueous gel, from ingredients

Figure 5 shows that for the type A synthesis, the originally very
low Si/Al ratio (PIGE) progressively increases and reaches, at the
end of the process, values close to that of the starting ingredients. This is consistent with the assumption that Si-richer
ZSM-5 crystallites grow from progressive dissolution of the Al-
rich gel phase. The latter, if isolated and dried in the beginning of the crystallization (after 48 h), must be essentially Al-
rich, as most of the soluble polysilicate fraction present in a
Pr$_4$NOH-silica sol is not retained by filtration ($\underline{11,65,72}$).
EDX measurements show the same trend, except that the Si/Al ratios
are always higher. Indeed, the EDX technique only probes the upper
layer of the solid sample which is supposed to contain less Al.
As the gel progressively dissolves to yield Si-richer zeolitic
phases, the EDX and PIGE Si/Al ratios become closer. However,
even for a 100 % crystalline phase, less Al is still probed by
EDX. This suggests that the A-synthesis ZSM-5 crystals must still
contain some Al-rich amorphous phase, deeply embedded within the
large particles, as to partially escape the EDX probing. For the
same reason, the XPS also continuously probes the outer Si-rich
layer of the growing particles. However, at the end of the crystallization, more Al is detected on the outer rim than in the

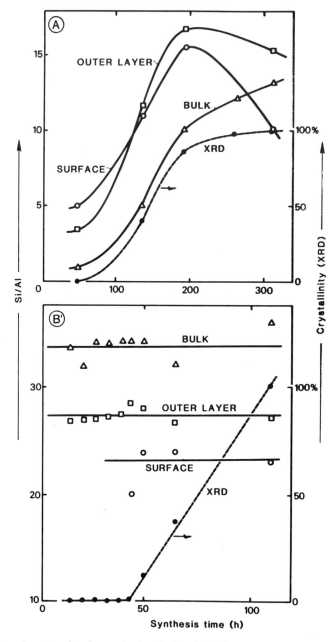

Figure 5.  Variation of the bulk (PIGE), outer layer (EDX)
and surface (XPS) Si/Al ratios and XRD crystallinities, of
intermediate phases, as a function of synthesis time, for
procedures A and B'.

bulk (73). This important finding is strongly in favour of the formation of crystallites having more Al in their outer layer. For synthesis B', both PIGE and EDX Si/Al ratios remain remarkably constant throughout the whole crystallization process. This indicates that both the growing crystallites and their gel precursor must continuously keep the same composition. This is consistent with the direct hydrogel transformation mechanism. In that case, however, EDX and XPS systematically detect slightly more Al near the surface, suggesting that in the beginning of the B' process, more bulky Si-rich inhomogeneous domains are present within the gel. Through internal rearrangements, they will transform progressively in ZSM-5 nuclei. The latter will grow with a rate proportional to their Si-content and yield zeolite crystallites with different sizes but with an homogeneous Al distribution within each phase.

The A and B' crystallization procedures are schematically illustrated in Figure 6.

Influence of pH on the crystallization of ZSM-5 Three identical B'-type portions from the same batch have been heated at 130°C at three different pH values. Their synthesis characteristics and various physical parameters are compared in Table IV. Figure 7 illustrates their morphological differences.

Table IV. Variation of different parameters as a function of pH (synthesis B')

| Sample | pH before/after heating | | Cryst. time (h) | Si/Al (PIGE) | Crystals size (μm) | morphol. | amount (mass%) | Amorph. phase (mass%) |
|---|---|---|---|---|---|---|---|---|
| B'$_1$ | 6.9 | 8.4 | >64 | 13.2 | 5x10 | twins | 30 | 70 |
| B'$_2$ | 9.0 | 10.0 | 44 | 30.0 | 1-5 / 15-20 | polycryst. aggreg. | 3 / 97 | - |
| B'$_3$ | 11.05 | 10.9 | 15 | 27.9 | 2 | " | 100 | - |

Low pH values lead to long crystallization times and incomplete gel transformations. At high pH values, pure ZSM-5 crystals are obtained after a very short time. They have the same morphology as those grown at pH 9 but appear smaller and homogeneously distributed. At low pH (which, in this case, is proportional to the $OH^-/SiO_2$ ratio), the dissolution process would be largely absent and well defined isolated crystals are expected to form (37), probably from more homogeneous Si-rich gel domains, intermixed with aluminosilicate-$Al(OH)_3$ inhomogeneous phases, which remain amorphous. The low Si/Al ratio confirms these proposals and suggests that the hydrogel transformation is not the only mechanism through which ZSM-5 is formed. At higher $OH^-/SiO_2$ ratios, polycristalline

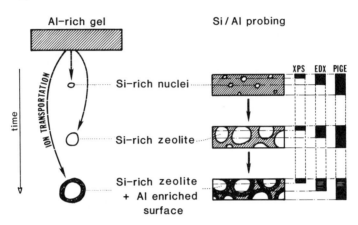

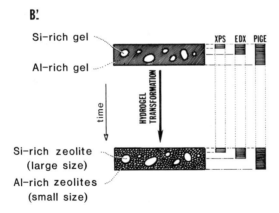

Figure 6. Schematic representation of the Si/Al variations in the synthesis A and B' mixtures during their crystallization.

pH = 7                    pH = 9

pH = 11

20 μm

Figure 7. SEM micrographs of ZSM-5 crystallites synthesized at different pH values (synthesis B', 130°C, 4 days).

aggregates are formed in high yield. Their wide size distribution
confirms the presence of Si-Al inhomogeneous distribution within
the gel. At very high $OH^-/SiO_2$, crystal growth and dissolution
would be competing and an equilibrium, producing smaller crystals,
would result (37). Dissolution will cause a diffuse surface orientation of the structure directing $Pr_4N^+$ units and impede any crystal formation.

Influence of the nature of the liquid phase on the crystallization
of ZSM-5    The aqueous part of a classical B'-type gel was replaced by the equivalent volume of pure glycerol. Crystals obtained
after hydrothermal heating of the aqueous and glycerol gels are
compared in Figure 8 and Table V.

Table V.   Variation of different parameters as a function of the
           nature of the liquid phase of the B'-type synthesis.

| Sample | B'4 | B'5 |
|---|---|---|
| Liquid phase | aqueous | glycerol |
| Material | 100 % ZSM-5 | 100 % ZSM-5 |
| Aver. size (μm) | 3.4 | 9 |
| Size distribution (μm) | (1-7.5) | (8.5-9.5) |

Larger sized and more homogeneous particles are formed in glycerol
medium. The increase is viscosity probably impedes the dissolu-
tion-growth competitive process and the growth predominates. On
the other hand, the high affinity of glycerol towards water results in important supersaturation of the sol which favours a rapid nucleation and growth. As a result, numerous and large crystals will be formed.

Role of alkali and $NH_4^+$ cations in the crystallization of ZSM-5
Introduced in an aqueous (alumino) silicate gel (sol), the bare
alkali cations will behave in various ways : firstly, they will interact with water dipoles and increase the (super) saturation of
the sol. Secondly, once hydrated, they will interact with the aluminosilicate anions with, as a result, the precipitation of the so
formed gel (salting-out effect). Thirdly, if sufficiently small,
they also can order the structural subunits precursors to nucleation species of various zeolites (template function-fulfilled by
hydrated $Na^+$ in the case of ZSM-5 (11,48)).
      Small bare cations ($Li^+$, $Na^+$) will readily order water molecu-
les in forming regular structural entities with aluminosilicate
species (structure-forming effect). Larger cations ($K^+$, $NH_4^+$, $Rb^+$,
$Cs^+$) interact less with $H_2O$ and even disrupt its regular structure
stabilized through H-bonding (structure-breaking effect). The very large $Pr_4N^+$ with shielded charge will tend to build stable water
clathrates (74). They are also structure-forming towards water.

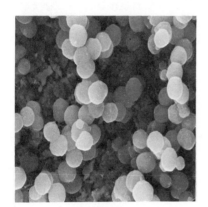 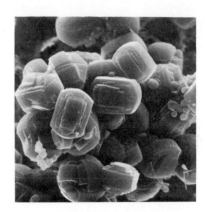

aqueous solution                    glycerol

20 µm

Figure 8. SEM micrographs of ZSM-5 obtained in aqueous solution and in glycerol (synthesis B', 130°C, 3 days).

Six ZSM-5 samples were synthesized from gels containing the whole series of alkali cations, added in form of chloride, as described in procedure B'. Synthesis data, principal properties and analyses are summarized in Tables VI and VII.

Table VI.   Synthesis data and properties of various (M)ZSM-5 samples (M = alkali cation) 'adapted from ref (26), by permission)

| Zeolite | Cryst. time (days) | Aver. size (µm) | H$_2$O content (mol./u.c.) | H$_2$O content (mol./M$^I$) | Pr$_4$N$^+$ content (mol./u.c.) |
|---------|------|------|------|------|------|
| (Li)ZSM-5 | 3 | 1.7 | 10.5 | 13.0 | 3.4 |
| (Na)ZSM-5 | 3.5 | 4.5 | 11.8 | 14.8 | 3.6 |
| (K)ZSM-5 | 3.5 | 18 | 8.2 | 7.5 | 3.7 |
| (Rb)ZSM-5 | 4 | 22 | 8.6 | 7.8 | 3.4 |
| (Cs)ZSM-5 | 7 | 25 | 5.6 | 2.8 | 3.6 |
| (NH$_4$)ZSM-5 | 93 | (see text) | 2.9 | 1.7 | 3.9 |

Table VII.   "Surface" (EDX) and bulk (PIGE) analyses of various (M)ZSM-5 samples (adapted from ref (26), by permission)

| Zeolite | Si/Al (PIGE) | Si/Al (EDX) | Al/Na (PIGE) | Al/M (EDX) | Al/u.c. (PIGE) | Na/u.c. (PIGE) | M/u.c. (EDX) | (Na+M)/ u.c. |
|---------|------|------|------|------|------|------|------|------|
| (Li)ZSM-5 | 32.4 | 31.0 | 14.2 | 3.5 | 2.87 | 0.07 | 0.81 | 0.88 |
| (Na)ZSM-5 | 36.1 | 33.5 | 3.2 | 2.4 | 2.56 | 0.80 | – | 0.80 |
| (K)ZSM-5 | 38.5 | 28.2 | 13.9 | 2.2 | 2.43 | 0.17 | 1.10 | 1.27 |
| (Rb)ZSM-5 | 47.5 | 35.0 | 9.1 | 1.8 | 1.98 | 0.22 | 1.10 | 1.32 |
| (Cs)ZSM-5 | 45.2 | 35.5 | 10.0 | 1.1 | 2.08 | 0.21 | 2.00 | 2.21 |
| (NH$_4$)ZSM-5 | 41.2 | 39.0 | 70.0 | 1.4 | 2.27 | 0.03 | 1.68 | 1.71 |

These data show that most of the physical parameters and properties which characterize each (M)ZSM-5 sample, show regular variations as a function of the nature of the cation, i.e. essentially of its size.

Except the particular behaviour of NH$_4^+$ which is envisaged separately, starting from Li to Cs, the following trends are observed : crystallization times and average particles sizes increase while the size distribution becomes less homogeneous (Figure 9). The water and Al contents decrease while the amount of M cations per unit cell of zeolite increases. For Li and Na, the morphology consists of clusters of polycrystalline aggregates. Better outlined single crystals are observed for K and Rb and additional pronounced twinning appears for (Cs)ZSM-5 (Figure 10).

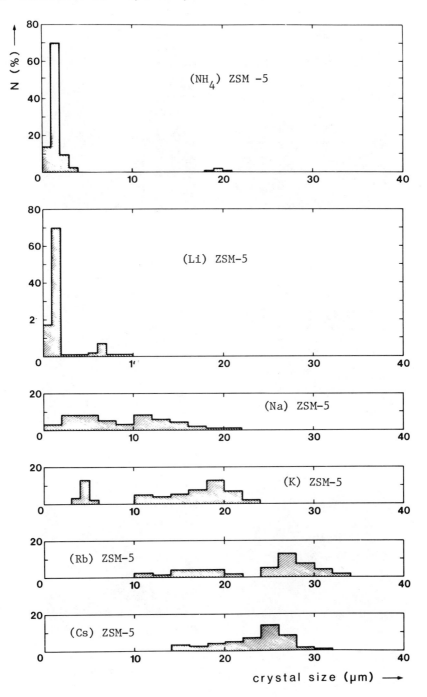

Figure 9. Particle size distribution for the various
(M)ZSM-5 zeolites, as computed from SEM (Reproduction with
permission, from ref (26), Elsevier, Sci. Publ. Co.).

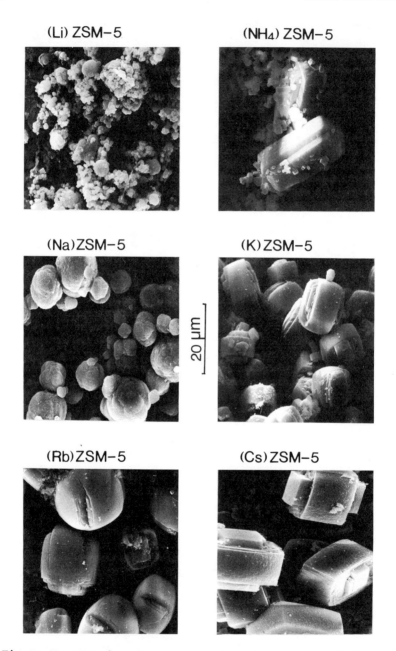

Figure 10. SEM micrographs comparing sizes and morphologies of the various (M)ZSM-5 zeolites (Reproduced with permission, from ref (26), Elsevier, Sci. Publ. Co.).

The various chemical analyses yield the following interesting conclusions :

(a) Small sized crystallites are equally probed by EDX and PIGE (Si/Al ratios, Table VII). The bigger the average size, the greater the difference between "surface" and bulk ratios.

(b) EDX shows that the average "surface" of a pellet composed of various sized crystallites, is always enriched in Al with respect to the bulk. On the other hand,(Si/Al) ratios determined by EDX on individual crystallites indicate that in all cases except $NH_4^+$, small crystallites contain more Al than the large ones (Figure 11). Combining these two informations, it appears that Al must be distributed homogeneously within the crystallites, although each crystallite must have an Al concentration inversely proportional to its size. Such a conclusion substantially confirms that (M)ZSM-5 are produced within the gel by structural rearrangements, as expected for a B'-type procedure.

(c) When M (from MCl) is present in the gel, very few $Na^+$ (from Na silicate) is incorporated in the zeolite lattice (Table VII). A possible explanation is to assume that the "free" $M^I$ ions (among which $Na^+$ from NaCl) interact with the aluminosilicate gel (and eventually direct its structure towards a zeolite framework by template effect), more readily than the $Na^+$ which were originally associated with the silicate anions.

(d) (Al/M) ratio, greater than unity for Li and Na zeolites, almost reaches unity for (Cs) ZSM-5. It is concluded that in the first case, $Pr_4N^+$ entities act as both structure directing units and exchange cations, while in the second case, they are essentially present within the zeolite lattice as $Pr_4NOH$. Further studies are in progress to elucidate the actual structure of $Pr_4N^+$ ions in the (M)ZSM-5 zeolites (75).

(e) (Al/M) ratio is constant for various crystal sizes (Figure 11), indicating that the incorporation of the alkali is governed by the Al concentration.

The general behaviour of the alkali cations in presence of aluminosilicate gels and their influence on zeolite nucleation and crystallization rates has been discussed in detail elsewhere (26). Their specific role in the formation of the various (M)ZSM-5 zeolites can be depicted on the basis of the above described observations.

Effect of $Cs^+$, a structure-breaking type cation   In acidic medium, the silica anions will interact preferentially with the smallest positively charged entities which appear to be the hydrated $Cs^+$ ions. Consequently, a small number of nuclei (because silicate-$Pr_4N^+$ interactions are less preferred), essentially Si-rich (because silicate-$Al(H_2O)_n^{3+}$ interactions are less favoured) and containing very little Na (because silicate-$Na(H_2O)_x^+$ interactions are

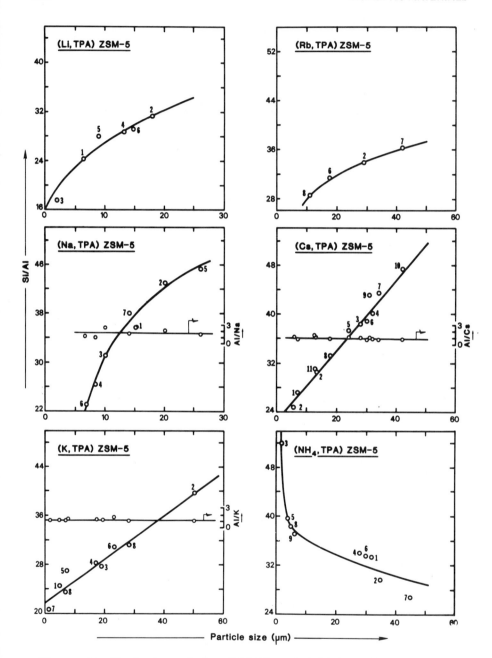

Figure 11. Variation of the (Si/Al) and (Al/M) ratios, as measured by EDX, as a function of crystal size, for various (M)ZSM-5 zeolites.

impeded), will be formed. Moreover, because of their structure-breaking behaviour towards water, the $Cs^+$ cations will desorganize the clathrated water structure of the silicate anions, preventing their "precipitation" at the $Pr_4N^+$ centers where the templating effect occurs.

In basic medium, negative Al species appear and condense with the remaining silicate entities, as to yield an Al-rich aluminosilicate gel. A few Al-richer nuclei can then form and yield a small amount of Al-richer ZSM-5 particles at the end of the process.

<u>Effect of $Na^+$, a structure-forming cation</u> Hydrated $Na^+$ cations have a large size and a more delocalized charge than the $Cs^+$ ions, so that the interaction between the silicate complex anions and the $Pr_4N^+$ cations is more favoured in presence of $Na(H_2O)_x^+$. Even the hydrated Al cationic species may interact at this stage of the process. In addition, the $Na^+$ ions will favour the water ordering in the aluminosilicate species that appear at higher pH values. Consequently, the hydrous gel will undergo nucleation rapidly and a large number of centers are formed. Smaller crystallites are obtained due to the relatively important Al concentration, either present in the original nulcei or incorporated during growth. Their particular morphology consists of medium-size (3-15 μm) units, still showing crystal faces, which however are less developed and poorly outlined (Figure 3, synthesis B). They always appear to be covered with a multitude of very small (0.1 μm) crystallites (Figure 12). Such a morphology suggests that the growth of the nuclei initially starts from the silicate species available. Once the gel is enriched in Al, it will yield, through a secondary nucleation, these small platelet-like crystallites, which either agglomerate separately or cover the primary Si-rich crystals, sometimes so heavily that the polycrystalline aggregates so formed develop a spherulitic shape (Figure 10).

<u>The particular behaviour of $NH_4^+$ ions</u> Because of their small size, $NH_4^+$ ions develop an important charge and behave essentially as structure-breaking towards water. As in the case of $Cs^+$ ions, few nuclei are formed and yield very large well defined crystals. A number of unusually large (up to 25 x 45 μm) crystals appear indeed, in agreement with previous observations (10,12,13,30). Combined EDX and PIGE analyses reveal that they must have a Si-rich core and an Al-enriched outer rim, as it was observed for the large single crystals grown from syntheses of type A. Indeed, because of the very poor salting-out effect of the $NH_4^+$ ions, gelation will be impeded and a liquid phase ion transportation mechanism can occur.

On the other hand, we also observed the formation of numerous small hexagonal ZSM-5 platelets which appear, after a long heating period, either well separated or sprinkled onto the primary large crystals (Figure 13). They contain more Si than the outer rim of the big crystals (EDX). Their formation through a secondary

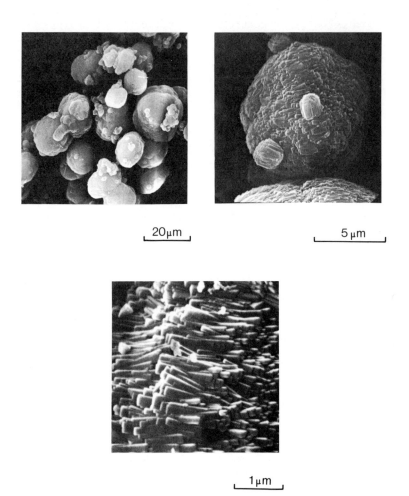

Figure 12. SEM micrographs showing the polycrystalline aggregate-type morphology obtained for (Na)ZSM-5.

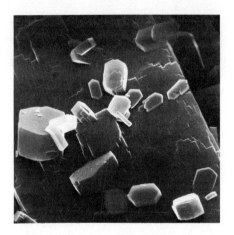

2μm

Figure 13. Secondary crystallization of small hexagonal
crystallites of (NH$_4$)ZSM-5 on the original single crystal
of the same (Reproduced with permission, from ref (26),
Elsevier Sci. Publ. Co.).

nucleation is explained as follows. Because of the preferential aluminate-$NH_4^+$ interactions, which predominate over silicate-$NH_4^+$ interactions (26), the Al-rich surface of the big crystallites will have tendency to adsorb the excess of the $NH_4^+$ ions and the growth process is stopped. The resulting solution still contains $Pr_4N^+$ cations and Al-depleted aluminosilicate anions, while the concentration of $NH_4^+$ ions is now very low. This favours the secondary formation of a large amount of new Si-rich crystallites. Their growth is stopped as soon as the solution is exhausted from ingredients, which explains their average small size.

Specific structure-directing effects of some organic bases or cations  When in the procedure B' $Pr_4N^+$ is replaced by other organics, various pentasil-type zeolitic precursors are formed. It appears that specific zeolites are formed only when quaternary ammonium salts are used, their nature (structure) being essentially dependent on the length of the alkyl chains : pure ZSM-8, ZSM-5 and ZSM-11 are obtained respectively with $Et_4N^+$, $Pr_4N^+$ and $Bu_4N^+$ cations. TG data indicate that the latter fill nearly completely the zeolitic channel system (Table VIII).

Table VIII.   Characteristics of some pentasil zeolites obtained from synthesis B' using various organic bases or cations (adapted from ref (25), by permission).

| Zeolite | Si/Al (PIGE) | Chemical nature of the organic occluded (NMR) | %ZSM-5 present (XRD) | Mol.of orga- nics/u.c. (TG) | Theor. channel length (nm) | %filled pore volume (TG) |
|---|---|---|---|---|---|---|
| $(Et_4N^+)$ZSM-8 | 31.3 | $Et_4N^+$ | – | (2.23)* | * | ($\sim$90)* |
| $(Pr_4N^+)$ZSM-5 | 36.1 | $Pr_4N^+$ | 100 | 3.64 | 8.8 | 98.8 |
| $(Bu_4N^+)$ZSM-11 | 43.0 | $Bu_4N^+$ | 0 | 2.62 | 8.0 | 94.7 |
| $(Pr_3N)$ZSM-11/5 | 34.0 | $Pr_3N$ | 10 | 1.72 | 8.08 | 38.1 |
| $(Bu_3N)$ZSM-5/11 | 29.4 | $Bu_3N$ | 60 | 1.61 | 8.48 | 41.2 |
| $(Pr_3N+PrBr)$ZSM-5 | 32.9 | $Pr_3N$ | 92 | 2.57 | 8.72 | 52.8 |

*  Theoretical channel length estimated $\simeq$ 4.65 nm, which should accomodate 2.45 mol of $Et_4N^+$, if 100 % filling is reached (25).

When tripropylamine or tributylamine is used instead of the corresponding $Alk_4N^+$ salt, ZSM-5/11 mixed phases (intergrowths ?) are formed, suggesting that the $Alk_3N$ species are less efficient in directing a specific structure. Unexpectedly, $Bu_3N$ yields essentially a ZSM-5-rich phase while ZSM-11-rich phases are preferentially obtained with $Pr_3N$ (XRD data). When an organic molecule acts as template towards (alumino)silicate species to form an ordered zeolitic framework, the latter is supposed to organize itself around the host organic species in such a way that a complete filling of the inorganic framework is achieved.  The nature of the zeolite formed will then essentially depend on the pore filling and not on the necessary location of the organic molecules at the (four) channel intersections of the so formed unit cell.  Experimental data have suggested that for ZSM-5 and ZSM-11 the most energetically favourable filling is obtained when $Pr_4N^+$ and $Bu_4N^+$ ions respectively achieve an  end-to-end configuration with no other steric constraints (28,76).  This also corresponds to about 4 $Pr_4N^+$ ions per unit cell of ZSM-5, while, because of the large size of the $Bu_4N^+$ units, only about 2.6 $Bu_4N^+$ units occupy (completely) one ZSM-11 unit cell on the average (28).  Bearing such a "complete pore filling" model in mind, one can also calculate the maximum space filled by four $Pr_3N$ and four $Bu_3N$ when they direct the formation of a hypothetical zeolite framework, (one molecule per intersection of a fictitious pentasil-type zeolite) (25).  The fictitious unit cell space length obtained for 4 $Pr_3N$ is 7.16 nm and for 4 $Bu_3N$, 8.68 nm, indicating that the best filling with $Pr_3N$ is achieved for ZSM-11 (total pore volume = 8.0 nm) and with $Bu_3N$, for ZSM-5 (total pore volume = 8.8 nm).  The actual composition of the mixed phases is close to that respectively predicted by our calculations, assuming a complete pore filling (Table VIII).  However, the actual percentage of the space filled, as experimentally evaluated by TG (Table VIII), does not correspond to the theoretical predictions.  This suggests that the complete filling is achieved only at the early stages of the nucleation-crystallization processes, which are critical to determine the nature of the zeolitic material.  Once a particular structure is achieved, it can grow through stereospecific spatial replications without necessarily incorporating the organic molecules.

When $Pr_3N$ and n-propyl bromide are used, the zeolite which is formed appears to be a pure ZSM-5 phase.  However, [13]C-NMR experiments indicate that the occluded organic species are $Pr_3N$ and not $Pr_4N^+$.  However, $Pr_4N^+$ must form because they specifically direct ZSM-5 structures.  $Pr_3N$ alone would have yielded  a ZSM-11-rich intergrowth.  It is then assumed that $Pr_4N^+$, formed in small amounts (traces) by reaction of $Pr_3N$ with propylbromide, must operate as structure-directing agent only during the nucleation stage.  Once ZSM-5 nuclei are formed, they can grow by progressively incorporating the most abundant $Pr_3N$ species, which remain in solution.

## General Conclusions

The existence of two distinct mechanisms describing the synthesis of ZSM-5 type materials has been proposed and experimentally confirmed.  The control of primary synthesis variables such as the source of silica and the composition of the reaction mixtures enables us to prepare ZSM-5 particles with specific properties desired for various catalytic applications : large single crystals having a Si-rich core and an Al-enriched outer rim, smaller and homogeneous particles or even very small "X-ray amorphous" ZSM-5 microcrystallites embedded in Al-rich aluminosilicate gel phases. The further investigation of the influence of some other secondary synthesis variables (pH, viscosity, or alkali cations) on the nucleation and crystallization processes, enabled us to confirm, to extend and to generalize the proposed mechanisms and to delineate more specific operating conditions which would yield ZSM-5 crystallites with controlled size, morphology, homogeneity and chemical composition. The clathrating and templating role of the $Pr_4N^+$ cations was recognized and emphasized. The chemical nature of the structure-directing organic base or cation appears to be an essential factor affecting the structure of the zeolitic material which is formed. Either the formation of pure pentasil zeolites or their intermediate intergrowths could be predicted and obtained from specific organic templates. The structure-directing effect of the latter operates only at the nucleation stages of the synthesis. The nature of novel potential zeolitic phases can be better explained by simplified models involving their channel filling by various templates, than by assuming the specific location of the organic molecules at the channel intersections.

## Acknowledgments

The authors wish to thank Dr. J. Rostrup-Nielsen (Haldor Topsøe A.S.) for his continuous interest in this work. They are also indebted to Elsevier Scientific Publishing Co. for having granted the permission to publish original data.

## Literature Cited

1.  Barrer, R. M.; Denny, P. J. J. Chem. Soc. 1961, 971.
2.  Kerr, G. T.; Kokotailo, G. T. J. Am. Chem. Soc. 1961, 83, 4675.
3.  Barrer, R. M. "Hydrothermal Chemistry of Zeolites"; Academic : London, 1982, p. 162.
4.  Kokotailo, G. T.; Meier, W. M. in "The Properties and Applications of Zeolites"; Townsend, R. P., Ed.; The Chemical Society : London, 1980, p. 133.
5.  Olson, D. H.; Kokotailo, G. T.; Lawton, S. L.; Meier, W. M. J. Phys. Chem. 1981, 85, 2238.
6.  Argauer, R. J.; Landolt, G. R. U.S. Patent 3 702 886, 1972.
7.  Erdem, A.; Sand, L. B. J. Catal. 1979, 60, 241.
8.  Erdem, A.; Sand, L. B. Proc 5th Intern. Conf. Zeolites, 1980, p. 64.
9.  Lecluse, V.; Sand, L. B. Rec. Progr. Rept.-5th Intern. Conf. Zeolites, 1981, p. 41.
10. Bibby, D. M.; Milestone, N. B.; Aldridge, L. P. Nature 1980, 285, 30.
11. Derouane, E. G.; Detremmerie, S.; Gabelica, Z.; Blom, N. Appl. Catal. 1981, 1, 101.

12.  Von Ballmoos, R. Ph.D. Thesis, Zurich University, Zurich, 1981.
13.  Von Ballmoos, R.; Meier, W.M. Nature, 1981, 289, 782.
14.  Chao, K. J.; Tasi, T. C.; Chen, M. S.; Wang, I. J. Chem. Soc. Faraday Trans. I 1981, 77, 547.
15.  Nakamoto, M.; Takahashi, H. Chem. Lett. 1981, 169.
16.  Hagiwara, H.; Kiyozumi, Y.; Kurita, M.; Sato, T.; Shimada, H.; Suzuki, K.; Shin, S.; Nishijima, A.; Todo, S. Chem. Lett. 1981, 1653.
17.  Nakamoto, H.; Takahashi, H. Chem. Lett. 1981, 1739.
18.  Howden, M. G.,"The role of tetrapropylammonium template in the synthesis of ZSM-5", CSIR Report CENG 413, Pretoria, 1982.
19.  Lyman, C. E.; Betteridge, P. W.; Moran, E. F. in "Intrazeolite Chemistry"; Stucky, G. D.; Dwyer, F. G., Eds.; ACS SYMPOSIUM SERIES N° 218, American Chemical Society : Washington, D.C., 1983, p. 199.
20.  Coudurier, G.; Naccache, C.; Vedrine, J. C. J. Chem. Soc. Chem. Commun. 1982, 1413.
21.  Mostowicz, R.; Sand, L. B. Zeolites 1982, 2, 143.
22.  Cavell, K. J.; Masters, A. F.; Wilshier, K. G. Zeolites 1982, 2,244.
23.  Derouane, E. G.; B.Nagy, J.; Gabelica, Z.; Blom, N. Zeolites 1982, 2, 299.
24.  Kulkarni, S. B.; Shiralkar, V. P.; Kotasthane, A. N.; Borade, R. B.; Ratnasamy, P. Zeolites 1982, 2, 313.
25.  Gabelica, Z.; Derouane, E. G.; Blom, N. Appl. Catal. 1983, 5, 109.
26.  Gabelica, Z.; Blom, N.; Derouane, E. G. Appl. Catal. 1983, 5, 227.
27.  Hughes, A. E.; Wilshier, K. G.; Sexton, B. A.; Smart, P. J. Catal. 1983, 80, 221.
28.  B.Nagy, J.; Gabelica, Z.; Derouane, E. G. Zeolites 1983, 3, 43.
29.  Nastro, A.; Sand, L. B. Zeolites 1983, 3, 57.
30.  Ghamami, M.; Sand, L. B. Zeolites 1983, 3, 155.
31.  Auroux, A.; Dexpert, H.; Leclercq, C.; Vedrine, J. C. Appl. Catal. 1983, 6, 95.
32.  Gabelica, Z.; B.Nagy, J.; Debras, G.; Derouane, E. G. Proc. 6th Intern. Conf. Zeolites, 1983, in press.
33.  Gabelica, Z.; B.Nagy, J.; Debras, G. J. Catal. 1983, in press.
34.  Barrer, R. M. Zeolites 1981, 1, 130.
35.  Chen, N. Y.; Miale, J. N.; Reagan, N. Y. U.S. Patent 4 112 056, 1978.
36.  Barrer, R. M. Ref. 3, p. 155.
37.  Rollmann, L. D.; Valyocsik, E. W. Eur. Patent 21 674 and 21 675, 1981.
38.  Rollmann, L. D. in "Zeolites : Science and Technology"; Ribeiro, F.R. et al., Eds; M. Nijhoff : Den Haag, 1983, in press.

39. Breck, D. W. in "Molecular Sieve Zeolites", Gould, R. F
    Ed.; ADVANCES IN CHEMISTRY SERIES N° 101, American Chemical
    Society : Washington, D. C.,1971, p. 1.
40. Flanigen, E. M. in "Molecular Sieves"; Meier, W. M.;
    Uytterhoeven, J. B.,Eds.; ADVANCES IN CHEMISTRY SERIES N° 121,
    American Chemical Society : Washington, D. ., 1973, p. 119.
41. Barrer, R. M. Chem. Ind. (London) 1968, 1857.
42. Breck, D. W. "Zeolite Molecular Sieves : Structure, Chemis-
    try and Use", Wiley : New York, 1974, chap. 4.
43. Robson, H. Chemtech. 1978, 8, 176.
44. Dwyer, F. G.; Cormier Jr. W.E.; Chu, P. German Patent
    2 836 076, 1978.
45. Grose, R. W.; Flanigen  E. M. U.S. Patent 4 257 885, 1981.
46. Flanigen  E. M.; Bennett, J. M.; Grose, R. W.; Cohen, J. P.;
    Patton, R. L.; Kirchner, R. M.; Smith, J. V. Nature 1978,
    271, 512.
47. Rollmann, L. D. in "Inorganic Compounds with Unusual Proper-
    ties"; King, R. B., Ed.; American Chemical Society : New
    York, 1979, vol. II, p. 387.
48. Flanigen, E. M. Pure Appl. Chem. 1980, 52, 2191.
49. B.Nagy, J.; Gabelica, Z.; Debras, G.; Bodart, P.; Derouane,
    E. G.; Jacobs, P. A. J. Molec. Catal. 1983, 20, 327.
50. Gabelica, Z.; B.Nagy, J.; Bodart, P.; Debras, G.; Derouane,
    E.G.; Jacobs, P. A. in "Zeolites : Science and Technology";
    Ribeiro, F. R. et al., Eds; M. Nijhoff : Den Haag, 1983,
    in press.
51. Chu, P. U.S. Patent 3 709 979, 1973.
52. Rubin, M. K.; Plank, C. J.; Rosinski, E. J.; Dwyer, F. G.
    Eur. Patent 14 059, 1980.
53. Chen, N. Y.; U.S. Patent 3 700 585, 1972.
54. Gabelica, Z.; Derouane, E. G.; Blom, N. Appl. Catal. 1983,
    5, 109; references 16 to 29 cited thereïn.
55. Limova, T. V.; Amirov, S. T.; Meged', N. F.; Mamedov, Kh. S;
    Izv. Akad. Nauk. SSSR, Neorg. Mater. (English transl.) 1980,
    16, 1593.
56. Kokotailo, G. T. Eur. Patent 18 090, 1980 and U.S. Patent
    4 289 607, 1981.
57. Debras, G.; Derouane, E. G.; Gilson, J. P.; Gabelica, Z.;
    Demortier, G. Zeolites  1983, 3, 37.
58. B.Nagy, J.; Gabelica, Z.; Derouane, E. G.; Jacobs, P. A.
    Chem. Lett. 1982, 2003.
59. Derouane, E. G.; Gilson, J. P.; Gabelica, Z.; Mousty-
    Desbuquoit, C.; Verbist, J. J. Catal. 1981, 71, 447.
60. Gabelica, Z.; Gilson, J. P.; Debras, G.; Derouane, E. G.;
    Proc. 7th Intern. Conf. Thermal. Anal.; Miller, B., Ed.;
    Wiley-Heyden : New York, 1982, vol. II, p. 1203.
61. B.Nagy, J.; Gabelica, Z.; Derouane, E. G. to be published.
62. Guth, J. L. and Sand, L. B. Rec. Progr. Rept. - 5 th Intern.
    Conf. Zeolites 1981, p. 206.

63. Iler, R. K. "The chemistry of silica"; Wiley : New York,1979.
64. Iler, R. K. in "Soluble Silicates"; Falcone Jr, J. S.; Ed.; ACS SYMPOSIUM SERIES N° 194, American Chemical Society : Washington, D.C., 1982, p. 95.
65. Andersson, K. R.; Dent Glasser, L. S.; Smith, D. N. in "Soluble Silicates"; Falcone Jr, J. S., Eds.; ACS SYMPOSIUM SERIES N° 194, American Chemical Society : Washington, D.C., 1982, p. 115.
66. Jacobs, P. A.; Derouane, E. G., Weitkamp, J. J. Chem. Soc. Chem. Commun. 1981, 591.
67. Howden, M. G. "Preparation and evaluation of an amorphous silica-aluminia catalyst synthesized in the presence of tetrapropylammonium hydroxides", CSIR Report CENG 441, Pretoria, 1982.
68. Doelle, H. J.; Heering, J.; Riekert, L. J. Catal. 1981, 71, 27.
69. Suib, S. L.; Stucky, G. D.; Blattner, R. J.; J. Catal. 1980, 65, 174.
70. Dwyer, J.; Fitch, F. R.; Machado, F.; Qin, G.; Smyth, S. M.; Vickerman, J. C. J. Chem. Soc. Chem. Commun. 1981, 422.
71. Dwyer, J.; Fitch, F.R.; Qin, G.; Vickerman, J. C. J. Phys. Chem. 1982, 86, 4574.
72. Guth, J. L.; Caullet, P.; Jacques, P.; Wey, R. Bull. Soc. Chim. Fr. 1980, 121.
73. Mousty-Desbuquoit, C. unpublished results.
74. Jencks, W. P. "Catalysis in Chemistry and Enzymology", Mc.Graw Hill : New York, 1969.
75. Gabelica, Z.; Debras, G.; B.Nagy, J.; Bloom, P. J. unpublished results.
76. Jacobs, P. A.; Beyer, H. K.; Valyon, J. Zeolites, 1981, 1, 161.

RECEIVED November 8, 1983

# Combined Physical Techniques in the Characterization of Zeolite ZSM-5 and ZSM-11 Acidity and Basicity

JACQUES C. VEDRINE, ALINE AUROUX, and GISÈLE COUDURIER

Institut de Recherches sur la Catalyse, C.N.R.S., 2, avenue Albert Einstein, 69626—Villeurbanne, France

Acidic pentasil zeolite ZSM-11 and ZSM-5 present fascinating properties related to the size of the channels and to the acidic characteristics of the material. Moreover the Al distribution along the material particles was shown to depend on the synthesis procedure and to be more or less regular resulting in differences in the strength and the distribution of the acid sites. Samples differing by their Al content, particle size, acidification or chemical treatment procedures were prepared and characterized by X-ray diffraction, TEM, EDX-STEM, XPS and n-hexane adsorption capacity. Their acidic properties were determined using ir spectroscopy (OH groups, $NH_3$ adsorption), microcalorimetric measurements of the differential heat of $NH_3$ adsorption at different temperatures vs pulses of $NH_3$ and ESR. By ESR, electron acceptor (acid) or donor (basic) properties were characterized by means of the charge transfer complexes formed with organic molecules able to give or accept one electron resulting in paramagnetic radical ions such as $C_6H_6^+$ and $SO_2^-$ or $C_6H_4(NO_2)_2^-$ respectively. It is found that the combination of several techniques is necessary to obtain information about the nature, strength and concentration of sites. Acid sites were found to be very strong for ZSM-5 and to a lesser extent for ZSM-11 samples. No electron donor (basic) sites could be evidenced.

0097–6156/84/0248–0253$06.25/0

Acidic properties of zeolitic materials have widely been investigated in the past fifteen years (1), particularly for Y-type
zeolites. The presence of strong acid sites, as it is usually the
case for acid zeolites, results in the presence of only weak basic
sites, if any. Therefore, up to now, majority of the studies has
dealt with the characterization of acid rather than basic properties. The acid sites (Brönsted : $H^+$, Lewis : Al) and basic
sites ($O^{2-}$, $OH^-$) may be characterized directly by using physical
techniques or indirectly by the adsorption of a basic (ammonia,
pyridine, n-butylamine etc) or an acidic ($CO_2$, acetic acid, $SO_2$,
hydrochloric acid, etc) probe. In the case of small pore zeolite,
as in the present work, the methods involving large probe molecules as Hammett's indicators (2,3) have obviously to be rejected.
The main techniques available can be summarized as follows :

i.      infra red spectroscopy (3,4) of hydroxyl groups (3500-3800
        $cm^{-1}$ region), of adsorbed molecule probes and of their
        further thermal desorption. The technique is the most
        widely used since it is very fruitful and allows to
        differentiate the Brönsted and the Lewis sites.

ii.     optical UV technique (5) of adsorbed appropriate molecules
        The technique is very sensitive but of narrower appli-
        cations than ir spectroscopy particularly because quan-
        titative determination is very limited and overlapping of
        broad peaks occurs precluding any precise characterization
        of different sites.

iii.    nuclear magnetic resonance technique (6) applied to the
        measurement of the proton jump frequency, even in the
        absence of adsorbate. This frequency increases with a
        higher proton mobility and characterizes the occupancy
        factor of the proton at the various oxygen atoms.

iv.     electron spin resonance of radical ions (anions or
        cations) (7), as dinitrobenzene, tetracyanoethylene or
        perylene, anthracene, benzene, respectively which probe the
        electron donor (basic) and electron acceptor (acid)
        properties of a surface.

v.      calorimetric determination of the differential heat (8) of
        adsorption of a probe molecule ($NH_3$, pyridine,
        benzene,n-butylamine, etc) at a given temperature. This
        heat is related to the ability of the sites to react with
        the probe molecule, i.e. to its basic or acidic character.
        Note, that the concentration and strength of sites but not
        their nature may be obtained.

vi.     temperature programmed desorption (DTA, TPD or TG) (9) of
        the adsorbed probe molecule. Information about the strength
        of the sites and about the number of sites, assuming one
        adsorbed molecule per site, may be obtained. However, the
        exact nature of the sites remains unknown while the nature
        of the desorbed molecules is assumed to be identical to

      that of the starting probe, which is not always true.

vii.    catalytic test reactions which are postulated to occur only
for a given acid or basic strength of sites depending on
the reaction and on the reactants chosen (10).

viii.   chemical titration of protons transfered to aqueous
solution where the material is introduced.

In the present work ir, esr and microcalorimetry techniques were
used to characterize the acid-base properties of acid ZSM-5 and
ZSM-11 samples. Complementary studies by TEM, EDX-STEM and XPS
were also carried out to determine the size and shape of zeolite
particles and the Al distribution within a particle. Catalytic
properties for methanol conversion were also determined.

## EXPERIMENTAL PART

ZSM-5 and ZSM-11 samples were prepared as previously described
(11) using tetrapropylammonium hydroxide and tetrabutyl ammonium
bromide, respectively. The nature and crystallinity of the
materials were verified by X ray diffraction, ir spectroscopy of
lattice vibrational bands (12), n-hexane adsorption capacity at
room temperature and constraint index (13) measurements. All
samples correspond to highly crystalline ZSM-5 or ZSM-11 materials. The chemical compositions of the samples as determined from
chemical analysis of Al and Na contents, are given in table 1.
The ir measurements were carried out with a Perkin Elmer 580
spectrometer and fused silica cell with KBr windows allowing to
outgass the zeolitic wafer at a desired temperature and to
introduce and further outgass a probe molecule without contact
with air.
The esr experiments were performed in usual silica 4 mm i.d. tubes
using a Varian E9 spectrometer monitored in X-band mode. Quantitative determination of radical ion concentration was obtained by
comparison with a Varian strong Pitch standard sample.
A heat-flow calorimeter of Tian-Calvet type from Setaram maintained at a desired temperature, from room temperature up to 400°C,
was used in connection with a volumetric apparatus equipped with a
Mc Leod gauge. Sample weights were typically 100 mg and ammonia
doses 0.1 cm$^3$ NTP.
Transmission electron microscopy pictures were taken using a JEOL
100 CX microscope. For some samples lateral micro-analysis of thin
sections of zeolite was carried out using a HB-5 VG microscope
equipped with EDX accessory at IFP (11).
ESCA experiments were carried out with a HP 5950A spectrometer of
the "Centre Commun ESCA de l'Université de Lyon". The Al signal
being small because of the low Al content, its accumulation was
necessary for one hour. Al2p peak was the only Al peak to be
analyzed since overlapping of Si phonon peak with Al2s peak takes
place. Smoothing of the peaks, substraction of the background and
determination of the surface, A, of the peaks were carried out
with the computer. The relative concentration n for element 1 and
2 (Si and Al for instance) was calculated using the approximate
relation :

$$\frac{n_1}{n_2} \simeq \frac{A_1}{A_2} \cdot \frac{\sigma_2}{\sigma_1} \frac{E_{k2}^{1/2}}{E_{k1}^{1/2}}$$

where $\sigma$ is the electron cross section tabulated by Scofield (14) and $E_k$ the kinetic energy value of the peaks analyzed for elements 1 and 2.

EXPERIMENTAL RESULTS

Infrared experiments

The self supporting wafers of zeolites (3 to 4 mg.cm$^{-2}$) presented an i.r. absorption continuum depending on the particle size. The most intense absorptions were obtained for the samples 5 and 6 with particle size of ca. 6 µm and a spherulitic shape.

For all the samples outgassed at 400°C two ir bands were observed in the 3750-3600 cm$^{-1}$ region, corresponding to two types of hydroxyl groups. The higher frequency band was found at 3720 cm$^{-1}$ (half band width = 55 cm$^{-1}$) for the ZSM-5 samples. It was restored after NH$_3$ adsorption by outgassing at room temperature. This band has been assigned to non-acidic terminal silanols and therefore its intensity may be related to the crystal size and/or the presence of structural defects. Indeed for highly crystalline materials with large crystal size this band is absent (15) while for well crystallized samples 1 and 2 with particle size of 1 µm, it is weak (fig. 1).

For all our ZSM-11 samples, an intense and narrow band was observed at 3740 cm$^{-1}$ (half band width = 20 - 30 cm$^{-1}$) (fig. 1). It could be thought at first glance that it corresponded to amorphous silica material. However a detailed high resolution and microdiffraction analysis (11) conjointly with i.r. framework spectra and n-hexane adsorption capacity studies showed that the materials were zeolitic in nature, but mainly composed of aggregates of tiny particles (5 to 10 nm in diameter). Sample 5 was even more pecular since it presented such an aggregate as a core with emerging needles (11) giving a spherulite-type shape (overall diameter = 6 µm). Microdiffraction study of the needles showed that they are single-crystal in nature with a very nice pattern of dots (11). The overall particles looked as golf balls in TEM pictures. Moreover a NaOH 6N treatment at 80°C resulting in sample 6 was shown by TEM to have dissolved the above needles and obviously any amorphous silica material leaving the aggregated core (diameter = 6 µm). The i.r. spectrum still exhibited a narrow band at 3740 cm$^{-1}$ more intense than for sample 5. It may thus be concluded that the band at 3740 cm$^{-1}$ in ZSM-11 samples as the one at 3720 cm$^{-1}$ in ZSM-5 was due to terminal silanol groups, obviously in large amount because of the presence of tiny particles in the aggregated core, and not due to extrazeolitic material (15). The low frequency band was found at 3602 cm$^{-1}$ (half band width = 30 cm$^{-1}$) for ZSM-5 samples and 3612 cm$^{-1}$ (half band width : 32-40 cm$^{-1}$) for ZSM-11 samples. Its intensity depended on the Al content of the zeolites as already observed (15). It was assigned to

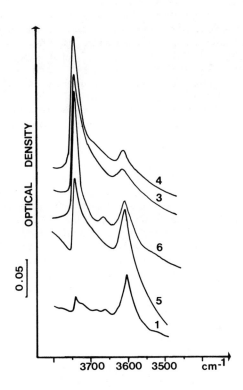

Figure 1 : ir spectra hydroxyl groups of the different samples : 1 (3.6 mg cm$^{-2}$), 3 (3.2 mg cm$^{-2}$), 4 (3.6 mg cm$^{-2}$), 5 (4.3 mg cm$^{-2}$) and 6 (3.7 mg cm$^{-2}$).

acidic hydroxyl groups of the zeolites. $NH_3$ adsorption and further desorption at increasing temperatures allowed to follow the acid strength of the hydroxyl groups (16). In fig. 2 the intensity of the $\nu NH$ vibration (3390 cm$^{-1}$) is reported against the outgassing temperatures. Considering the absolute values of the optical density (o.d.) and of their variations with outgassing temperature it can be concluded that ZSM-11 sample 3 has lesser and weaker acid sites than the ZSM-5 samples 1 and 2. The number of acidic sites is also related to the intensity of the low frequency OH band (o.d. = 0.081 and 0.050 respectively for 4 mg cm$^{-2}$ of samples 1 and 3) and to the Al content. The acid strength of the sites also decreased with the increase in frequency of this OH vibration : 3602 cm$^{-1}$ for ZSM-5, 3612 cm$^{-1}$ for ZSM-11. This result, that ZSM-5 is more acidic than ZSM-11, is in agreement with other findings by Jacobs et al (17) using OH group band shift upon benzene adsorption.

By heating the samples at high temperatures (> 600°C) irreversible modification of the material was observed to occur (18). As a matter of fact it was observed that the low frequency OH band disappeared upon outgassing at 800°C whereas the intensity of the 3740 cm$^{-1}$ band considerably decreased. In the mean time, a new band appeared at 1380 cm$^{-1}$ which may be assigned to an aluminic acid type species (19). Further rehydration did not restore the 3612 cm$^{-1}$ band but generated a band at 3700 cm$^{-1}$ that may be due to hydroxyl groups bound to the aluminic species. One has suggested previously (18) that such high temperature dehydration results in a partial dealumination of the zeolitic framework and therefore to a lower acidity.

## Calorimetry experiments

Differential heats of $NH_3$ adsorption were measured for the samples outgassed at different temperatures ranging from 400 to 800°C. Ammonia was chosen as a basic probe because its size is small, which may limitate diffusion effects in small pore zeolite materials. The variations of the differential heats of adsorption are plotted in fig. 3 as a function of the successive pulses of $NH_3$ introduced, i.e. of $NH_3$ coverage. The adsorption temperature was changed from 150 up to 400°C. The main experimental features can be summarized as follows :

i.   some samples, particularly those acidified by $NH_4Cl$ and with larger particle size, gave rise to a curve with a maximum, whose shape greatly depended on the adsorption temperature (fig. 3).

ii.  when the $NH_3$ adsorption temperature increased there was competition between the formation of ammonium ions between $NH_3$ and a Brönsted site and their decomposition (fig. 3). Upon $NH_3$ adsorption a drop of pressure occured almost instantaneously due to strong chemisorption. However for weaker chemisorption the pressure did not drop to zero instantaneously. Such a change in behaviour, which is presently considered as a change in chemisorption strength type, is shown by an arrow in the figures dealing with

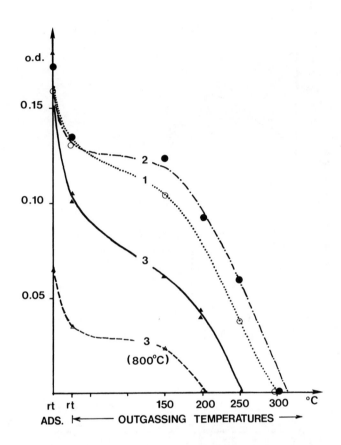

Figure 2 : Variations of optical density of vNH band
(3390 cm$^{-1}$) for ZSM-5 (samples 1 and 2) and ZSM-11 (sample
3) vs the outgassing temperature. Wafers of 4.0 mg. cm$^{-2}$.
The dashed curve 3 corresponds to sample 3 outgassed at 800°C
rehydrated at room temperature and dehydrated again at 400°C.

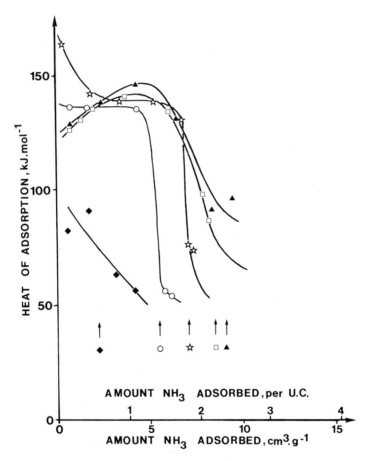

Figure 3 : Variations with coverage of the differential
heats of adsorption of ammonia on H−ZSM−5 (sample 1) measured at 150°C, (▲), 200°C, (□), 250°C, ☆ 300°C (0) and
400°C (◆). The sample was outgassed at 400°C prior NH₃ adsorption. The meaning of the arrows is explained in the text.

calorimetry. For instance for samples outgassed at 400°C one
then gets. 2.4, 2.2, 1.9, 1.4 and 0.6 $NH_3$ molecules per unit
cell for adsorption at 150, 200, 250, 300 and 400°C respectively. These numbers have to be compared with the number
of protons $H^+$ equal to 3.2 per u.c. and are always smaller.
This presumably arises from the presence of "weak" Brönsted
acid sites in the sample. Therefore it appears that the
calorimetry technique allows to determine the number of
"strong" Brönsted sites assuming that outgassing at 400°C
before $NH_3$ adsorption gives rises to a maximum in $H^+$ i.e.,
that no dehydroxylation had occured. This is a crude approximation (20) but can be considered as valid in a rough
approximation and for comparison of different samples.
Moreover, temperature of $NH_3$ adsorption as low as 150°C
seemed to be reliable for calorimetric experiments. Room
temperature adsorption did not allow to differentiate strong
and weak acid sites. It also appeared, as previously observed, that the number of strong acid sites was lower that
the number of protons calculated from chemical analysis,
which may be interpreted as due to a weak acidity of part of
these protons.

iii. if the samples were calcined at 800°C before $NH_3$ adsorption,
a curve with a maximum was still observed ; there are less
acid sites but they are stronger. If such samples were
further rehydrated at room temperature and outgassed again at
400°C, the starting curve was not obtained (fig. 2). It can
then be concluded that such a heat treatment has irreversibly
modified the material. From XPS data of Si : Al ratio
measurement and from i.r. experiment of pyridine adsorption,
it was previously (18) suggested that Al from the lattice was
extracted upon calcination resulting in an aluminic acid
compound within the cavities and consequently in additional
steric constraints. Ir study, as described above, also showed
that dehydroxylation was not reversible for ZSM-11 sample.

iv. the shape of the calorimetric curves with a maximum was
previously (20-21) asssigned to an unusual conjunction of
three phenomena : (i) immobile adsorption, (ii) mass transfer
limitation and (iii) preferential location of the most
energetic acid sites in the internal pores of the zeolite.
The latter point could correspond to an heterogeneous
distribution of Al within the zeolite grain as it was
observed later on (11, 22 - 24). Mass transfer limitation
obviously played a role since the shape with a maximum
disappeared when the $NH_3$ adsorption temperature increased
(fig. 3), i.e. when diffusion rate increased. This conclusion
was also supported by the experiment for which calorimetric
cell was disconnected from the line, heated outside the
calorimeter for 2 h at 250°C between each $NH_3$ pulse to
facilitate the diffusion of adsorbed $NH_3$. The maximum in the
curve was not observed anymore as described in ref. 21 when
the samples were partly deactivated by methanol conversion
reaction or by introduction of a phosphorous compound ;

stronger but inner acid sites were then accessible to the
first $NH_3$ molecules introduced. The maximum in the curve
disappeared whereas higher heat of adsorption values were
obtained as it can be seen in figs. 3 and 5 in ref. 24. It
followed that for curves with a maximum the highest value of
the $NH_3$ adsorption heat did not characterize really the
strongest acid sites as it could be thougt at first glance.
As a matter of fact there was a competition for ammonia to
neutralize the strongest acid sites which are located inside
the zeolite particles and to freely diffuse within the
channel to reach these sites. A kind of average value was
then obtained which depended on adsorption temperature, size
of the particles and channels, the time between successive
pulses if slow diffusion occured. Nevertheless the number of
strong acid sites, determined   as discussed above, remains
reliable data from the experiments. When a maximum in the
calorimetry curve was not obtained it can be either because
the particle size of the zeolite was small or more often
because the strongest acid sites were located in the outer
layers of the grains. Such a phenomenon is clearly seen in
fig. 4 for the different ZSM-11 samples. Note also that
hydrochloric acid treatment partly dealuminated surface
layers of the zeolite particles as evidenced by XPS (table 1)
and resulted in stronger surface acide sites, i.e. in no
maximum in the calorimetry curve.

v.     another striking feature of calorimetry measurements is worth
while mentioning. Sample 1 was used respectively for methanol, $C_2H_4$ and $C_2H_4$ + $CH_3OH$ conversion reactions for 20
min. The calorimetry results are plotted in fig. 5 for
samples outgassed after reaction. It clearly appeared that no
strong acidic sites were then accessible to $NH_3$ if $C_2H_4$
conversion was performed while acid sites were still accessible after $CH_3OH$ or $C_2H_4$ + $CH_3OH$ reactions. This was
presumably due to the formation of linear polymeric residues
when $C_2H_4$ was used alone which then filled the channels. This
also explained the low activity of ZSM-5 zeolite for $C_2H_4$
conversion reaction (25, 26). Such a low activity was
considered by some authors to rule out the possibility that
$C_2H_4$ was an intermediate species in $CH_3OH$ conversion reaction
(27). However, the calorimetric curve using a mixture of $C_2H_4$
and $CH_3OH$ showed clearly in connection with catalytic results
that methanol acted as an alkylating agent and that strong
acid sites were still available after reaction. Therefore in
$CH_3OH$ conversion $C_2H_4$ may well be an intermediate compounds,
the presence of $CH_3OH$ in the feed led the reaction to proceed
to higher hydrocarbons rather than to deactivate the material
by the formation of linear hydrocarbon residues which fill
the channels. Methanol conversion studies at low conversion
level (< 1 %) and low temperature (250°C) clearly showed that
$C_2$ to $C_4$ olefins were all primary products. Such an ex-
périment indicated that the rattle mechanism, as proposed by
Guisnet et al. (27), was valid on ZSM-5 samples as represented
as follows on p. 266 (28).

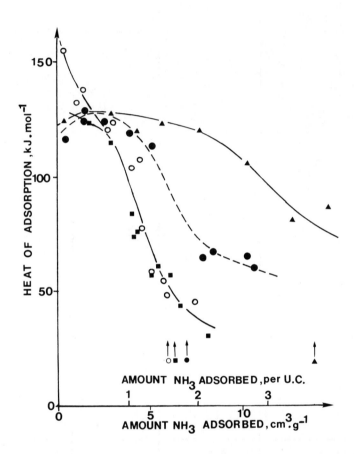

Figure 4 : Variations with coverage of the differential heats of NH₃ adsorption measured at 143°C on H-ZSM-11 samples outgassed at 400°C. Samples 3 (O) 4 (■), 5 (●) and 6 (▲).

Table 1 : Main characteristics of the different ZSM-5 and ZSM-11 samples.

| Samples* | 1 | 2 | 3 | 4 | 5 | 6 |
|---|---|---|---|---|---|---|
| treatment | no | no | no | no | no | NaOH, 20 % wt |
| acidification at 80°C | HCl, N/2 | NH$_4^+$, M/2 | HCl, N/2 | NH$_4^+$, M/2 | NH$_4^+$, M/2 | NH$_4^+$, M/2 |
| zeolite type | ZSM-5 | ZSM-5 | ZSM-11 | ZSM-11 | ZSM-11 | ZSM-11 |
| **Particles** | | | | | | |
| shape | platelets | platelets | spheroids | spheroids | spherulite | spheroids |
| size | 0.5-2µm | 0.5-2µm | 0.6 µm | 0.6 µm | 6 µm | 6 µm |
| structure | single crystal type | single crystal type | aggregates of tiny particles ($\phi \approx$ 5-10nm) | aggregates of tiny particles ($\phi \approx$ 5-10nm) | core of aggregates + emerging needles (0.2 - 0.4 µm) | aggregate of tiny particles ($\phi \approx$ 5-10nm) |
| **Composition (per u.c.)** | | | | | | |
| Al | 3.4 | 4.8 | 1.8 | 2.5 | 2.2 | 5.9 |
| Na | 0.2 | 0.3 | 0.1 | 0.3 | 0.0 | 0.2 |
| H | 3.2 | 4.5 | 1.7 | 2.2 | 2.2 | 5.7 |
| **Si : Al (atoms ratios)** | | | | | | |
| Chemical analysis | 27 | 19 | 53 | 37 | 43 | 15 |
| XPS data | 30 | 23 | 24 | 21 | 70 | 16 |
| EDX-STEM | 20 - 50 | 20 - 30 | 25 - 45 | - | core : 15-25 needles:30-70 | 15 - 25 |

* Samples 1 and 2, 3 and 4, 5 and 6 originate from the same of the three batches respectively.

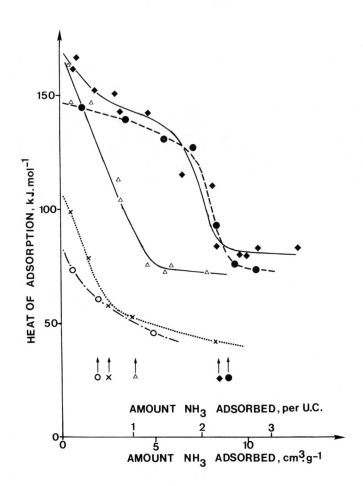

Figure 5 : Variations with coverage of the differential
heats of NH$_3$ adsorption at 143°C on H–ZSM–5 sample not mentionned in table 1

● : starting sample outgassed at 400°C

○ : 20 min. C$_2$H$_4$ conversion reaction followed by outgassing
at 400°C

x : 20 min. C$_2$H$_4$ + CH$_3$OH reaction followed by outgassing at
150°C

◆ : 20 min. C$_2$H$_4$ + CH$_3$OH reaction followed by outgassing at
400°C

△ : 20 min. CH$_3$OH reaction followed by outgassing at 400°C

Gas phase :    $Me_2O$         MeOEt         $C_2H_4$         MeOPr         $C_3H_6$
               or MeOH
                  $\uparrow\downarrow$              $\uparrow\downarrow$           $-H^+\downarrow+H^+$     $\uparrow\downarrow$    $-H^+\uparrow\downarrow+H^+$

Adsorbed
intermediate    $CH_3O-Z$  $\overset{\rightarrow}{\leftarrow}$  MeOEt  $\rightleftarrows$  $C_2H_5^+$   MeOPr  $\rightleftarrows$  $C_3H_7^+$
                                                        $+ C_1$

                                          $- C_1$

vi. the acidification procedure and calcination or chemical
treatment may also be important in the distribution of acid
sites within the channels. Particularly, if protons stemming
from tetrapropyl ammonium ion decomposition were exchanged by
$Na^+$ or neutralized by $NH_3$ gas before acidification by
exchange by ammonium salt or by HCl, the material might
behave differently (29). The difference in calorimetric
curves when comparing the two acidification procedures, with
or without neutralizing $H^+$ from TPA ion decomposition, is
clearly seen in fig. 6 while noticeable differences in
catalytic properties were observed (16,29). More and slightly
heavier aromatics were formed (see for instance table 2 in
ref. 29) which could be interpreted as due to an higher
acidity. The calorimetric curves did not support such a
conclusion since either they were very similar or corresponded to slightly less acidic sites. Jacobs et al. (15) have
suggested that exchange by Na ions of protons formed by TPA
ion decomposition may limit the hydrolysis of framework
aluminium under treatment at high temperature. Such hydrolysis may then explain the calorimetry shape with a
maximum more marked in figure 6 since diffusion limitation
may be enhanced. However it is worth while noting that the
number of strong acid sites as characterized by calorimetry
was not modified significantly. As calorimetry did not allow
to determine the nature of the acid sites, one cannot go
further in the interpretation although the material was
obviously slightly modified. However, it turns out that the
role of $Na^+$ exchange or $NH_3$ adsorption prior to acidification
modifies the further exchange of the starting Na ions, in
other words the location of acid sites.

## ESR experiments

It is now well established that when a surface presents electron
donor or electron acceptor sites, it is possible to ionize
molecules of relatively high electron affinity (> 2 eV) or low
ionization potential values, resulting in paramagnetic radical
ions. For instance anthracene and perylene are easily positively
ionized on alumina (7) (IP = 7.2 and 6.8 eV respectively). The
adsorption at room temperature of benzenic solution of perylene,
anthracene and napthalene on H-ZSM-5 and H-ZSM-11 samples heated
up to 800°C prior to adsorption did not give rise to the formation
of the corresponding radical cation. For samples outgassed at high

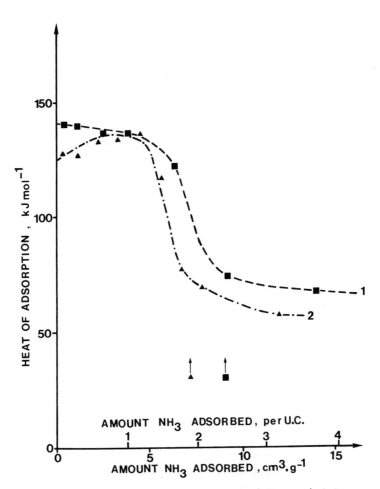

Figure 6a: Variations with coverage of differential heats of NH$_3$ adsorption at 143°C on H-ZSM-5 samples outgassed at 400°C not mentionned in table 1.
Key -- 1 : acidification via NH$_4$Cl (usual treatment) (■)
2 : introduction of NH$_3$ prior to acidification via NH$_4$Cl (▲)

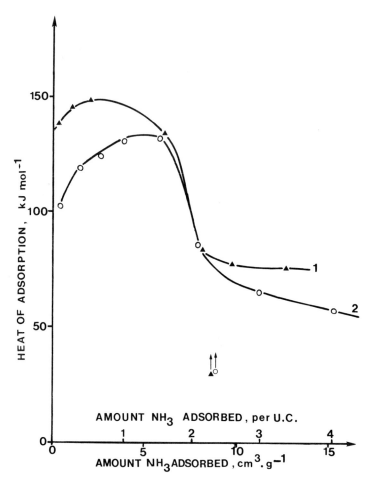

Figure 6b:  Same as Figure 6a.

Key -- 1 : acidification via $NH_4Cl$ (usual treatment) (▲)
      2 : exchange by $Na^+$ of $H^+$ from TPA ion decomposition
          followed by acidification via $NH_4Cl$ (O)..

temperature (T > 600°C) a seven hyperfine line spectrum was observed at room temperature with 4.4 Gs splitting (30) (fig. 7). The spectrum can be unambiguously assigned to $C_6H_6^+$ species (30) from the solvent. The intensity of the ESR signal varied with time (30) but its shape did not change. At variance, $C_6H_6^+$ species obtained on acid mordenite was observed to be transformed with time into a dimer, sandwich-like, $(C_6H_6)_2^+$ which exhibited a thirteen hyperfine line structure with a 2.2 Gs splitting. As perylene was not ionized at contrary to benzene although its ionization potential is much smaller than that of benzene (IP = 6.8 eV against 9.25 eV) it can be concluded that strong electron acceptor (acid type) sites were located within the zeolite channels and that the exterior of the zeolite particles did not exhibit electron acceptor properties high enough to ionize perylene. The ESR signal intensity increased with time, reached a maximum and then decreased again as shown in ref. 30. Note also that NO gas adsorbed at room temperature on the zeolite outgassed above 400°C gave rise to an ESR NO spectrum (Fig. 7) with a hyperfine structure attributable to interaction with Al (18). This shows that NO was adsorbed on Lewis acid sites. The intensity of the ESR spectrum increased sharply with outgassing temperature which corresponded to the increasing formation of Lewis acid sites (18) by a well known dehydroxylation procedure.

In order to characterize electron acceptor (basic type) properties of the samples, tetracyano ethylene compound, known to be easily ionizable in TCNE⁻ radical anion, was introduced at room temperature in the samples outgassed at different temperatures up to 800°C. No ESR signal was observed. As steric hindrance could preclude the experiment, smaller molecules as $SO_2$ and p-dinitro benzene were also introduced. Then too, no ESR spectrum could be detected although the ESR technique is extraordinarly sensitive. It may thus be concluded that the ZSM-5 and ZSM-11 materials did not exhibit electron donor (basic) properties as detectable by ESR.

## CONCLUSION

The present work allows to draw the following conclusions :

i.    H-ZSM-5 presents very strong acid sites, higher than H-mordenite and H-Y zeolites, with a heterogeneous strength distribution.

ii.   the heterogeneous acid strength distribution may be assigned to an heterogeneous distribution of framework Al within the zeolite particles, i.e. along the channels.

iii.  the H-ZSM-11 samples are very strongly acid but slightly less than the H-ZSM-5 samples. The acid OH groups have an i.r. frequency equal to 3612 and 3602 cm$^{-1}$ respectively.

iv.   calorimetry of $NH_3$ adsorption is an excellent method to characterize the strength and the amount of strong acid sites. However no information about the nature of the sites is obtained. Moreover $NH_3$ diffusion may give average heat of adsorption rather than the actual distribution in strength particularly for large zeolite particles. Inversely dif

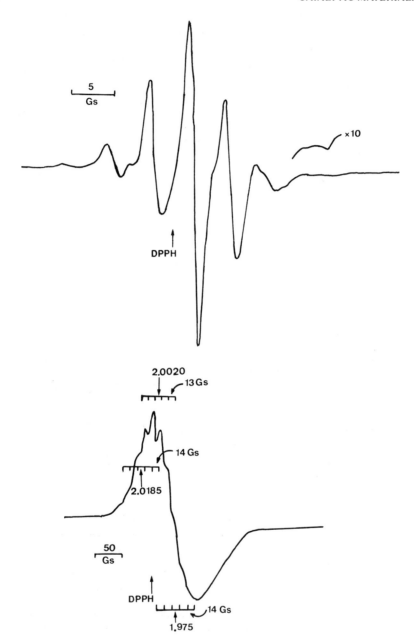

Figure 7 : ESR spectra : <u>upper</u> : room temperature spectrum of $C_6H_6$ adsorbed on sample 1 outgassed at 600°C prior to adsorption.

　　　　　　　　　　<u>bottom</u> : liquid nitrogen spectrum of NO (5 torr) adsorbed at room temperature on sample 1 outgassed at 600°C prior to absorption.

fusion limitation may be used to demonstrate that outer
layer acid sites are weaker than inner sites and indirectly
that acidity is heterogeneously distributed all along a
zeolite particle, i.e. lattice aluminium atoms as well.

v.   conjunction of calorimetry and ir techniques is necessary to
obtain information about the nature and amount of strong
acid sites but unfortunately complete characterization is
not reached because of limitations inherent to each method.

vi.   electron donor properties of the zeolite, related to basic
properties, were not detected, which indicates that if there
were basic sites they should be very weak.

vii.   methanol or olefin conversion reactions were shown to
involve first the acid sites located in the outerlayers of
the zeolitic particles, before the inner sites could act
catalytically. It follows that deactivation due to carbonaceous residues occurs primarily at outer layers of the
material, while the tridimensional channel network allows
ZSM-5 or ZSM-11 zeolite samples to be still active up to
complete annealing of the inner active sites.

viii.   synthesis conditions, acidification reactant, chemical
treatment before acidification procedure and dehydration
temperature were shown to affect the acid site strength,
nature and location within the particle. Methanol conversion
reaction was shown to be not very sensitive to such minor
modifications of the material, although alkylation of
aromatics with methanol and presumably other acid-type
reactions, more sensitive to the appropriate channel size or
more dependent of a given acid strength, may be very much
influenced by such modifications. Such a property may be
generalized to any other zeolitic material and opens a wide
field of further applications.

## Literature Cited

1.   Forni, L., Catal. Rev., (1973), 8, 65.
     Benesi, H.A. and Winquist, B.H.C., Adv. Catal., (1979), 27,
     97.
     Barthomeuf, D. in Molec. Sieves II,Ed. Katzer J.R., A.C.S.
     Symposium Ser. (1977), 40, 453 and J. Phys. Chem. (1979), 83,
     249.
     Jacobs, P.A., Catal. Rev., (1982), 24, 415.
2.   Hirschler A.E., J. Catal., (1963), 2, 248 and (1968), 11, 274.
3.   Beaumont, R. and Barthomeuf, D., J. Catal., (1972), 26, 218,
     and (1972), 27, 45.
     Moscou, L. and Mone, R., J. Catal., (1973), 30, 417.
4.   Rabo, J.A., Pickert, P.E., Stamires, D.N. and Boyle, J.E.,
     Proc. Intern. Congr. Catalysis 2nd, Paris, Edit. Dunod (1960),
     p. 2055.
     Norton C.J., Ibid., (1960), p. 2073.
     Bielanski, A. and Dakta, J. Bull. Acad. Sci. Polon. (1974) 22,
     341 and Datka, J. Ibid (1974), 27, 975.

5.  Kiselev, A.V., Kitiashuili, D.G. and Lygin, V.I., Kin. i
    Kat., (1971), 12, 1075.
    Kageyama, Y., Yotsuyanagi, T. and Aomura, K., J. Catal.,
    (1975), 36, 1.
    Zhdanov, S.P. and Kotov, E.I., Adv. Chem. Ser. (1973), 121,
    240.
6.  Freude, D., Oehme, W., Schmiedel, H. and Staudte, B., J.
    Catal., (1974), 32, 137.
    Derouane, E.G., Fraissard, J., Fripiat, J.J. and Stone,
    W.E.E., Catal. Rev. (1972), 7, 121.
    Mestdagh, M.N., Stone, W.E.E., Fripiat, J.J., J. Catal.,
    (1975), 38, 358 and J.C.S., Faraday Trans. I., (1976), 1, 154.
7.  Naccache, C., Kodratoff, Y., Pink, R.C. and Imelik, B., J.
    Chim. Phys., (1966), 63, 341.
    Kodratoff, Y., Naccache, C. and Imelik, B., J. Chim. Phys.,
    (1968), 65, 562.
    Hirschler, A.E., Neikam, W.C., Barmby D.S. and James, R.L., J.
    Catal. (1965), 4, 628.
8.  Brueva, T.P., Klyachko-Gurvich, A.K. and Rubinstein, A.M.,
    Isz. Akad. Nauk, SSSR, Ser. Khim., (1972), 2807, (1974), 1254,
    (1975), 939.
    Tsutsumi, K., Koh, H.O., Hagiwara, S., and Takahashi, H. Bull.
    Chem. Soc. Japan, (1975), 48, 3576.
    Masuda, T., Taniguchi, H., Tsutsumi, K. and Takahashi, H.,
    Bull. Chem. Soc. Japan, (1978), 51, 1, 65.
9.  Steinberg, K.H., Bremer, H. and Falke, P., Z. Chem., (1974),
    14, 110.
    Topsoe, N.Y., Pedersen, K., and Derouane, E.G., J. Catal.,
    (1981), 70, 41.
10. Steinber, K.H., Bremer, H., Hofmann, F., Minachef, Kh. M.,
    Dhitriev, R.V., and Detyuk, A.N., Z. Anorg. Allg. Chem.,
    (1974), 404, 129 and 142.
    Meylen, C.F., and Jacobs, P.A., Adv. Chem. Ser., (1973), 121,
    490.
    Cormerais, F.X., Chen, Y.S., Kern, M., Gnep, N.S., Perot, G.
    and Guisnet, M., J. Chem. Research S. (1981), 290.
    Olson, D.H., Haag, W.O., and Lago, R.M., J. Catal., (1980),
    61, 390.
11. Auroux, A., Dexpert, M., Leclercq, C. and Védrine, J.C., Appl.
    Catal., (1983), 6, 95.
12. Coudurier, G., Naccache, C., and Védrine, J.C., J.C.S. Chem.
    Commun, (1982), 1413.
13. Frilette, V.J., Haag, W.D. and Lago, R.M., J. Catal., (1981),
    67, 218.
14. Scofield, J.H., J. Electron. Spectrosc., (1976), 8, 129.
15. Jacobs, P.A. and Von Ballmoos, R., J. Phys. Chem., (1982), 86,
    3050.
16. Védrine, J.C., Auroux, A., Coudurier, G., Engelhard, P.,
    Gallez, P., and Szabo, G., Proceed. 6th Intern. Zeol.
    Confer., Publ. Butterworths Scient. Press, Reno, July 1983.

17. Jacobs, P.A., Martens, J.A., Weitkamp, J. and Beyern, K., Faraday Discussion,, (1981), 72, 353.
18. Védrine, J.C., Auroux, A., Bolis, V., Dejaifve, P., Naccache, C., Wierzchowski, P., Derouane, E.G., Nagy, J.B., Gilson, J.P., Van Hoff, J.H.C., Van Den Berg, J.P. and Wolthuizen, J., J. Catal., (1979), 59, 248.
19. Védrine, J.C., Abou Kais, A., Massardier, J., and Dalmai-Imelik, G., J. Catal., (1973), 29, 120.
20. Auroux, A., Bolis, V., Wierzchowski, P., Gravelle, P.C. and Védrine, J.C., J.C.S., Faraday Trans. II, (1979), 75, 2544.
21. Auroux, A., Wierzchowski, P. and Gravelle, P.C., Thermochim. Acta (1979), 32, 165.
22. Von Ballmoos, R. and Meier, W.M., Nature, (1981), 289, 782.
23. Derouane, E.G., Detremmerie, S., Gabelica, Z., and Blom, N., Appl. Catal. (1981), 1, 201
24. Derouane, E.G., Gilson, J.P., Gabelica, Z., Mousty-Desbuquoit, C. and Verbist, J., J. Catal., (1981), 71, 447.
25. Védrine, J.C., Dejaifve, P., Naccache, C. and Derouane, E.G., Proceed 7th Intern. Cong. Catalysis, Edit. by Seiyama and Tanabe (Kodanska Ltd), Tokyo, (1981), 724.
26. Anderson, J.R., Foger, K., Mole, T., Rajadhyaksha, R.A. and Sanders, J.V., J. Catal. (1979), 58, 114.
    Van Den Berg, J.P., Wolthuizen, J.P. and Van Hooff, J.H.C., J. Catal., (1983), 80, 139.
27. Ahn, B.J., Armando, J., Perot, G. and Guisnet, M., C.R. Acad. Sci., Paris, Ser. C, (1979), 288, 245.
    Cormerais, F.X., Perot, G., Chevalier, F. and Guisnet, M., J. Chem. Research(S) (1980), 362.
28. Dejaifve, P., Derouane, E.G., Ducarme, V. and Védrine, J.C., to be published.
29. Derouane, E.G., Dejaifve, P., Gabelica, Z., and Védrine, J.C., Discussion Faraday (1981), 72, 331.
30. Wierzchowski, P., Garbowski, E.D., and Védrine, J.C., J. Chim. Phys., (1981), 78, 41.
31. Edlung, O., Kinell, P.O., Lund, A., and Shimizu, A., J. Chem. Phys., (1967), 46, 3679.
    Corio, P.L. and Shih, S., J. Catal., (1970), 19, 126.

RECEIVED October 3, 1983

# Structure–Selectivity Relationship in Xylene Isomerization and Selective Toluene Disproportionation

D. H. OLSON and W. O. HAAG

Mobil Research and Development Corporation, Princeton, NJ 08540

As a result of steric constraints imposed by the
channel structure of ZSM-5, new or improved aromatics conversion processes have emerged. They
show greater product selectivities and reaction
paths that are shifted significantly from those
obtained with constraint-free catalysts. In
xylene isomerization, a high selectivity for isomerization versus disproportionation is shown to
be related to zeolite structure rather than
composition. The disproportionation of toluene
to benzene and xylene can be directed to produce
para-xylene in high selectivity by proper catalyst
modification. The para-xylene selectivity can be
quantitatively described in terms of three key
catalyst properties, i.e., activity, crystal size,
and diffusivity, supporting the diffusion model
of para-selectivity.

Intermediate pore zeolites typified by ZSM-5 (1) show unique
shape-selectivities. This has led to the development and
commercial use of several novel processes in the petroleum and
petrochemical industry (2-4). This paper describes the
selectivity characteristics of two different aromatics conversion
processes: Xylene Isomerization and Selective Toluene
Disproportionation (STDP). In these two reactions, two different
principles (5,6) are responsible for their high selectivity: a
restricted transition state in the first, and mass transfer
limitation in the second.

## Xylene Isomerization

Prior to the introduction of ZSM-5-based xylene isomerization
processes, most of the commercial units operated with a

0097-6156/84/0248-0275$09.25/0
© 1984 American Chemical Society

dual-functional catalyst containing a hydrogenation component —
usually platinum — and an acid catalyst. With such catalysts,
the isomerization occurs in the presence of hydrogen via
hydrogenated intermediates, e.g.:

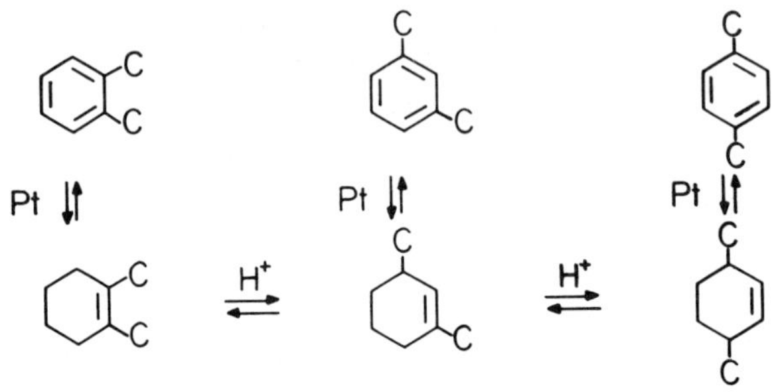

in addition to direct methyl migration around the aromatic ring.
Commercial $C_8$-aromatic streams contain considerable amounts of
ethylbenzene, 15-20% when derived from reformate, and 35-55% in
pyrolysis gasoline from ethylene crackers. In xylene
isomerization with dual-functional catalysts, ethylbenzene is
converted in part to additional xylenes by the same mechanism as
shown for xylene isomerization, e.g.:

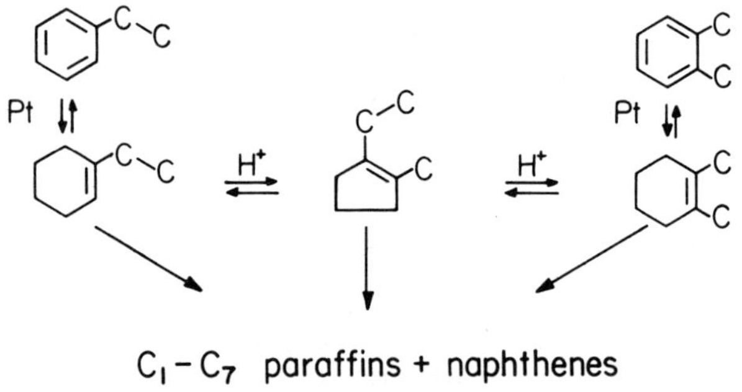

$$C_1 - C_7 \text{ paraffins + naphthenes}$$

However, the susceptibility to hydrocracking of the non-aromatic
intermediates leads to considerable formation of light gases and
naphthenes that reduce the xylene yield and the hydrogen purity.

Early attempts to utilize the high acid activity of faujasite zeolite catalysts for direct xylene isomerization suffered from low selectivity. Considerable improvement was obtained first by using a large pore zeolite (7) catalyst and subsequently in several process modifications that use ZSM-5 as catalyst (2). In the following we will show how these selectivity differences can be related to structural differences of the various zeolites.

The acid catalyzed isomerization of xylene

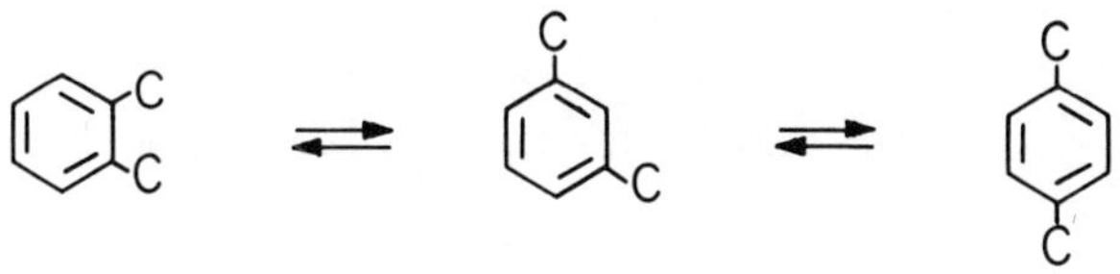

is accompanied by xylene disproportionation, e.g.:

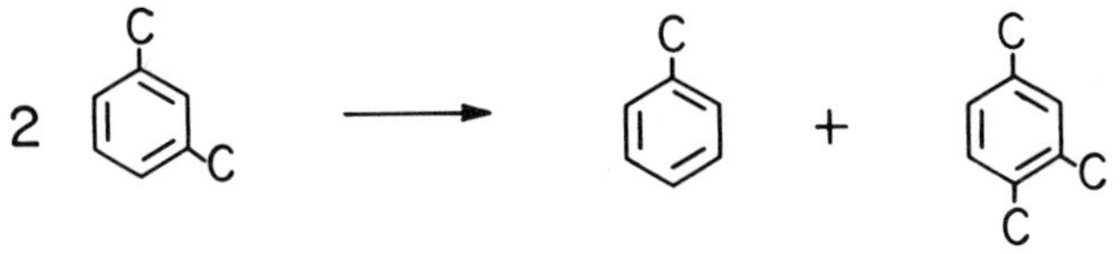

This side reaction leads to undesirable losses of xylenes. With REHY zeolite as catalyst, disproportionation occurs at a rate comparable to that of isomerization of m-xylene (8), e.g., 14% disproportionation at 16% isomerization. In fact, the product, trimethylbenzene, is postulated as an important intermediate in isomerization (8).

By contrast, under the same conditions, ZSM-5 produces orders of magnitude less disproportionation product, as shown in Figure 1.

We have examined the rate constants for disproportionation and isomerization for a variety of zeolites, using a commercial-type feed containing 70% m-xylene and 30% o-xylene in a fixed-bed flow reactor. The results, listed in Table I, show the exceptionally low disproportionation/isomerization selectivity of ZSM-5 relative to synthetic faujasite. Synthetic mordenite and ZSM-4 have intermediate selectivities.

It has been suggested that the reason for this difference is the different site density. According to this proposal, the large concentration of acid sites in synthetic faujasite (ca. 5 meq/g) favors the bimolecular disproportionation reaction relative to the monomolecular isomerization. By contrast, ZSM-5 has a low acid site concentration, typically less than 0.5 meq/g.

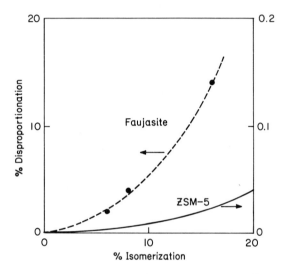

Figure 1.   Comparison of the relative disproportionation versus isomerization selectivities of HZSM-5 and synthetic faujasite (8).   Feed: m-xylene.   Temperature: 300°C.

Table I.  Selectivity in Xylene Isomerization
Feed: 70% m-/30% o-Xylene, 316°C
Pressure: 28 bar

|  | $SiO_2/Al_2O_3$ | $\dfrac{k_{disproportionation}}{k_{isomerization}}$ |
|---|---|---|
| HY | 5 | 0.050 |
| Synthetic Mordenite | 15 | 0.014 |
| ZSM-4 | 7 | 0.010 |
| ZSM-5 | 70 | 0.001 |

This argument, however, is unlikely on theoretical grounds. Both disproportionation and isomerization rates should depend linearly on the number of acid sites. Experimental findings have confirmed this. Also, the data in Table I show no correlation of the selectivity with $SiO_2/Al_2O_3$ ratio.

The best correlation of the observed isomerization selectivities was found in terms of the diameter of the intracrystalline cavity, determined from the known crystal structure (9) of these zeolites, as shown in Figure 2. While faujasite, mordenite and ZSM-4 all have 12-membered ring ports and hence should be similar in their diffusion properties, they differ considerably in the size of their largest intracrystalline cavity; both mordenite and ZSM-4 have essentially straight channels, whereas faujasite has a large cavity at the intersection of the three-dimensional channel system.

The correlation between selectivity and intracrystalline free space can be readily accounted for in terms of the mechanisms of the reactions involved. The acid-catalyzed xylene isomerization occurs via 1,2-methyl shifts in protonated xylenes (Figure 3). A mechanism via two transalkylation steps as proposed for synthetic faujasite (8) can be ruled out in view of the strictly consecutive nature of the isomerization sequence o   m   p and the low activity for disproportionation. Disproportionation involves a large diphenylmethane-type intermediate* (Figure 4). It is suggested that this intermediate can form readily in the large intracrystalline cavity (diameter ~1.3 nm) of faujasite, but is sterically inhibited in the smaller pores of mordenite and ZSM-4 (d ~0.8 nm) and especially of ZSM-5 (d ~0.6 nm). Thus, transition state selectivity rather than shape selective diffusion are responsible for the high xylene isomerization selectivity of ZSM-5.

*Methyl substituted diphenylmethanes are present in trace amounts in the reaction product with ZSM-5 catalyst, and in larger quantities with ZSM-4 catalyst.

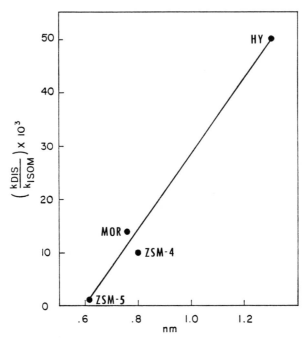

Figure 2.  Effect of intracrystalline cavity diameter of
several zeolites on selectivity in xylene isomerization.

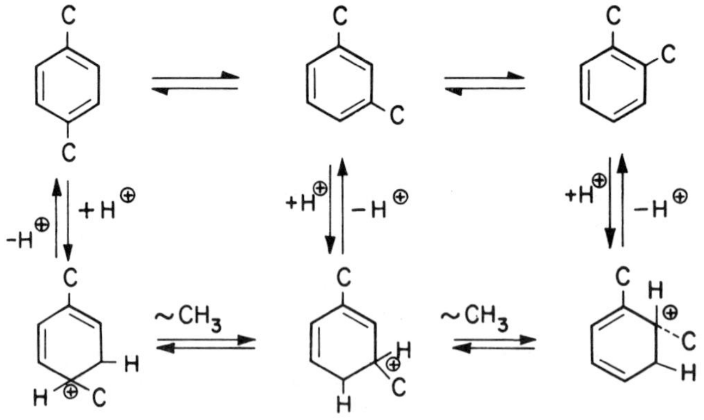

Figure 3.  Acid catalyzed xylene isomerization mechanism.

## Isomerization of Xylene Containing Ethylbenzene

Production of p-xylene via p-xylene removal, i.e., by crystallization or adsorption, and re-equilibration of the para-depleted stream requires recycle operation. Ethylbenzene in the feed must therefore be converted to lower or higher boiling products during the xylene isomerization step, otherwise it would build up in the recycle stream. With dual-functional catalysts, ethylbenzene is converted partly to xylenes and is partly hydrocracked. With mono-functional acid ZSM-5, ethylbenzene is converted at low temperature via transalkylation, and at higher temperature via transalkylation and dealkylation. In both cases, benzene of nitration grade purity is produced as a valuable by-product.

We have determined the relative rate constants for the various transalkylation reactions with a variety of zeolite catalysts (10) at 250-280°C. As indicated in Figure 5, the reaction designations (E,E), (E,X), (X,E) and (X,X) are chosen to indicate first the donor, then the acceptor of a transferred alkyl group, where E = ethylbenzene and X = xylene. Thus, (E,X) signifies the transfer of an ethyl group from ethylbenzene to a xylene molecule, while in reaction (X,E) a methyl group from xylene is transferred to ethylbenzene. The relative rate constants for these reactions were obtained from plots of the various transalkylation products vs conversion as illustrated in Figure 6, and using a kinetic model based on second-order mass action competition.

$$\frac{d \text{ Diethylbenzene}}{dt} = k_{E,E} [E]^2$$

$$\frac{d \text{ Ethylxylene}}{dt} = k_{E,X} [E][X] \qquad \text{etc.}$$

The validity of the second-order model was verified with feeds containing varying ratios of ethylbenzene and xylene.

The data are summarized in Table II. They have been normalized to $k_{X,X} \equiv 1$ for each zeolite catalyst. In general it is seen that the transfer of an ethyl group (E,E;E,X) occurs faster than that of a methyl group (X,E;X,X). This is in agreement with the indicated mechanism for transalkylation (Figure 4) which involves a benzylic carbenium ion intermediate. In the case of methyl transfer, this is a primary cation, $Ph-CH_2^+$, whereas during ethyl transfer it is a more stable secondary cation, $\Phi-\overset{+}{C}H-CH_3$, which is easier to form. It is also apparent, that ethylbenzene is a better acceptor than xylene. We suggest that this is largely a consequence of the larger steric requirement of the bulky diphenylmethane intermediate for alkyl transfer to xylene vs to ethylbenzene.

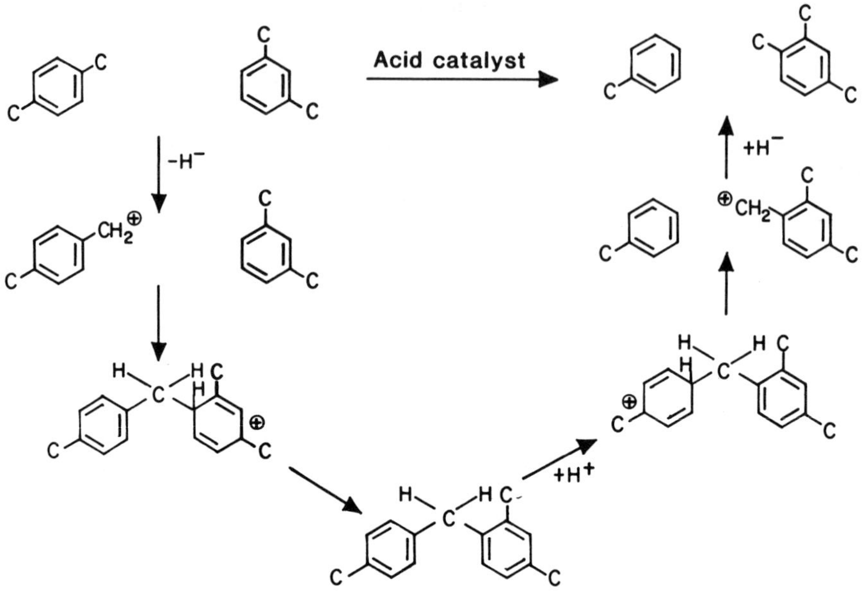

Figure 4.  Acid catalyzed xylene disproportionation
mechanism.

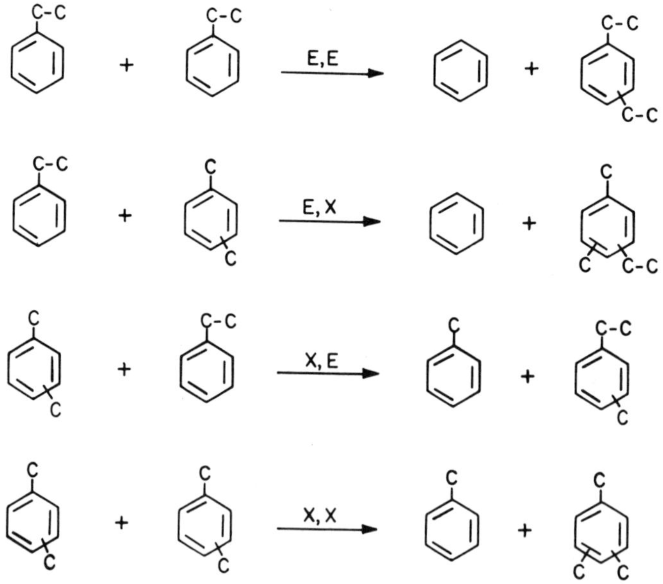

Figure 5.  Transalkylation reactions in the ethylbenzene-
xylene system.

Table II. Transalkylation Kinetics Over Various
Zeolite Catalysts
250–280°C, 28 bar, WHSV = 2–20

| | Relative Rate Constants | | |
|---|---|---|---|
| Reaction | ZSM–4 | Mordenite | ZSM–5 |
| E,E | 10.4 | 20.8 | 125.0 |
| E,X | 2.2 | 3.6 | 16.8 |
| X,E | 1.3 | 1.5 | 3.6 |
| X,X | 1.0 | 1.0 | 1.0 |
| **Ethyl vs Methyl Transfer** | | | |
| $k_{E,E}/k_{X,E}$ | 8.0 | 13.9 | 34.7 |
| $k_{E,X}/k_{X,X}$ | 2.2 | 3.6 | 16.8 |
| **Ethylbenzene vs Xylene** | | | |
| $k_{E,E}/k_{E,X}$ | 4.7 | 5.8 | 7.4 |
| $k_{X,E}/k_{X,X}$ | 1.3 | 1.5 | 3.6 |

The effect of different zeolite structures and pore systems is also reflected in the data of Table II. With the intermediate pore ZSM–5, xylene is apparently much less reactive than ethylbenzene, both as an alkyl donor and acceptor, than it is with the large pore zeolites, ZSM–4 and synthetic mordenite. This may be partly the result of increased steric crowding in the transition state of transalkylation. Another contributory factor to the increased selectivity in ZSM–5 is the higher diffusion rate of ethylbenzene vs m–/o–xylene in ZSM–5 and hence a higher steady state concentration ratio [EB]/[xyl] in the zeolite interior than in the outside phase. Diffusional restriction for xylenes vs ethylbenzene may also be indicated by the better selectivity of synthetic mordenite vs ZSM–4, since the former had a larger crystal size.

In commercial xylene isomerization, it is desirable that the necessary ethylbenzene conversion is accompanied by a minimum conversion (transalkylation) of xylenes, since the latter constitutes a downgrading to less valuable products. The ability of ZSM–5 to convert ethylbenzene via transalkylation in high selectivity, as shown in Table II, leads to high ultimate p-xylene yields in a commercial process. With a simulated commercial feed containing 85% m– and o–xylene and 15% ethylbenzene, we have obtained the data shown in Table III. It is seen that for a given ethylbenzene conversion, the xylene loss

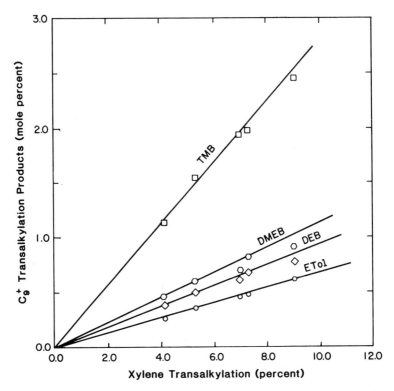

Figure 6. Transalkylation of an ethylbenzene-xylene feed over HZSM-4. TMB = trimethylbenzene, DMEB = dimethylethylbenzene, DEB = diethylbenzene, and ETol = ethyltoluene. Feed: 16% EB, 62% m-xylene, 22% o-xylene. Temperature: 282°C. Pressure: 29 bar. WHSV: 8-20.

is quite small with ZSM-5, twice as large with synthetic mordenite, and four times larger with ZSM-4. For a faujasite-type catalyst, the xylene loss is even greater. These data were obtained on a once-through basis. In commercial recycle operation, the absolute selectivity values can be further optimized by varying the recycle ratio.

Table III.   Selectivity in Xylene Isomerization
Feed:   15% Ethylbenzene, 85% Xylene (63% m, 22% o)

| Catalyst | % Xyl Transalkylated / % EB Transalkylated |
|----------|--------------------------------------------|
| ZSM-4 | .36 |
| Mordenite | .19 |
| ZSM-5 | .09 |

As mentioned earlier, at higher temperature the selective conversion of ethylbenzene is further enhanced by opening an additional pathway, i.e., dealkylation, that yields increased amounts of benzene of high purity:

$$C_6H_5-C_2H_5 \rightleftharpoons C_6H_6 + CH_2=CH_2$$

$$\underset{H_2/\text{Metal}}{\xrightarrow{\hspace{5cm}}} C-C$$

The reaction is rendered irreversible by hydrogenating the ethylene with a selective hydrogenation catalyst.

Toluene Disproportionation (TDP)

At temperatures above 450°C ZSM-5 is a very effective catalyst for the disproportionation of toluene. A process has been developed and put into commercial practice (2). The thermodynamic equilibrium composition (11) is listed in Figure 7. The product obtained with ZSM-5 contains less of the highly substituted aromatics, as a result of diffusion and transition-state inhibition, such that the process can be approximated by the equation:

$$2 \quad \text{Toluene} \longrightarrow \text{Benzene} + \text{Xylene}$$

The xylenes are produced in an equilibrium mixture containing 24% p-, 54% m-, and 22% o-xylene (11). This is readily understandable. The transalkylation occurs via an electrophilic substitution of toluene by a benzyl cation. In the absence of steric constraints, p- and o-xylene are expected as predominant

primary products. For example, the benzylation of toluene with
benzyl chloride produces 55% p-, 41% o-, and 4% m-methyldiphenyl
methane (12). Indeed, with rare earth X zeolite as catalyst, the
xylene produced from toluene at low conversion (<10%) contains
>50% p-xylene (13). For this catalyst, we estimate the ratio
$k_I/k_D$ <10. However, with ZSM-5 as catalyst, an equilibrium
mixture of xylene is expected at all significant conversions
since any kinetically controlled primary xylene product, e.g.,
p-xylene, would equilibrate rapidly in a consecutive reaction:

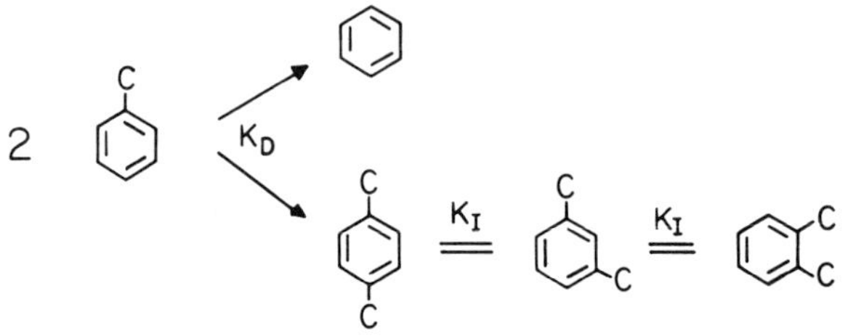

We find from separate experiments with ZSM-5 catalyst that the
intrinsic rate constant for isomerization, $k_I$, is much faster
than that for disproportionation, $k_D$; at 482°C, $k_I/k_D \geqslant 7000$,
i.e. $k_I/k_D$ is much faster than it is for the synthetic faujasite
catalyst.

## Selective Toluene Disproportionation (STDP)

It has been found that the disproportionation of toluene over
ZSM-5 catalyst can be directed such that p-xylene is the
predominant xylene isomer (14-17). This reaction, designated
STDP, is one of several in which disubstituted aromatics rich in
the para isomer are produced. Others are the alkylation of
toluene with methanol to produce p-xylene (15,18) and with
ethylene to produce p-ethyltoluene (19,20), as well as the
aromatization of olefins (20), paraffins (20) and of methanol
(21).

As is apparent from the previous discussion on toluene
disproportionation, the observation of high p-selectivity in STDP
requires a dramatic change in selectivity. First, the primary
product must be directed to be highly para-selective. Secondly,
the subsequent isomerization of the primary p-xylene product must
be selectively inhibited:

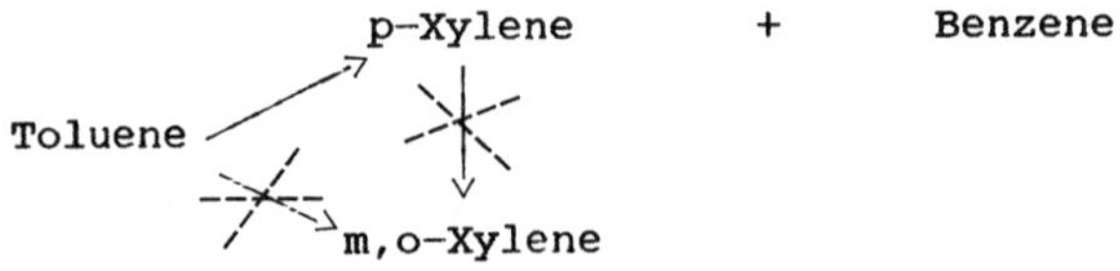

Various ways to modify ZSM-5 catalyst in order to induce para-selectivity have been described. They include an increase in crystal size (15,17,20) and treatment of the zeolite with a variety of modifying agents such as compounds of phosphorus (15,18), magnesium (15), boron (16), silicon (21), antimony (20), and with coke (14,18). Possible explanations of how these modifications may account for the observed selectivity changes have been presented (17) and a mathematical theory has been developed (22). A general description of the effect of diffusion on selectivity in simple parallel reactions has been given by Weisz (23).

In this paper we present a quantitative, correlative model for STDP based on simple, measured catalyst properties.

We find that the degree of para-selectivity obtainable depends uniquely on the activity and diffusion characteristics of the catalyst, independent of how these properties are obtained. While we will discuss these relationships with regard to STDP, the principles involved are generally applicable to those reactions over ZSM-5 where dialkylaromatic products are formed.

## Model for STDP

The general characteristics of toluene disproportionation are summarized by the data presented in Figure 8. With standard HZSM-5 catalyst, as indicated by the lowest curve, the xylenes produced contain essentially an equilibrium concentration of the para isomer (24%) and exceed it only slightly at low conversion. The other curves result from a variety of HZSM-5 catalysts modified in different ways and to different degrees. It is apparent that a wide range of para-selectivities can be obtained. At increasing toluene conversions, the para-selectivity decreases for all catalysts.

The reaction scheme to be considered is shown in Figure 9. Toluene diffuses into the zeolite with a diffusivity $D_T$. It undergoes disproportionation to benzene and either p-, m-, or o-xylene with a total rate constant $k_D$. The initial product distribution ($P_i$, $M_i$, $O_i$) is not known. In the absence of steric constraints, it is expected to be rich in the p- and o-isomers for mechanistic reasons, as discussed above. Under the steric constraints of the channels of ZSM-5, the initial formation of p-xylene should be further favored. Isomerization within the zeolite crystal will interconvert the initial product towards a sorbed equilibrium mixture.

The primary product ($P_o$, $M_o$, $O_o$), the first observable product outside the zeolite at conversions approaching zero percent, is determined by the initial isomer distribution, the rate of interconversion via isomerization ($k_I$), the respective diffusivities ($D_p$, $D_m$, $D_o$) and the length of the diffusion path, characterized by $r$.

At the longer contact times required to give practical toluene conversion, the primary product on reentering the zeolite

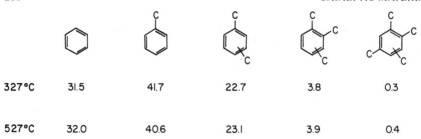

| | | | | |
|---|---|---|---|---|
| 327°C | 31.5 | 41.7 | 22.7 | 3.8 | 0.3 |
| 527°C | 32.0 | 40.6 | 23.1 | 3.9 | 0.4 |

Figure 7.   Thermodynamic equilibrium, in mole percent, for
toluene disproportionation at 327°C and 523°C.

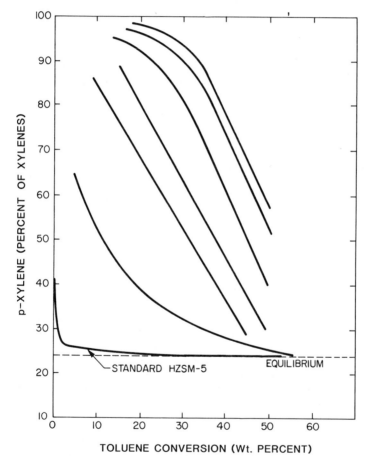

Figure 8.   STDP over various HZSM-5 catalysts. Temperature:
550°C.   Pressure: 41 bar.   H$_2$/Tol: 4, variable
contact time.

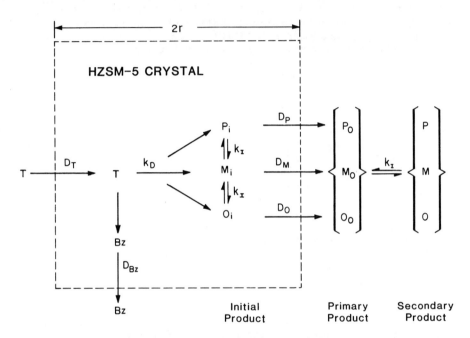

Figure 9.   Model for STDP.   Bz = benzene, T = toluene,
P,M,O = para-, meta-, and ortho-xylene.

crystals undergoes consecutive isomerization towards the
thermodynamic equilibrium mixture (Figure 8). The secondary
product is thus a function of the degree of conversion.

## Requirements for High Para-Selectivity

The primary product will be rich in the para isomer if initial m-
and o-xylene diffuse out of the zeolite crystal at a lower rate
($D_{m,o}/r^2$) than that of their conversion to p-xylene ($k_I$) and the
latter's diffusion ($D_p/r^2$). Conversion of the para-rich primary
product to secondary product low in p-xylene is minimized when
the actual, observed rate of isomerization ($k_I$)$_{obs}$ is lower than
the rate of toluene disproportionation ($k_D$).
     The following conditions need to be fulfilled:

High p-selectivity in primary product:

$$1. \quad D_p \gg D_{m,o}$$

$$2. \quad k_I \gg \frac{D_{m,o}}{r^2}$$

$$3. \quad k_D \ll \frac{D_T}{r^2}$$

Minimum secondary isomerization:

$$4. \quad \left(\frac{k_i}{k_D}\right)_{observed} \leqslant 1$$

Condition 1: Direct determination of diffusivities in ZSM-5 via
sorption rate measurements showed $D_p/D_o > 10^3$ and and $D_o \approx D_m$.

Condition 2: The para-selectivity is increased by an increase in
$k_I$ and in the crystal size ($r$) and by a reduction in the
diffusivity. The applicability of this relationship will be
illustrated below.

Condition 3: It is desirable that the toluene disproportionation
reaction ($k_D$) is not diffusion limited. Otherwise, the effective
crystal size would be smaller than the actual size ($r$), adversely
impacting on Condition 2. Experimentally we find that the
observed toluene disproportionation rates are similar for equal
activity HZSM-5's having $D/r^2$ values differing by four orders of
magnitude. Thus, this condition is met.

Condition 4: As mentioned above, the intrinsic ratio $k_I/k_D$
$\geqslant 7000$. Thus, a severe selective inhibition of the isomerization

rate is necessary. Since the catalytic sites for isomerization and disproportionation are the same, this cannot be accomplished by changing the nature of the sites, but only via the imposition of diffusion barriers that will reduce the observed xylene isomerization $(k_I)_{obs}$. This is fulfilled if:

$$k_I \geqslant \frac{D_{m,o}}{r^2}$$

Thus, the requirement to avoid consecutive isomerization is the same as that leading to a para-rich primary product (Condition 2).

It is seen that high para-selectivity should result from mass transport inhibition for formation of the undesirable m-/o-xylenes.

## Effect of Crystal Size

The effect of crystal size, 2r, in STDP is demonstrated in Figure 10. These data for three zeolites having similar activity, but with crystal sizes differing by nearly two orders of magnitude, show a significant increase in para-xylene selectivity with increasing crystal size. The primary product selectivity is enhanced and secondary isomerization is retarded.

## Effect of Diffusivity

In view of the difficulty of measuring the diffusivity of o-xylene at the reaction temperature, 482°C, we have used the diffusivity determined at 120°C. For a series of ZSM-5 catalysts, the two D-values should be proportional to each other. Para-xylene selectivities at constant toluene conversion for catalysts prepared from the same zeolite preparation (constant r) with two different modifiers are shown in Figure 11. The large effect of the modifier on diffusivity, and on para-selectivity, is apparent.

## Effect of $r^2/D$

In order to compare a number of different zeolite preparations we have found it convenient to determine not the diffusivity of o-xylene per se, but to characterize the samples by measuring the time $(t_{0.3})$ it takes to sorb 30% of the quantity sorbed at infinite time. The characteristic diffusion time, $t_{0.3}$, is a direct measure of the critical mass transfer property $r^2/D$:

$$t_{0.3} \propto r^2/D$$

Thus, the required value of $r^2/D$ is obtained from a single experiment; it also eliminates the effort and errors associated

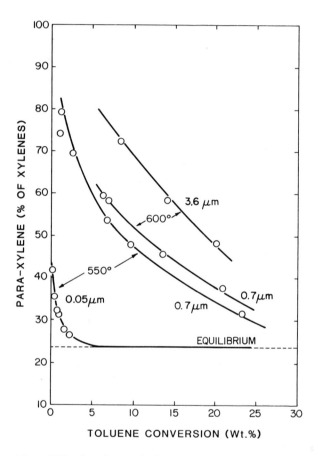

Figure 10. Effect of crystal size, 2r, on p-xylene selectivity in toluene disproportionation by HZSM-5.

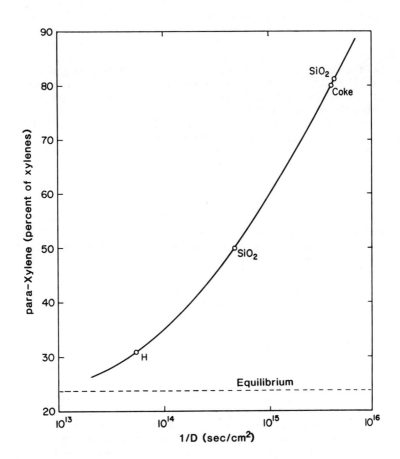

Figure 11.   Effect of diffusivity on p-xylene selectivity.
Toluene disproportionation at 550°C, 20% conversion;
o-xylene diffusivity at 120°C.

with crystal size determinations. With respect to the latter, additional complications include the fact that crystals of ZSM-5 are rarely equidimensional and its pore structure suggests that $D_x \neq D_y \neq D_z$. As an approximation we have assumed a flat plate diffusion model and used the diffusion equation solution given by Crank (24). That the straight channels in ZSM-5 are normally parallel to the smallest crystal dimension supports this approximation in lieu of a much more complex analysis of diffusion.

Para-selectivity for a wide variety of ZSM-5 preparations of comparable activity are shown in Figure 12. These data include results for unmodified HZSM-5's of varying crystal size as well as chemically modified HZSM-5's. Since the activity of these catalysts is nearly identical, these data clearly establish the major role of diffusion in the para-xylene content of the xylenes produced in TDP. We have examined in more detail the effect of the concentration of one of these chemical modifiers, MgO.

## Magnesium Modified ZSM-5

Both para-xylene selectivity and $r^2/D$ ($t_{0.3}$) increase smoothly with MgO level for a series of large crystal, Mg modified HZSM-5 catalysts, and again para-xylene selectivity increases with $t_{0.3}$ (Figure 13, Table IV). However, these catalysts appear to be significantly different from the catalysts just discussed, defining a separate functional dependence on $r^2/D$ ($t_{0.3}$). These differences will be shown to be attributable to differences in acid activity of this series of catalysts.

## Effect of Activity

Over zeolite catalysts both of the reactions of interest, toluene disproportionation and xylene isomerization, are catalyzed by acid sites. Further, the cracking rate of n-hexane, the $\alpha$-value, has been reported to be a measure of the active sites and hence of the acid activity (25), and has been found to be linearly related to other acid catalyzed reactions (26). In our study, both TDP and hexane cracking rate have been used to assess catalyst acidity. The effect of MgO level on activity as measured by both of these tests is shown in Figure 14. The acid activity dramatically decreases with MgO level implying that some ionic magnesium species, e.g. $Mg^{++}$ or $MgOH^+$, have exchanged for protons in the zeolite. Thus, the lower para-selectivity of the MgO-modified catalysts (Figure 13) is a direct consequence of their reduced activity (see Condition 2) which partly negates the beneficial effect of MgO incorporation on decreasing diffusivity (Table IV).

The beneficial effect of catalyst activity, k, can also be seen from the temperature dependence of the para-selectivity. Comparing the selectivity for the same, unmodified ZSM-5 catalyst at 550°C and 600°C in Figure 10 shows the advantage of the

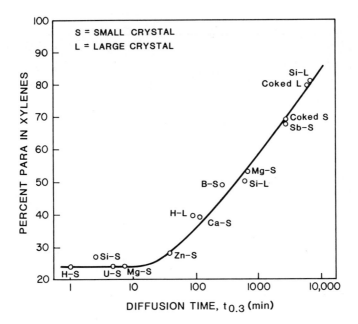

Figure 12. Relationship between the diffusion parameter, $t_{0.3}$, and p-xylene selectivity in toluene disproportion-ation. Temperature: 550°C. Pressure: 41 bar. Conversion: 20%. $t_{0.3}$: time to reach 30% of amount sorbed at infinite time.

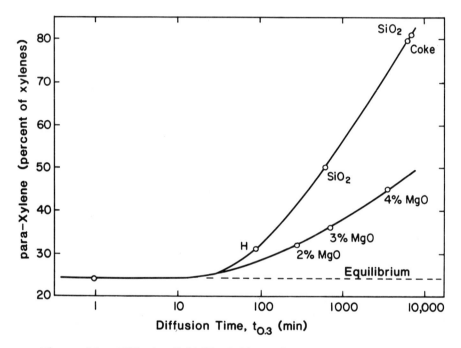

Figure 13. Effect of diffusivity and MgO content on
p-xylene selectivity. Toluene disproportionation at 550°C,
20% conversion; o-xylene diffusion time at 120°C.

Table IV.    Effect of MgO Content on the Properties of MgO-Modified HZSM-5 Catalysts

| MgO Level (wt %) | 0 | 1 | 2 | 3 | 4 | 4.6 | 5 |
|---|---|---|---|---|---|---|---|
| Sorption (n-C$_6$, mg/g)[a] | 105.0 | 103.7 | 100.6 | 101.4 | 97.6 | 91.9 | 92.7 |
| Percent Decrease | -- | 1.3 | 4.2 | 3.9 | 7.0 | 12.5 | 11.8 |
| Loss ($\mu l$/g) | 0.0 | 2.1 | 6.7 | 6.2 | 11.2 | 20.0 | 18.9 |
| Volume of MgO ($\mu l$/g)[b] | 0.0 | 2.9 | 5.8 | 8.6 | 11.7 | 13.4 | 14.7 |
| Diffusion Properties[c] | | | | | | | |
| $t_{0.3}$ (min.) | 80 | 155 | 390 | 700 | 3710 | 29700 | 29700 |
| Activity | | | | | | | |
| Toluene Conv. (wt %)[d] | 30.0 | -- | 15.0 | 5.5 | 3.9 | 3.0 | 1.0 |
| Alpha (hexane cracking) | 270.0 | -- | -- | -- | 36.0 | 29.0 | -- |

a.  Measured at 90°C, P(n-hexane) = 83 mm, calculated per gram of MgO-free zeolite.
b.  Calculated from content and density of MgO.
c.  Time to reach 30% of amount sorbed at infinite time, measured at 120°C,
    P(o-xylene) = 3.8 mm.
d.  TDP measured at 577°C, WHSV = 50, H$_2$/HC = 6, P = 26.5 bar; hexane cracking
    measured at 538°C.

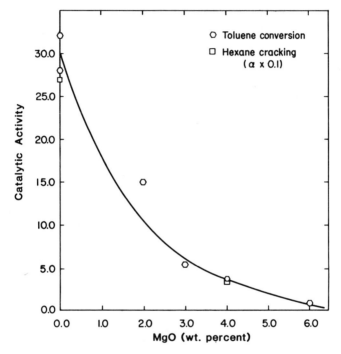

Figure 14.  Effect of MgO content on the activity of MgO–
HZSM-5 catalysts.  Toluene conversion at 577°C,
WHSV = 50, $H_2$/HC = 6, pressure = 26.5 bar; hexane cracking
at 538°C.

higher temperature. Similar data for Mg-modified ZSM-5 were presented previously (16). We have determined the activation energies for xylene isomerization (ca. 30 kcal/mol) and for o-xylene diffusivity (9 kcal/mole). Since the former is considerably higher, raising the temperature will significantly increase the Thiele modulus $\phi^2 = r^2k/D$ and the mass transfer limitation.

## The Parameter $\alpha \cdot t_{0.3}$

Rigorous treatment of the para-selectivity requires a knowledge of the intrinsic value of the rate constant for all the reactions involved and of the absolute value of the crystal size and of the diffusivity, all under reaction conditions. These values are obtainable only with considerable difficulty and effort. As has been mentioned, the 30 percent sorption time for o-xylene at 120°C, $t_{0.3}$, is proportional to the actual values, $r^2/D$. Likewise, the value for hexane cracking at 538°C, the $\alpha$-value, is proportional to the intrinsic value for toluene disproportionation and xylene isomerization, and is much easier to obtain since it is not affected by crystal size or diffusive alteration (5).

Thus, the critical value determining para-selectivity $r^2k/D$ can be replaced by two readily measured quantities, $\alpha$ and $t_{0.3}$.

$$(k \times r^2/D) \propto (\alpha \times t_{0.3})$$

If the data in Figure 13 are replotted against $\alpha \cdot t_{0.3}$ as a pseudo Thiele-modulus, a single curve describing all the catalysts (Figure 15) results. Thus, the para-selectivity of a catalyst can be readily predicted from this empirical correlation and a knowledge of two basic catalyst properties, activity and diffusion time. Furthermore, these data are in full agreement with the model advanced above, which describes para-selectivity in terms of a classical diffusion-reaction interplay.

## Model for the Catalyst

As pointed out above, the reduced activity of the MgO modified catalysts indicates exchange of a magnesium species for some of the protonic centers in the zeolite. However, this exchange would account for less than ~1/4 of the added MgO. Measurements of loss in n-hexane sorption capacity are in good agreement with the volume of MgO added, indicating that most of the MgO is incorporated into the zeolite channels (Table IV). Further, we find that log $(D/r^2)$ is a linear function of the percent of pore volume filled by MgO (Figure 16). Over the range examined, $D/r^2$ decreases by one order of magnitude for each five percent of pore volume filled by a blocking agent.

Similar measurements on a coke modified (4% coke), 0.7$\mu$m HZSM-5 revealed less than 2% loss in sorption capacity, and no

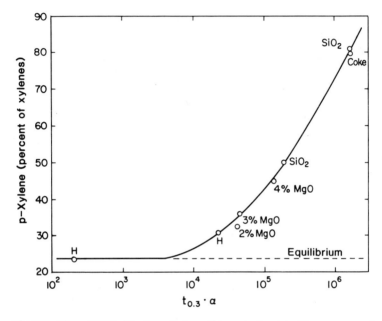

Figure 15. Dependence of p-xylene selectivity on pseudo Thiele modulus. Toluene disproportionation at 550°C, 20% conversion.

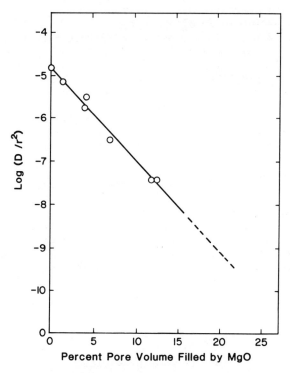

Figure 16. Effect of pore filling on diffusivity of o-xylene at 120°C in HZSM-5.

change in activity, whereas $r^2/D$ increased by nearly two orders
of magnitude. Figure 17 shows a model for coke modified HZSM-5.
This shows the coke forming an external surface layer blocking
off a large fraction of the entrance ports to the zeolite channel
system.

These data demonstrate two different principles for altering
the diffusivity of a catalyst:
With $D = D_o \rho . \tau$,

where $D$ = effective diffusivity,
$D_o$ = intrinsic diffusivity,
$\rho$ = porosity = fraction of catalyst that is void,
$\tau$ = tortuosity,

It is seen that the intracrystalline MgO induces pore blockage in
a fraction of the pore system and alters the porosity as well as
$D_o$, and/or $\tau$ with the latter factors contributing most to the
reduced diffusivity. In contrast, the coke modifier appears to
affect mainly the surface-to-volume ratio and suggests that the
effective surface area, number of available entrance ports, is
reduced by two orders of magnitude.

Conclusion

We have shown that the high selectivity of ZSM-5 in xylene
isomerization relative to larger pore acid catalysts is a result
of its pore size. It is large enough to admit the three xylenes
and to allow their interconversion to an equilibrium mixture; it
also catalyzes the transalkylation and dealkylation of
ethylbenzene (EB), a necessary requirement for commercial
feed; but it selectively retards transalkylation of xylenes, an
undesired side reaction.

$$o\text{-xylene} \rightleftharpoons m\text{-xylene} \rightleftharpoons p\text{-xylene}$$

$$EB + EB \longrightarrow BZ + DIEB$$

$$EB \longrightarrow BZ + C_2^=$$

$$\text{Xylene} + \text{Xylene} \xrightarrow{\quad\times\quad} \text{Toluene} + TMB$$

It has also been shown that the selectivity features of
para-selective catalysts can be readily understood from an
interplay of catalytic reaction with mass transfer. This
interaction is described by classical diffusion-reaction
equations. Two catalyst properties, diffusion time and intrinsic
activity, are sufficient to characterize the shape selectivity of
a catalyst, both its primary product distribution and products at
higher degrees of conversion. In the correlative model, the
diffusion time used is that for o-xylene adsorption at

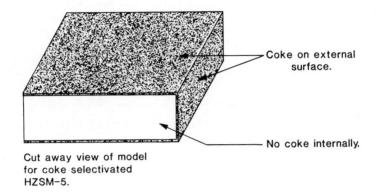

Coke on external surface.

No coke internally.

Cut away view of model for coke selectivated HZSM-5.

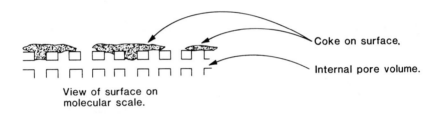

Coke on surface.

Internal pore volume.

View of surface on molecular scale.

Figure 17.   Schematic model for coke selectivated HZSM-5.

120°C; the intrinsic activity is characterized by hexane cracking at 538°C.

A quantitative model requires knowledge of the diffusivity under reaction conditions and of the intrinsic activities for toluene disproportionation and xylene isomerization. While these are not easily obtained, the methodology has been worked out for the case of paraffin and olefin cracking (5). So far, we have obtained an approximate value for the diffusivity, D, of o-xylene at operation conditions from the rate of sorptive o-xylene uptake at lower temperature and extrapolation to 482°C (Table V).

Table V.   Diffusion Properties of Selected HZSM-5 Catalysts

| Crystal Size (2r, $\mu$m) | 0.05 | 0.7 | 0.7 |
|---|---|---|---|
| Modification | None | None | Coked |
| $(r^2/D)^{120°C}$ (s) | <1 | 67,000 | $5.0 \times 10^6$ |
| $D^{120}$ (cm$^2$/s) | $<5 \times 10^{-15}$ | $1.8 \times 10^{-14}$ | $2.4 \times 10^{-16}$ |
| $(r^2/D^{482°C}$ (s)$^a$ | ~1 | 270 | 10,000 |
| $D^{482}$ (cm$^2$/s) | $(5 \times 10^{-12})$ | $4.6 \times 10^{-12}$ | $6.1 \times 10^{-14}$ |

a.   Computed using 9 kcal/mole activation energy for o-xylene diffusion.

We have also determined the intrinsic rate constant $k_{intr}$ for TDP and have made an estimate of the intrinsic rate constant for xylene isomerization. These data are collected in Table VI, together with calculated values of the Thiele parameter $\phi^2 = k' r^2/D$ and the effectiveness factor, $\eta$:

$$\eta = k_{observed}/k_{intrinsic},$$

$$\eta = \tanh \phi/\phi, \text{ for flat plate geometry.}$$

Also listed are the characteristic times for diffusion, i.e., Einstein time $\tau_E = r^2/2D$, and for reaction, $\tau_k = 1/k$.

It is seen that TDP is indeed not diffusion limited ($\eta = 1$), in agreement with condition 3, with coked large crystal catalyst being just on the borderline. Xylene isomerization is severely inhibited with both catalysts used. The ratio of rate constants $k_I/k_D$ shows that the intrinsic ratio of 7000 is reduced via diffusion control to an apparent value of 2 in the para-selective large crystal catalyst, in agreement with the observed, greatly reduced p-xylene isomerization in a consecutive reaction (Condition 4). Inspection of the $\tau_E/\tau_k$ ratios reveals that a m-xylene molecule formed from toluene in a small crystal is 200 times as likely to isomerize as it is to diffuse out of the crystal. For the larger crystal size catalyst with reduced diffusivity, the corresponding value is very much larger, about

Table VI.  Estimated Reaction Parameters for Toluene Disproportionation (TDP) and Xylene Isomerization (ISOM) at 482°C

| Reaction | $k_{intrinsic}$ (s⁻¹) | Small Crystal | | | Large Crystal, Coked | | |
|---|---|---|---|---|---|---|---|
| | | $r^2/D$(s) | $kr^2/D$ | $\eta$ | $r^2/D$(s) | $kr^2/D$ | $\eta$ |
| TDP | 0.14 | $10^{-4}$ | $10^{-5}$ | 1 | 1 | $10^{-1}$ | .97 |
| ISOM | $10^3$ | 0.4 | $4 \times 10^2$ | $5 \times 10^{-2}$ | $10^4$ | $10^7$ | $3 \times 10^{-4}$ |

| $k_{ISOM}/k_{TDP}$ | Intrinsic | Observed (Small Crystal) | Observed (Large Crystal) |
|---|---|---|---|
| | 7000 | 360 | 2 |

| Reaction | $\tau_k$ (s) | Small Crystal | | Large Crystal, Coked | |
|---|---|---|---|---|---|
| | | $\tau_E$ (s) | $\tau_E/\tau_k$ | $\tau_E$ (s) | $\tau_E/\tau_k$ |
| TDP | 7 | $5 \times 10^{-5}$ | $10^{-5}$ | 0.5 | 0.07 |
| ISOM | $10^{-3}$ | 0.2 | $2 \times 10^2$ | $5 \times 10^3$ | $5 \times 10^6$ |

$10^7$ (Condition 2). Thus, the postulates of the para-selectivity model are well supported by the pertinent reaction/ diffusion parameters.

Experimental

The zeolites and catalysts used in this study were prepared as described previously (1,16,18,20). The ortho-xylene sorption rate data, obtained on a computer-controlled Du Pont 951 TGA, were measured at 120°C and P(o-xylene) = 3.8 torr. The isomerization and disproportionation data were obtained using a 0.3 inch diameter stainless steel, downflow reactor and the products analyzed with an online sampling GC system. The hexane cracking test, alpha test, has been described earlier (25,27,28).

Acknowledgments

We wish to acknowledge helpful discussions with P.B. Weisz, J.R. Katzer and P.G. Rodewald, R.M. Lago for the hexane cracking measurements, F.G. Dwyer for supplying some of the zeolites used in this study, and N.H. Goeke for his experimental assistance.

Literature Cited

1.  Argauer, R.J.; Landolt, G.R. U.S. Patent 3,702,886, 1972.
2.  Meisel, S.L.; McCullough, J.P.; Lechthaler, C.H.; Weisz, P.B. "Recent Advances in the Production of Fuels and Chemicals Over Zeolite Catalysts"; Leo Friend Symposium 174, American Chemical Society; Chicago, Illinois, August 1977.
3.  Meisel, S.L.; McCullough, J.P.; Lechthaler, C.H.; Weisz, P.B. Chemtech. 1976, 6, 86.
4.  Chen, N.Y.; Garwood, W.E.; Haag, W.O.; Schwartz, A.B. "Shape Selective Conversion Over Intermediate Pore Size Zeolite Catalysts"; Am. Inst. of Chem. Eng. 72, An. Meeting; San Francisco, California, November 1979.
5.  Haag, W.O.; Lago, R.M.; Weisz, P.B. Faraday Disc./1982, 72, 317.
6.  Csicsery, S.M. J. Catalysis 1970, 19, 394; ibid, 1971, 23, 124.
7.  Chutoransky, P. Jr.; Dwyer, F.G. ADVANCES IN CHEMISTRY SERIES No. 121, 1973, 540.
8.  Lanewalda, M.A.; Bolton, A. J. Org. Chem. 1969, 34, 3107.
9.  Meier, W.M.; Olson, D.H. "Atlas of Zeolite Structure Types"; Structure Commission of the Intl. Zeolite Assn., Polycrystal Book Service, Pittsburgh, PA, 1978.
10. Haag, W.O.; Olson, D.H. U.S. Patent 3,856,871, December 24, 1974.

11.   Stull, D.R.; Westrum, E.F., Jr. Sinke, G.C. "The
      Chemical Thermodynamics of Organic Compounds"; John
      Wiley and Sons Inc., 1969.
12.   Olah, G.A.; Tashiro, M.; Kobayashi,
      S. J. Amer. Chem. Soc. 1970, 70, 6369.
13.   Haag, W.O.; Wise, J.J., unpublished laboratory results.
14.   (a) Kaeding, W.W. U.S. Patent 4,029,716, June 14,
      1977; (b) Haag, W.O.; Olson, D.H. U.S. Patent 4,097,543,
      June 27, 1978.
15.   Chen, N.Y.; Kaeding, W.W.; Dwyer, F.G. J. Amer. Chem.
      Soc. 1979, 101, 6783.
16.   Kaeding, W.W.; Chu, C.; Young, L.B.; Butter,
      S.A. J. Catal. 1981, 69, 392.
17.   Young, L.B.; Butter, S.A.; Kaeding, W.W. J. Catal.
      1982, 76, 418.
18.   Kaeding, W.W.; Chu, C.; Young, L.B.; Winstein,
      B.; Butter, S.A. J. Catal. 1981, 67, 159.
19.   (a) Kaeding, W.W.; Young, L.B.; Prapas, A.B. Chemtech.
      1982, 12, 556; (b) Kaeding, W.W.; Young, L.B.
      U.S. Patent 4,086,287, April 25, 1978.
20.   Haag, W.O.; Olson, D.H.  U.S. Patent 4,117,026,
      September 26, 1978.
21.   Rogewald, P.G.  U.S. Patents 4,060,568, November 29,
      1977;  4,090,981, May 23, 1978;  4,127,616, November 28,
      1978;  4,145,315, March 20, 1979.
22.   Wei, J. J. Catal. 1982, 76, 433.
23.   Weisz, P.B. J. Pure and Appl. Chem.1980, 52, 2091;
      ibid., "New Horizons in Catalysis", Proceed. of the
      Seventh Intl. Congress on Catalysis, Tokyo, Japan, June
      30—July 4, 1980, Elsevier Scientific Publ. Co., NY,
      p. 3.
24.   Crank, J. "The Mathematics of Diffusion", Oxford
      University Press, Ely House, London, 1967.
25.   Olson, D.H.; Haag, W.O.; Lago, R.M.  J. Catal. 1980,
      61, 390.
26.   Haag, W.O.; Lago, R.M., unpublished laboratory data.
27.   Weisz, P.B.; Miale, J.N. J. Catal. 1965, 4, 527.
28.   Miale, J.N.; Chen, N.Y.; Weisz, P.B. J. Catal. 1966, 6,
      278.

RECEIVED September 20, 1983

# MICROSCOPY AND OTHER
# NOVEL METHODS

# Analytical Electron Microscopy of Heterogeneous Catalyst Particles

C. E. LYMAN

Central Research and Development Department, E. I. du Pont de Nemours & Company, Experimental Station, Wilmington, DE 19898

Analytical electron microscopy of individual catalyst particles provides much more information than just particle size and shape. The scanning transmission electron microscope (STEM) with analytical facilities allows chemical analysis and electron diffraction patterns to be obtained from areas on the order of 10nm in diameter. In this paper, examples of high spatial resolution chemical analysis by x-ray emission spectroscopy are drawn from supported Pd, bismuth and ferric molybdates, and ZSM-5 zeolite.

The origins of analytical electron microscopy go back only about 15 years when the first x-ray spectra were obtained from submicron diameter areas of thin specimens in an electron microscope [1]. Characterization of catalyst materials using AEM is even more recent[2,3] but is currently a very active research area in several industrial and academic laboratories. The primary advantage of this technique for catalyst research is that it is the only technique that can yield chemical and structural information from individual submicron catalyst particles.

## Microanalysis of Catalysts

Bulk techniques (wet chemistry, x-ray diffraction, IR, NMR, etc.) analyze milligram to gram-sized samples. Surface analysis techniques (Auger electron spectroscopy, XPS, SIMS, ISS, etc.) analyze only the first few monolayers of areas usually larger than a square micrometer. The electron microprobe and the scanning electron microscope employing an x-ray spectrometer attachment can analyze regions down to about 1μm in diameter. Although these instruments can produce submicron electron beams, the spatial resolution for x-ray analysis is always on the order of 1μm or greater because of electron penetration and scattering beneath the specimen surface. Figure 1 shows how this limitation

0097–6156/84/0248–0311$06.25/0
© 1984 American Chemical Society

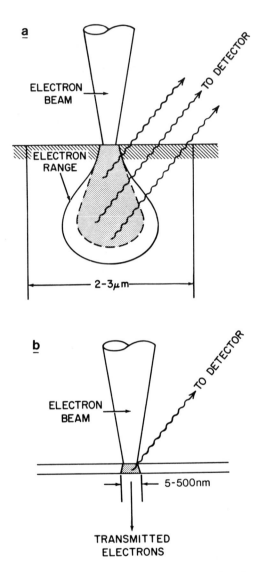

Fig. 1:   Spatial resolution of x-ray analysis from a) bulk
          microprobe specimens and b) thin specimens
          suitable for analytical electron microscopy.

can be removed for the case of thin specimens. When the electron
diffusion zone beneath the surface is removed, the spatial
resolution for x-ray analysis becomes a function of the specimen
thickness, beam size, average specimen atomic number, and the
electron accelerating voltage[4]. Currently, under optimum
conditions, a spatial resolution for x-ray microanalysis on the
order of 2-10nm can be achieved with a scanning transmission
electron microscope (STEM) employing a field-emission gun. The
field-emission gun[5] is required for very high resolution
analysis because more electrons can be focussed into a 1-10nm
diameter electron beam with this electron source than with any
other. Another advantage in analyzing thin specimens versus bulk
specimens is that the absorption and fluorescence corrections,
required for quantitative analysis of bulk specimens, often can
be neglected.

Although x-ray microanalysis in the STEM is the most
developed form of analytical electron microscopy, many other
types of  information can be obtained when an electron beam
interacts with a thin specimen. Figure 2 shows the various
signals generated as electrons traverse a thin specimen. The
following information about heterogeneous catalysts can be
obtained from these signals:

1. Transmitted electrons
   - shape and size of phases by mass-thickness contrast
     (bright-field images)
   - structure images of thin crystals by phase contrast
     (in combination with diffracted beams)
   - pore sizes and shapes

2. Bragg-diffracted beams
   - defect structure of crystalline phases
   - determination of amorphous versus crystalline phases
   - identification of phases
   - distribution of phases relative to one another
     (dark-field images)

3. Electron energy loss spectra (EELS)
   - analysis of light elements in very thin specimens
     (e.g., carbon in coke formations)
   - fingerprint-type identification of phases using
     low-loss spectra similar to optical spectra around
     1-10eV
   - distribution of light elements within phases
     (EELS element distribution mapping)

4. Backscattered electrons
   - phase distribution by atomic number contrast
   - crystallographic contrast
     (electron channeling patterns)
   - topographical images of catalyst surfaces

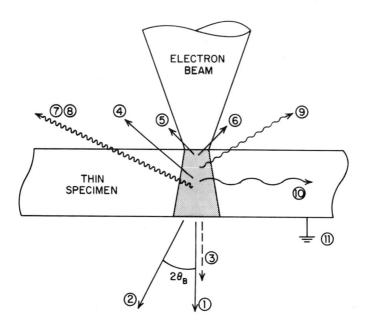

1. Transmitted electrons
2. Bragg-diffracted beams
3. Electron energy loss electrons
4. Backscattered electrons
5. Secondary electrons
6. Auger electrons
7. Characteristic x-rays
8. Bremsstrahlung (continuous x-rays)
9. Cathodoluminescence (light)
10. Heat
11. Specimen current

Fig. 2:    Electron-beam interactions with a thin specimen.

5. Secondary electrons
    - topographical images of catalyst surfaces
      (scanning electron micrographs)

6. Auger electrons
    - surface analysis
      (information from 2nm depths)
    - elemental analysis
      (lower backgrounds with thin specimens)

7. Characteristic x-rays
    - chemical analysis for elements sodium to uranium
    - identification of phases
      (measurement of atomic ratios of elements)
    - distribution of elements within phases
      (x-ray element distribution mapping)

8. Bremsstrahlung
    - continuous x-radiation which interferes with analysis
      (not a useful signal at present)

9. Cathodoluminescence
    - identification of phases
      (optical emission spectra)
    - location of electron trapping defects
    - band gap measurements

10. Heat
    - subsurface imaging and crystallographic contrast
      (thermal wave microscopy)
    - effects due to elastic constant variations

11. Specimen current
    - images that are complementary to backscattered
      electron images
    - electron beam induced current (EBIC) images of
      semiconductor materials

Many of these signals can be used to form images on the STEM and
for analysis of small regions on the specimen. Currently,
signals 1 through 7 have been employed most often in the analysis
of catalysts, although catalyst applications of thin specimen
Auger analysis (signal 6) have not yet appeared. In this paper,
x-ray microanalysis (signal 7) in combination with bright-field
and dark-field images will be discussed exclusively.

## Quantitative Analysis

Analytical electron microscopy by x-ray emission spectroscopy can
be extremely useful as a qualitative analysis tool, e.g. to
determine which elements are present in 10nm diameter areas of
the specimen. However, the greatest impact of AEM comes from
quantitative chemical profiles across minute regions or features
in the specimen, information that usually cannot be obtained by
other means.
    Local thickness variations in a thin specimen complicate the
quantitative analysis of a single element in the absence of
precise knowledge of specimen thickness and without the ability
to compare the measured x-ray intensities with those of thin
standards. To avoid this difficulty, the x-ray intensity for the
element of interest can be divided either by the intensity of a
region of background between peaks as in the Hall method[8], or
by the intensity from another element as in the Cliff-Lorimer
method[9]. The former is largely used for biological analysis
while the latter has become the standard thin specimen
microanalysis method for materials science applications. The
Cliff-Lorimer method is expressed in the following equation:

$$\frac{C_A}{C_B} = k_{AB} \frac{I_A}{I_B} \tag{1}$$

where $I_A$ and $I_B$ are the measured x-ray intensities from elements
A and B. This x-ray intensity ratio $I_A/I_B$ is directly related to
the mass concentration ratio $C_A/C_B$ through the scaling factor,
$k_{AB}$, which varies with various instrumental parameters but is
independent of sample thickness and composition if the two x-ray
intensities, $I_A$ and $I_B$, are measured simultaneously. These
k-values can be either measured from known standards or
calculated[10]. Once the $k_{AB}$ factors are known for various
elements in a particular specimen, then weight fraction ratios or
atomic ratios can be obtained for these elements by multiplying
the intensity ratio by the k-value. However, this analysis
scheme assumes an infinitely thin specimen. To analyze practical
thin specimens, we must accept some error in the analysis if we
neglect the absorption and fluorescence corrections. A guide to
the relative error involved in neglecting the absorption
correction is found in the "thin film criteria" proposed by
Tixier and Philibert[11],

$$t < \frac{0.1}{\chi_A \rho} \tag{2}$$

and by Goldstein, et al.[10],

$$t < \frac{0.2}{(x_B - x_A)\rho} \tag{3}$$

where $x_A$ is the x-ray mass absorption coefficient for element A times the cosecant of the x-ray take-off angle between the specimen surface and the x-ray detector, and $\rho$ is the average density of the specimen. If the specimen thickness t is below the values given in Equations (2) and (3), then Equation (1) can be used without modification with an error of less than 10%. However, if the specimen is thicker than t as given by Equation (2), or if there is a large difference in mass absorption coefficients between elements A and B as indicated by Equation (3), then an absorption correction should be applied to avoid unacceptably large systematic errors. The absorption correction factor usually has the following form[10]:

$$ACF = [(\mu/\rho)^A_{sp}/(\mu/\rho)^B_{sp}] \frac{1-\exp[-(\mu/\rho)^B_{sp}\csc\alpha(\rho t)]}{1-\exp[-(\mu/\rho)^A_{sp}\csc\alpha(\rho t)]} \tag{4}$$

where $(\mu/\rho)^A_{sp}$ and $(\mu/\rho)^B_{sp}$ are the mass absorption coefficients for x-ray lines A and B absorbed by the specimen, $\alpha$ is the x-ray take-off angle, $\rho$ is the specimen density, and t is the specimen thickness. Note that $(\mu/\rho)^A_{sp}$ includes the effect of absorption by all the elements in the specimen even if they are not detectable with the x-ray detector. For example, in the case of zeolites, it is important to include the mass absorption coefficients of $AlK_\alpha$ and $SiK_\alpha$ absorbed by oxygen as well as by each other.

Characteristic fluorescence of x-rays of one element by the x-rays from another element in the specimen also can lead to errors in certain cases, e.g., $CrK_\alpha$ in an iron alloy. Although at times significant, this correction is usually much smaller than the absorption correction. The general formulation of the characteristic fluorescence correction factor is[12]:

$$FCF = [1 + \frac{I^A_f}{I_A}]^{-1} \tag{5}$$

where
$$\frac{I^A_f}{I_A} = C_B \omega_B \frac{r_A - 1}{r_A} \frac{A_A}{A_B} (\mu/\rho)^B_A \frac{E_{KA}}{E_{KB}}$$

$$\times \frac{\ln(E_o/E_{KB})}{\ln(E_o/E_{KA})}(t/2)\{0.923 - \ln[t(\mu/\rho)^B_{sp}]\}$$

where $I_f^A$ is the fluorescent intensity over and above the primary
intensity $I_A$, $\omega_B$ is the fluorescence yield for element B, $r_A$ is
the absorption jump ratio for element A, $(\mu/\rho)_{sp}^B$ and $(\mu/\rho)_B^A$ are
the mass absorption coefficients for x-rays from element B in
element A and the specimen, $A_A$ and $A_B$ are the atomic weights of
elements A and B, $E_o$ is the electron acceleration voltage, and
$E_{KA}$ and $E_{KB}$ are the critical excitation energies for the
characteristic K lines of elements A and B. The total correction
scheme for thin specimen microanalysis is thus:

$$\frac{C_A}{C_B} = k_{AB} \frac{I_A}{I_B} [ACF][FCF] \tag{6}$$

where the last two factors are given by Equations (4) and (5).
Note that Equations (4) and (5) depend on the specimen thickness
t. In specimens of irregular geometry, such as catalyst
particles, measurement of specimen thickness at each analysis
point can be a major difficulty in the practical application of
these correction factors.

In catalyst specimens, the major difficulties in
quantitative analysis are a) low x-ray signal, b) spurious
x-rays, c) electron beam damage, d) specimen drift, and e)
uncertainty of specimen thickness. Low x-ray signals are common
because the very thin specimens necessary to meet the thin-film
criteria (Equations 2 and 3) may contain only small amounts of
the elements to be analyzed. High brightness electron guns
(e.g., field emission) and long x-ray counting times must be used
to compensate for the small amount of an element. The second
difficulty is the spurious x-ray signal not originating from the
area under electron bombardment. Contributions to this signal
from the microscope itself are very low for the Vacuum Generators
STEM (<0.1% of the desired x-ray peak) used in this work.
However, electrons scattered by, and bremsstrahlung continuum
radiation generated in, the specimen area under examination can
give rise to x-rays from remote regions of the specimen which are
not related to the area under the electron beam. This is a
fundamental limitation of the AEM technique, but it is usually
not serious for the case of catalyst particles which are often
widely dispersed on a support film. Electron beam damage may be
the limiting factor in analysis of some catalyst materials such
as zeolites. Differential loss of mass from the specimen can
have serious effects on microanalysis. The necessity for low
beam currents in these cases causes reduced x-ray signals. At
very high spatial resolutions of analysis, drift of the specimen
in the microscope is likely to occur. Although for an ideal
specimen, the Vacuum Generators STEM meets a specification for
drift rate of less than 0.01nm per sec., this amount will lead to

1nm total drift in 100 sec. making drift an important limitation
in analyzing particles less than 10nm in diameter.  The final
difficulty is the uncertainty of specimen thickness measurements
in the application of the absorption and fluorescence correction
factors.

## Specimen Preparation Methods

Specimens for AEM should be on the order of 20-100nm thick and
should accurately represent the features which are to be
analyzed.  In general, these requirements are often difficult to
achieve simultaneously, and various specimen preparation methods
must be used to approach the ideal specimen.  For catalyst
specimens, three main specimen preparation methods can be used
depending on the catalyst material, the form of the catalyst, and
the information desired.  These are grinding and dispersing,
microtomy, and ion-beam thinning.
     Dispersion of catalyst particulates on a carbon support film
is the easiest method.  Most end-use catalyst forms, such as
pellets and extrudates, are far too large to examine directly in
a transmission electron microscope.  These materials must be
ground up with an agate or boron carbide mortar and pestle.  It
is important that elements from the grinding apparatus do not
contaminate the specimen.  The products of grinding have a size
range from less than 0.1μm to several micrometers.  By
ultrasonically suspending the powder in a solvent, such as
distilled water or alcohol, the finest suspended particles may be
selected.  A drop of the suspension may then be placed on a
carbon-coated microscope grid and allowed to dry.  If the
preparation is successful, many electron transparent particles
less than 0.1μm in size will be separated by several microns in a
typical field of view.  We have found two difficulties with this
method of specimen preparation.  First, some phases in the
specimen may be soluble in water or alcohol.  Dry dispersion is
then recommended.  Secondly, determination of where the analyzed
areas were located in the original catalyst pellet is difficult.
This latter difficulty can be partially overcome by making thin
specimens from the desired location on the pellet.
     Microtoming thin-sections of catalyst materials is a
demanding art requiring great patience.  The desired region for
analysis must be cut, scraped, or ground up from the catalyst
pellet.  The 100μm particles are then mounted in an epoxy resin
which is cured overnight.  The cured blocks are trimmed to the
proper shape for microtomy and then cut with a diamond knife
ultramicrotome.  Various thicknesses can be cut from 50nm to
500nm.  Thinner sections are best for high resolution electron
microscopy; whereas, thicker sections are often used to improve
the x-ray counting statistics by increasing the amount of the
element under the electron beam.  Thicker sections also help to
keep the sections of catalysts from popping out of the epoxy.

Difficulties in microtomy include the presence of Si, Cl, and
sometimes S in the embedding resin which may interfere with the
elements under analysis; failure to retain the particle within
the epoxy; and drift of the section with respect to the support
grid. Even when these problems are minimized, it requires
patience to survey many grids to find an area to analyze that
relates to the catalyst surface, pore structure, defect
structure, etc.

Ion-beam thinning is usually used for dense bulk specimens
where particular regions must be analyzed. It can be useful in
AEM for thinning the same single crystals used in surface
analysis to make direct comparisons with results from AES, XPS,
etc. Ion-beam thinning can also be useful in analysis of
interfaces and defects within bulk metallic catalysts such as Pt
and Pd and their alloys.

## Results and Discussion

Supported Palladium. Heavy metal particles on a light-element
support material offer the opportunity to image and analyze small
amounts of an element. Supported palladium is a common
hydrogenation catlayst. Figure 3a shows an annular dark-field
STEM image of Pd particles supported on carbon. Figure 3b shows
an x-ray emission spectrum for the 2nm particle marked as area 1
in Fig. 3a. This spectrum clearly demonstrates the capability to
detect as little as $4 \times 10^{-19}$ grams of Pd. Figure 3c shows the
spectrum from area 2 about 10nm away from the Pd particle. This
spectrum contains even smaller peaks for Pd, Cl, and S; however,
at least some of the Pd signal in this spectrum is due to
scattered electrons or bremsstrahlung exciting Pd x-rays from
nearby Pd particles. Electron scattering and bremsstrahlung
generation are also responsible for the Al signal from the
specimen grid. The Si peak is from an impurity in the carbon.
These data indicate the current level of qualitative
microanalysis with a field-emission analytical electron
microscope.

Bismuth Molybdates. Bismuth molybdates are used as selective
oxidation catalysts. Several phases containing Bi and/or Mo may
be mixed together to obtain desired catalytic properties. While
selected area electron diffraction patterns can identify
individual crystalline particles, diffraction techniques usually
require considerable time for developing film and analyzing
patterns. X-ray emission spectroscopy in the AEM can identify
individual phases containing two detectable elements within a few
minutes while the operator is at the microscope.

Three standard phases were analyzed: $Bi_2MoO_6$, $Bi_2Mo_2O_9$, and
$Bi_2Mo_3O_{12}$. Figure 4 shows a typical group of $Bi_2Mo_3O_{12}$ particles

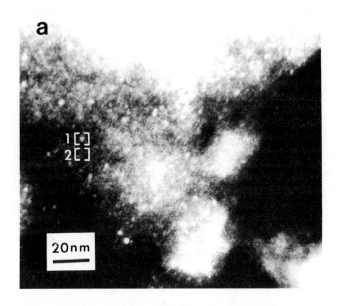

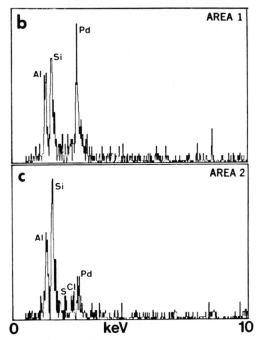

Fig. 3:   High spatial resolution analysis of Pd/carbon
          experimental catalyst: a) annular dark-field
          image showing areas analyzed, b) x-ray spectrum
          of 2nm Pd particle (area 1), c) x-ray spectrum of
          carbon support (area 2). (Counting time = 100s,
          full scale = 115 counts, Al grid).

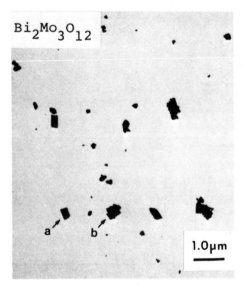

Fig. 4:    A typical distribution of $Bi_2Mo_3O_{12}$ particles on
           a carbon film.  Particles (a) and (b) are $MoO_3$
           and $Bi_2Mo_3O_{12}$, respectively.

dispersed on a carbon film. Particle A had a spectrum containing
no Bi indicating that this particle was $MoO_3$. Particle B had a
spectrum typical of $Bi_2Mo_3O_{12}$. X-ray spectra were collected by
scanning the electron beam in a raster adjusted to completely
cover a single particle. Spectra from about 30 particles of each
phase were collected. Only particles less than $0.8\mu m$ in size
were analyzed. This ensured that the "thin-film criteria",
Equations (2) and (3), would be met for these compounds. The
ratio of the $BiL_\alpha$ peak intensity to the $MoK_\alpha$ peak intensity was
converted to the true Bi/Mo atomic ratio for each particle by
multiplying by the bismuth-molybdenum Cliff-Lorimer k-value of
0.398 measured separately. Figure 5 shows histograms of the
numbers of particles at each Bi/Mo value. From this figure, one
can see that these three bismuth molybdate phases may be
identified on the basis of their measured Bi/Mo atomic ratio
alone. Particles with Bi/Mo ratios off the main distributions
for each phase were impurity phases or combinations of two phases
in a single particle.

$\underline{Fe_2(MoO_4)_3 - Cr_2(MoO_4)_3 \text{ Solid Solutions}}$. An experimental series
of Fe and Cr molybdates was prepared from pure $Fe_2(MoO_4)_3$ to
$Cr_2(MoO_4)_3$ [13]. A sample from this solid solution series was
prepared at each ten mole per cent of $Cr_2(MoO_4)_3$ in $Fe_2(MoO_4)_3$
including the end members. Chemical analysis was performed by
analytical electron microscopy and atomic absorption
spectroscopy. The resulting calibration curves shown in Figure 6
indicate that the relative error for AEM analysis is about the
same or less than that for the AAS. The major systematic error
in the AEM analysis can be explained by considering fluorescence
of the Fe K-series by the $CuK_\alpha$ line, from the support grid, at
low Fe levels. This systematic error could be removed by using
grids of a different material.
   The major disadvantages in application of the AEM method
relative to AAS is the time required for specimen preparation and
analysis. Care must be taken that the particles analyzed are
characteristic of the bulk material, that they are thin enough to
meet the "thin-film criteria" of Equations (2) and (3), and that
enough particles are analyzed to reduce the random error to
acceptable levels. The AEM analyses shown in Figure 6 took
several days to collect and involved considerble operator
attention.

<u>Zeolite ZSM-5</u>. Zeolite catalytic selectivity is usually related
to channel size, but for many reactions catalytic activity and
selectivity also are influenced by the number of acid sites that
contact reactant molecules. Since the acidity of a local region
within a zeolite crystal is directly related to the aluminum
concentration in that region[14], it is important to know the

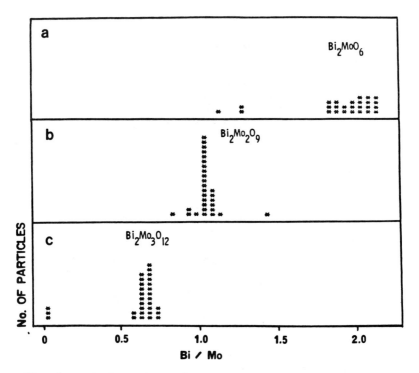

Fig. 5:   Number of particles measured versus Bi/Mo ratio
          for three bismuth molybdates:   a) $Bi_2MoO_6$, b)
          $Bi_2Mo_2O_9$, c) $Bi_2Mo_3O_{12}$.

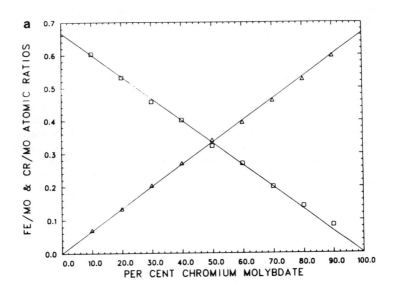

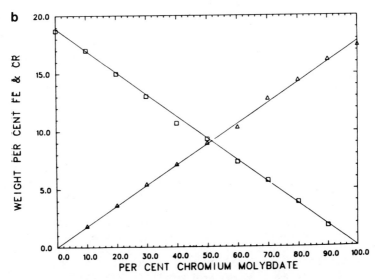

Fig. 6:  Chemical analyses of solid solutions of
$Cr_2(MoO_4)_3$ in $Fe_2(MoO_4)_3$ by a) analytical
electron microscopy and b) atomic absorption
spectroscopy.

actual chemical profile of Al across individual zeolite
particles. Bulk chemical analyses give only an average Si/Al
ratio. However, as shown here ZSM-5 may have different chemical
profiles depending on the synthesis method and the starting
Si/Al. Specimens of ZSM-5 were prepared following the general
procedures of Rollmann and Valyocsik[15]. Compositional
variations within Na-ZSM-5 zeolites are strongly dependent on the
synthesis conditions in a particular run. A series of synthesis
runs were carried out in a 1 gallon stirred titanium autoclave
for 24 hours at 160°C and autogeneous pressure. Bulk chemical
analyses of the products were determined by atomic absorption
spectroscopy (Al) and gravimetric analysis (Si). X-ray
diffraction of all the members of this series showed that they
possess the peaks and d-values characteristic of published data
on ZSM-5[16]. Figures 7a and 7b show surface SEM micrographs for
preparations with bulk Si/Al=9.3 and Si/Al=39.5 designated
ZSM-5(9) and ZSM-5(40), respectively. The experimental x-ray
intensity ratios were corrected to atomic ratios using an
aluminum-silicon Cliff-Lorimer k-factor of k=1.31 measured on the
mineral albite.

Compositional profiles across thin whole particles of
ZSM-5(9) showed two types of particles. Particles which appear
to be single crystals have more aluminum in the center of the
crystal than near the edges as shown in Figure 8a. Other
particles in the same preparation, shown in Figure 8b, exhibited
no compositional variations at a spatial resolution of analysis
of approximately 30nm. This second type of particle is composed
of many individual crystals about 30nm in size (Figure 9). These
small crystals stack together with nearly the same orientation
into a polycrystalline particle which has no apparent chemical
gradient. Each small crystal may actually have a compositional
gradient itself but this was difficult to measure since the
crystals were about the same size as the spatial resolution of
analysis.

Particles of ZSM-5(40) have a much different morphology as
shown in Figure 7b. A thin section of this type of particle can
be assumed to be taken from near the particle center if the
outline of a pillbox-like twin (Figure 7b) can be seen in the
thin section. Figure 10 shows the chemical profile of aluminum
across such a particle. In this case, more aluminum is present
near the surface than in the interior of the particle. At each
analysis point, the composition was sampled through the 80nm
section thickness so that an average through-section composition
was measured.

The significance of this work is that while bulk analysis
techniques would imply that ZSM-5 zeolites are of uniform
composition, they actually may exhibit gradients in Al
concentration across individual particles. This leads to the

Fig. 7:    Secondary electron SEM images of zeolites
           analyzed by AEM:    a) Na-ZSM-5 (bulk Si/Al=9.3),
           b) Na-ZSM-5 (bulk Si/Al=39.5).

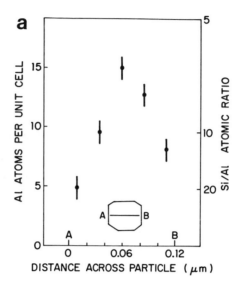

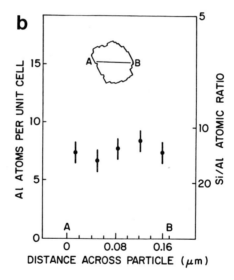

Fig. 8:    Compositional profiles across thin particles of
           ZSM-5(9).  Two types of particles were found:   a)
           single crystals and b) polycrystalline
           aggregates.  AB are loci of analysis points
           across the particles.  Reproduced with permission
           from Ref. 23, Copyright 1983, Elsevier Sequoia.

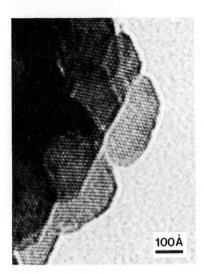

Fig. 9:    Polycrystalline particle of Na-ZSM-5(9) showing
           individual crystals.  Reproduced with permission
           from Ref. 24, Copyright 1983, American Chemical
           Society.

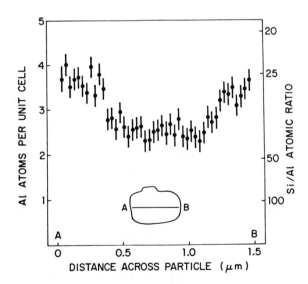

Fig. 10:   Compositional profile across a thin section of
           Na-ZSM-5(40).  46 separate analyses were taken
           along the line AB.  Reproduced with permission
           from Ref. 23, Copyright 1983, Elsevier Sequoia.

conclusion that the distribution of catalytically active sites is
not necessarily uniform throughout a ZSM-5 crystal. The activity
and selectivity for a reaction occurring in ZSM-5 has been shown
to be most favorable for a particular average separation of Al
sites[17]. There have been several somewhat conflicting reports
of distributions of Al in ZSM-5 crystals[18-22]. The present
work generally supports the conclusions of two of these
reports[19,20]. The other reports can be understood by assuming
that slight variations or variations in synthesis conditions can
change both particle morphology and Al distribution. The present
work represents the first time that chemical profiles in zeolites
have been measured directly in small particles in the size range
actually used in catalysis.

The average x-ray signal of a single unthinned particle can
be obtained by collecting x-rays from an area scan covering an
entire particle. This technique was used to analyze a series of
five Na-ZSM-5 samples with varying Si/Al. Figure 11 shows
histograms of the AEM analyses for five preparations of ZSM-5 in
which the bulk Si/Al was varied from 9.3 to 49.5. Note that for
the three preparations with the lowest Si/Al values (shown in
Figures 11a-c), the bulk and AEM analyses follow one another
closely. For the two specimens with higher Si/Al, Figures 11d
and 11e, the bulk Si/Al is higher than that measured by AEM.
Application of the absorption and fluorescence correction factors
of Equations (4) and (5) shift the average Si/Al for a 2μm
particle toward slightly lower Si/Al and further from the bulk
chemical analysis. However, since Figure 10 shows that for
typical 1-2μm particles of these preparations there is more Al
near the surface of the particle than the center, the average
x-ray intensity ratios for these inhomogeneous particles are
likely to indicate an annomalously low Si/Al.

From images of these particles, it was found that the
particles in Figures 11a, 11b, 11c and peak A of 11d were of
nearly the same morphology as those analyzed in Figures 8 and 9.
Peak B of Figure 11d was comprised entirely of the well-formed
ZSM-5 crystals shown in Figures 7b and 10. The differences in
particle morphology and Al distribution for particles in peak A
and peak B are likely to have considerable catalytic consequence.
Diffusion paths are shorter for the particles of peak A while the
Al distribution is quite different between the two types of
particles.

Conclusion

Analytical electron microscopy has been shown to be an effective
technique for the chemical analysis of catalyst particles. In
some cases AEM may be the only technique to provide chemical
profiles across small particles. Analysis of thin sections of

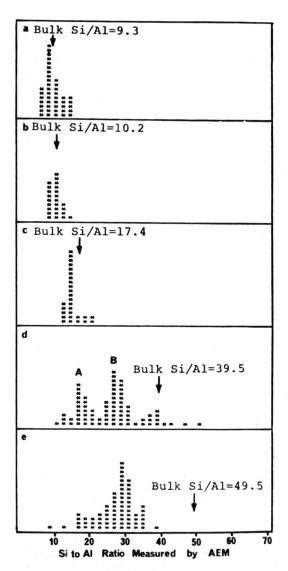

Fig. 11: Histograms showing the number of particles of Na-ZSM-5 versus Si/Al as measured by AEM for various bulk Si/Al ratios: a) bulk Si/Al=9.3, b) bulk Si/Al=10.2, c) bulk Si/Al = 17.4, d) bulk Si/Al=39.5, and e) bulk Si/Al=49.5.

whole particles is useful in understanding chemical inhomogeneities within single catalyst particles.

## Acknowledgments

The author thanks J. B. Michel, U. Chowdhry, W. T. A. Harrison,
and E. F. Moran for providing samples; J. B. Michel, P. W.
Betteridge, and W. T. A. Harrison for assistance in the electron
microscopy; B. F. Burgess and W. F. Farguhar for chemical
analysis; and R. D. Farlee for NMR spectroscopy.

## References

1.  Duncumb, P. J. Micros. (Paris) 1968, 7, 581.
2.  Delannay, F. Catal. Rev. Sci. Eng. 1980, 22, 141.
3.  Lyman, C. E. Microbeam Analysis 1981, Geiss, R. H. (ed.),
    San Francisco Press, 261.
4.  Introduction to Analytical Electron Microscopy, Hren, J. J.;
    Joy, D. C.; Goldstein, J. I. (eds.), Plenum Press, New York
    1979.
5.  Crewe, A. V.; Wall, J.; Langmore, J. Science 1970, 168,
    1338.
6.  Langmore, J. P.; Wall, J.; Isaacson, M. Optik 1973, 38, 335.
7.  Fiori, C. E.;  Newbury, D. E. Scanning Electron Microscopy/
    1978, Vol. 1, SEM Inc., AMF O'Hare, IL., 401.
8.  Hall, T. A. in Physical Techniques in Biological Research,
    Oster, G. (ed.) 2nd edition, Vol. IA, Academic Press, New
    York, 1971, 157.
9.  Cliff, G.; Lorimer, G. W. J. Micros. (Oxford) 1975, 103,
    203.
10. Goldstein, J. I.; Costley, J. L.; Lorimer, G. W.; Reed,
    S.J.B. Scanning Electron Microscopy/1977, Vol. I, ITT Res.
    Inst., Chicago, 315.
11. Philibert, J.; Tixier, R. Physical Aspects of Electron
    Microscopy and Microbeam Analysis, Siegel, B. M.; Beaman, D.
    R. (ed.), Wiley, New York 1975, 333.
12. Nockolds, C.; Nasir, M. J.; Cliff, G.; Lorimer, G. W.
    Electron Microscopy and Analysis, Inst. Phys. Conf. Ser. No.
    52, Inst. Physics, London and Bristol, 1979, 417.
13. Harrison, W. T. A.; Lyman, C. E.; Chowdhry, U. to be
    published.
14. Olson, D. H.; Haag, W. O.; Lago, R. M. J. Catal. 1981, 61,
    390.
15. Rollmann, L. D.;  Valyocsik, E. W. to be published in
    Inorganic Syntheses, 22, Holt, S. L. (ed.), Wiley, New York.
16. Argauer, R. J.; Landolt, G. R. U.S. Patent No. 3 702 886,
    1972 to Mobil Oil Corp.
17. Nayak, U. S.;  Chowdhary, V. R. Appl. Catal. 1982, 4, 333.

18. Suib, S. L.; Stucky, G. D.; Blattner, R. J. J. Catal. 1980,
    65, 174.
19. von Ballmoos, R.; Meier, W. M. Nature 1981, 389, 782.
20. Derouane, E. G.; Gilson, J. P.; Gabelica, Z.;
    Mousty-Desbugnoit, C.; Verbist, J. J. Catal. 1981, 71, 447.
21. Dwyer, J.; Fitch, F. R.; Qin, G.; Vickerman, J. C. J. Phys.
    Chem. 1982, 86, 4574.
22. Huges, A. E.; Wilshier, K. G.; Sexton, B. A.; Smart, P.
    J. Catal. 1983, 80, 221.
23. Lyman, C. E., J. Molecular Catal. to be published.
24. Lyman, C. E.; Betteridge, P. W.; Moran, E. F. ACS Symposium
    Series No. 218, Intrazeolite Chemistry, Stucky, G. D.;
    Dwyer, F. G. eds., American Chemical Society 1983, 199.

RECEIVED January 23, 1984

# Single-Particle Diffraction, Weak-Beam Dark-Field, and Topographic Images of Small Metallic Particles in Supported Catalysts

M. J. YACAMAN

Instituto de Fisica, Universidad Nacional Autonoma de Mexico, Apartado Postal 20-364, Deleg. Alvaro Obregon, 01000 Mexico, D.F.

In the present paper non-conventional TEM methods to characterize small metallic particles are presented. The topographic information on the particles shape can be combined with micro-diffraction (using STEM) data to obtain a full characterization of the particle. The case of gold particles evaporated on a NaCl substrate is used as example. The particle shapes observed are discussed. It is shown that many particles have a crystal structure which is different from the bulk (Fcc). The results are exteded for the case of the Pt/Graphite supported catalysts. Finally it is shown that by using refraction topographic images that rough supports can also be studied. This last method is used to study particles with size down to ∿ 5 Å in diameter which are not visible by other techniques.

The study of shape and crystal structure of small metallic particles is of prime importance in modern catalysis science. The relation between reactivity and structure is still not well known. The main problem in studying small metallic particles is that conventional techniques fail in the manometer diameter range. However it is possible to overcome these difficulties by the application of non-conventional methods. It is the purpose of this paper to review some of these methods and to present some results on the characterization of gold and platinum particles.

TEM Techniques
Topographic Images

This method is based on the use of the refracted portion of the electron beam (1-3). Refraction is very
sensitive to topography changes (Snell law) and can be
used to produce high resolution images of rough objects. Fig. 1 shows objective aperture arrangement in
forming such images. This method can be applied to the
study of particles without the need of removing the
substrate. A typical topographic image of a Pt/Graph-
ite catalyst is shown in Fig. 2. The sample in this
case was heated at $\sim$ 800° C in $H_2$. As it is well known
methane (4-5) is produced under these conditions. The
source of carbon is the substrate itself, when atoms
are removed by "channeling" of the Pt particles. By
combining these pictures with standard selected area
diffraction pattern it is possible to determine the
crystallographic direction of the channels. The channels were always found along {10.0} and {11.0} directions in the basal plane of the graphite. A further
important application of this technique is the observation of very small particles in rough supports such
as $\gamma$-$A\ell_2O_3$. This is of considerable importance in the
study of particle size effects. In normal observation
conditions particles are not easely distinguished from
the support. This is particularly conspicuous in the
small particle size range. ($\sim$ 10 Å).
        Fig. 3 shows a topographic image of a Pt/$\gamma$-$A\ell_2O_3$
catalyst. Contrast from particles is clearly separated
from the substrate topography. On the other hand pores
on the substrate are well defined. If the aperture includes some portion of the dark field spot then the
resolution for small particles is improved. Fig. 4
shows an image of a 100% dispersed catalyst (as measur
ed by chemisorption methods) in which particles of
about 5 Å can be seen.
        It should be mentioned that observation of very
small particles requires extremely clean operation con
ditions on the microscope. Images such as the one in
Fig. 4 will be obtained only if the microscope contamination is substantially reduced.

Weak Beam Dark Field

Dark field images can be used to determine the shape
of a small particle as shown by Yacaman et al. (6-7).
When an electron beam enters on a small particle the
presence of wedges will produce a splitting on the
diffracted spots as shown by Gomez et al (8). A

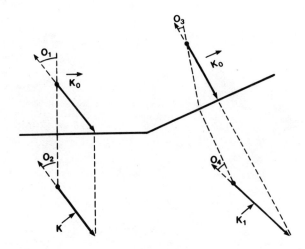

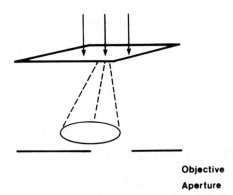

Figure 1. Diagram showing the effect of refraction on the electrons traveling a small particle.

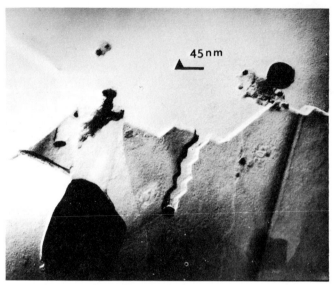

Figure 2. Example of a topographic image of
Pt/Graphite catalyst.

Figure 3. Topographic image of $\gamma$-$Al_2O_3$ catalyst.
Particles are distinguished from the substrate.

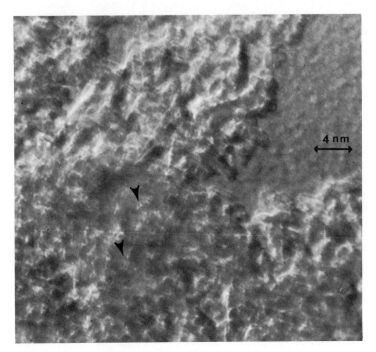

Figure 4. Topographic image showing ∿5Å particles in a 100% dispersed γ-Al₂O₃ catalyst.

wedge of slope m will alter the reciprocal lattice
vectors by:

$$\Delta g_i \ = \ g_i \pm \ \gamma^{(i)} m \hspace{3cm} (1)$$

where $\gamma^{(i)}$ are the eigenvalues of the scattering matrix (9). Although in general will be N values of in
most cases only two or three are significant. Therefore the split will only show few components. In Fig.
5 the splitting effect is shown for a square particle.
If an image is formed placing the aperture in a split
spot fringes will be observed that follow the shape
of the wedges. Such fringes are known as equal thickness fringes. The fringes corresponding to the square
particle in Fig. 6 indicate that this has shape of a
pyramid with four faces. Calculations of the contrast
using dynamical theory have shown that the faces are
{111} planes. A spliting of $\Delta g$ will produce fringes
spaced $1/\Delta g$ on the image. The larger the spliting the
finer the fringes and the more accurate the shape determination. This can be controled experimentally by
tilting away from the Bragg angle for the spot used.
This can be measured through the excitation error Sg.
In Kinematical approximation

$$\Delta g \ \underset{\sim}{} \ S_g m \hspace{3cm} (2)$$

In practice with this method fringes of $\sim 7$ Å spacing
can be obtained.

## Microdiffraction of Individual Particles

In a TEM with STEM attachment it is possible to obtain diffraction patterns from areas from 50 $\sim$ 200 Å.
That allows in most cases to obtain patterns from individual particles. In order to study the crystal
structure of the particle is more convinient to use a
non-convergent beam ($\sim 10^{-3}$ rad). This produces sharp
spots and avoids interference effects such as the ones
described by Roy et al. (9) that makes the interpretation of the data more complicated. Again in this case
the operation conditions must be as clean as possible.

## Characterization of Gold Evaporated Particles

In the present section we will present the shape characterization of different types of gold particles
which are present on an evaporated film grown on a
NaCl substrate. This is an interest model not only for
epitaxy studies but also because the same shapes are
found in real catalyst.

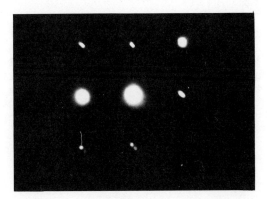

Figure 5. Micro-diffraction pattern of square gold particle showing splitting of the spots.

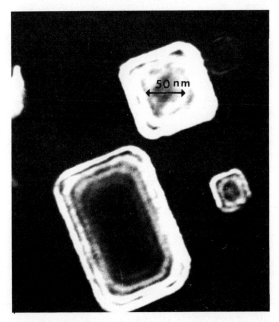

Figure 6. Weak beam image of square particle showing thickness fringes.

## Fcc Particles
## The Square Pyramid

This particle is a regular Fcc structure already described in Fig. 6. The base is a {100} plane and the
sides are {111} planes. This structure is in most
cases truncated on the top generating an extra {100}
plane. The truncation might be defined by a percentage R with respect to the non truncated figure (R=O).
In this structure the truncation might be up to
R = 60%.

## Octahedron and Cubo-Octahedron

These are several Fcc structures which are based on
the octahedron and are illustrated in Fig. 7. In fact
the square pyramid is a half-octahedron. The cubo-
octahedron structures are obtained from the octahedron
by truncations in {100} planes. It is interesting to
note that the polyhedra might have different faces in
contact with the substrate.

## Platelet Structures

In many cases the octahedron based structures appear
with a truncation R = 75%. In that case they can be
described as platelets. On the other hand the fact
that the growth rate of the various crystal faces
might be different generates irregular shaped platelets. A particulary common shape are the triangular
plates (shown in Fig. 8). These are the result of
truncating a single tetrahedron with {111} faces.

## Icosahedral and Decahedral Particles

A special group of particles that are often produced
are the icosahedral ($I_5$) and decahedral ($D_5$) structures shown in Fig. 9. These particles have a five-
fold symetry axis which is forbidden for infinite
crystals. Yang (10) has described these particles
using a non-Fcc model. The particles are composed by
five ($D_5$) and twenty ($I_5$) tetrahedral units in twin
relationship. However the units have a non-Fcc structure. The decahedral is composed by body-centered
orthorhombic units and the icosahedral by rhombohedral

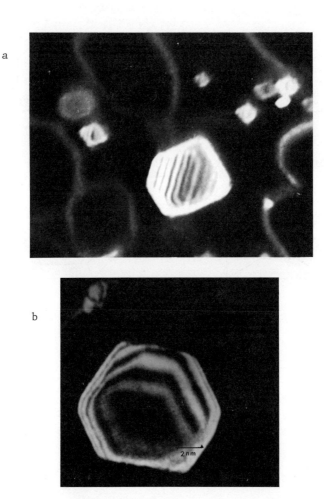

Figure 7. Weak beam images of a)octahedral and b)cubo-octahedral particles.

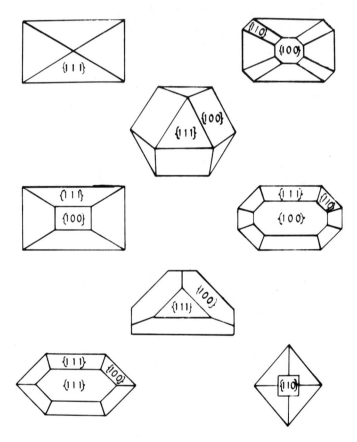

Figure 8. Platelet structures observed by weak beam dark field.

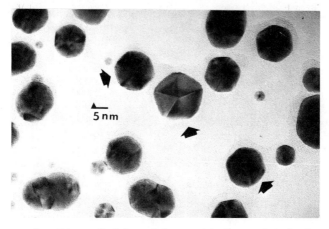

Figure 9. Five-fold gold particles which do not
have structure of the bulk.

units. This model has early confirmation in dark
field experiments (11) and has now full confirmation
from microdiffraction experiments (12).

   In Fig. 10 the micro-diffraction patterns of I5
and D5 particles are shown. These patterns are identical to the ones calculated by Yang et al. (11) based on the non-Fcc model. This type of particles is a
very important example of departures from the Fcc
bulk symmetry. In addition is also an example of poli
hedral particles formed by several units.

## Additional Non-Fcc Particles

Another kind of anomalous particles are observed in
evaporated gold. The diffraction pattern (Fig. 11)
shows six spots in an hexagonal array corresponding
to an interplanar distance $\sim$ 2.46 Å. This spots have
been observed before in larger particles or in continuous films (13-14). A number of explanations have
been offered in the literature to explain those anomalous spots. The present experiments, since the electron spot used is very small, rule out all the explanation based on double diffraction on twin boundaries.
Twin related explanations will require a bend foil
which will have regions in diffraction condition for
the twins. This condition will not follow in the present case. In another type of explanation Cherns (14)
has suggested that in the case of continuos foils,
the extra spots are due to monoatomic steps on the
surface (which break the ABC stacking sequence). This
explanation however will not explain the strong inten
sity of the extra spots in the case of a small particle. Another explanation is based on the surface reconstruction which has been observed in LEED studies
of noble metals (15).

   The extra spots can be explained in terms of the
hexagonal lattice contained in the Fcc structure.
However, intensity calculations suggest that the
breaking of the Fcc symmetry occurs in the whole particle and not just on the surface. This is probably
due to small displacements of the rows of atoms from
the normal position. This might be the result of
strain in the particle. This point is important for
catalytic activity since the surface array of atoms
will be different from the normal Fcc arrays.

## Characterization of Pt/Graphite Catalysis

An interesting catalyst is the platinum supported on
graphite. Fig. 12 shows a bright field image. The

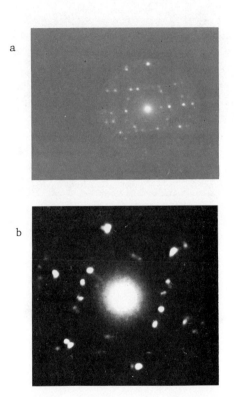

Figure 10. Micro-diffraction patterns of icosahedral particles in two orientations a) Five-fold
axis parallel to the electron beam;b) <112> direc
tion parallel to the electron beam.

Figure 11. Anomalous diffraction pattern of a
small metallic particle, showing 2.46 Å spots.

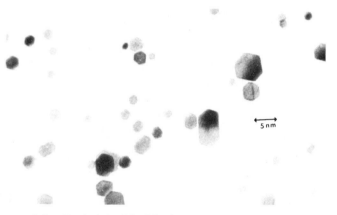

Figure 12. Bright field image of Pt particles
on a graphite support.

most dominant profile is the hexagonal. The dark field
profiles (7) indicate that the shapes correspond to
the octahedron and its truncations such as the cubo-
octahedron. In addition there are compound particles
and platelets. In this catalyst the microdiffraction
patterns have the advantage that spots from the graphite are also present. Therefore epitaxial information
can also be extracted. On the other hand graphite
spots can be used for internal calibration of the camera length and then for more accurate measurements of
lattice parameters of small particles.

   Particle shape and structure can change after a
chemical reaction. For instance the particles in the
Pt/Graphite were heated at ∿ 850°C during several periods. The particle in Fig. 2 corresponds to this case
It was found that the shapes of the particles changed.
A tendency to flat structures was apparent. Fig. 13
shows the typical shapes developed after heating. At
higher temperatures of about 950°C the particles became rounded with a very irregular shape as shown in
Fig. 14. In general the same trend observed in Au evap
orated particles were found in catalyst. In the Pt/C
system particles with non-Fcc diffraction patterns are
also observed. Icosahedral and decahedral particles
have been observed in $Rh/\gamma-Al_2O_3$ and $Rh/SiO_2$ catalysts
(16).

## Atomic Resolution in Small Particles

The modern methods of high resolution can be applied
to the study of small metallic particles. The most
usefull technique is perhaps the projected potential
images (17). Fig. 15 shows an image of a gold particle with icosahedral shape. Atomic resolution along
the {111} planes is observed. The continuity of the
planes along the boundary is clearly seen.

   Atom resolution images can be obtained in particles with diameter down to 100 Å. This technique can
be extremely important in particle characterization.
However in order to obtain usefull information from
atomic images, computer calculations are required for
proper image interpretation.

## Conclusion

The new TEM techniques can provide a full characterization of small particles. The combination of weak
beam images and microdiffraction information can ren-
der a very complete picture of the particle structure. In addition, refracted electron images can be

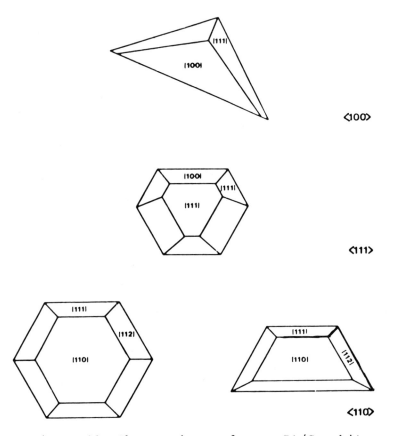

Figure 13. Shapes observed on a Pt/Graphite
catalyst after a methanation reaction at 850°C.

Figure 14. Pt particle
showing a rough shape after
a methanation reaction at
about 950 °C.

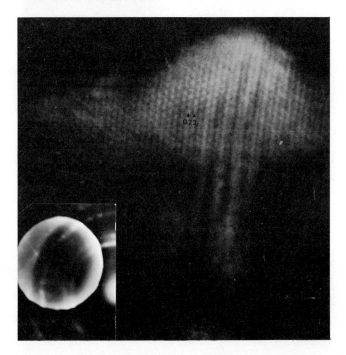

Figure 15. Dark field atom resolution on a gold
particle. {111} planes are observed.

used to characterize the substrate and to obtain reliable particle size distributions. At the present
time, these techniques are in the begining. It is
expected that in the future correlations between
structure and reactivity can be stablished.

## Literature Cited

1. Collis, A.; Maher, D. Ultramicroscopy 1975, 1, 97.
2. Gomez, A.; Hernandez, P.; Yacaman, M.J. Surf. and
   Interf. Anal. 1982, 4, 129
3. Saxton, W.O. in "Advances in Electron Physics",
   Suppl. 10 1978, Academic Press, London.
4. Baker, R.T.K.; Sherwood, R.D. J. of Catal. 1981,
   70, 1a 8.
5. Baker, R.T.K.; Sherwood, R.D.; Dewovane, E.O.
   J. of Catal. 1982, 75.
6. Yacaman, M.J.; Ocaña, T. Phys. Stat. Sol. 1977,
   A42, 571.
7. Yacaman, M.J.; Dominguez, J.M. J. Catal. 1980,
   64, 213.
8. Gomez, A.; Schabes, P.; Yacaman, M.J.; Ocaña, T.
   Phil. Mag 1983,
9. Roy, R.A.; Messier, R.; Cowley, J.M. Thin Solid
   Films 1981, 79, 207.
10. Yank, C.Y.

11. Yacaman, M.J.; Yang, C.Y.; Heinemann, K.; Poppa,
    H. J. of Cryst. Growth 1979, 47, 177.
12. Gomez, A.; Schabes, P.; Yacaman, M.J. Thin Sol.
    Films.
13. Krakow, W.; Ast, D.G. Surf. Sci. 1970, 58, 485.
14. Cherns, D. Phil. Mag. 1974, 30, 549.
15. Fedak, D.G.; Gjostein, N.A. Surf. Sci. 1967,
    8, 77.
16. Yacaman, M.J.; Romeu, D.; Fuentes, S.; Dominguez,
    J.M. J. Chim. Phys. 1981, 78, 861.
17. Cowley, J.M. Diffraction Physics, American
    Elsevier, 1981.

RECEIVED November 7, 1983

# The Use of Scanning Transmission Electron Microscopes to Study Surfaces and Small Particles

J. M. COWLEY

Department of Physics, Arizona State University, Tempe, AZ 85287

The scanning transmission electron microscope
provides unique capabilities for the study of
the surfaces of small particles. Special imaging
modes are possible to allow particular features
of the specimen to be emphasized e.g. the selective
imaging of high atomic number elements. Electron
beams of diameter less than 1 nm may be used to
obtain microdiffraction and microanalysis from very
small regions. By running such beams parallel to
the surface of small crystals it is possible to
probe the surface energies very effectively. These
methods have been applied to the study of small
gold and platinum particles, both unsupported and
supported on amorphous alumina or silica substrates,
or on single crystals.

For straightforward high resolution imaging of small particles or
of structural details on the surfaces of thin crystals, the
conventional transmission electron microscopy (TEM) instrument is
unsurpassed. The scanning transmission electron microscopy (STEM)
instrument can in principle provide comparable resolution and
information when used in equivalent modes but in practice the
picture quality and useful picture area often suffer from limitations inherent in the use of the scanning system. However, the
fact that in STEM the image is formed by scanning an electron beam
of small diameter over the specimen introduces many possibilities
for the use of special imaging modes and for the derivation of
other types of information concerning the specimen, not directly
accessible with TEM. These possibilities are particularly
valuable for many of the types of sample which are common in
catalysis research.

For small particles or surface features, the principle
difficulties arising in TEM relate to the necessary presence of
supporting material. In some cases of special single crystal
supporting films, the contribution of the support to the image can

0097-6156/84/0248-0353$06.00/0
© 1984 American Chemical Society

be minimized so that small particles (1,2) or surface structure
(3,4) can be observed with almost the same clarity as if they were
suspended in space without any support. However, most cases of
practical significance involve a supporting film which is
amorphous, microcrystalline or else crystalline but with internal
faults and surface features which obscure the required information.
The image contrast in bright field TEM of thin films depends on
the spatial variation of the phase change suffered by the transmitted electron wave, caused by the variations of the projection
of the electrostatic potential distribution in the sample. The
fluctuations of projected potential due to structural inhomogeneities of the support can equal or exceed those of a small
particle or surface feature, obscuring the desired information.

One important advantage of STEM is that it is relatively easy
to apply special detector configurations which allow specific
types of information to be selectively displayed in the image. As
suggested in figure 1, the electrons which have passed through the
specimen spread to form a diffraction pattern of the irradiated
region of the specimen on the detector plane. Any portion of this
diffraction pattern may be collected to form an image or any
portion of it may be passed through an aperture to the energy
analyser so that images or diffraction patterns may be formed with
electrons having lost any particular amount of energy in inelastic
scattering processes. The Z-contrast method of Crewe and
coworkers (5), employing the ratio of dark-field and inelastic
small angle scattering signals to emphasize the scattering from
high atomic number elements, has been applied effectively to the
inhancement of the contrast of small metal particles on various
supports (6).

It has been shown that heavy atom particles in a light atom
matrix may also be emphasized by using an annular detector of
large inner radius to collect electrons scattered to very high
angles (7). Variants of this method, designed to avoid confusion
with strong scattering from  microcrystalline regions of the
substrate include the use of multiple high angle detectors.

In this report we concentrate on the special applications of
the STEM instrument involving the use of a stationary incident
beam or a beam moved slowly in a controlled manner. If the
scanning of the incident beam is stopped, the beam will illuminate
a region of the specimen of diameter approximately equal to the
resolution limit of the STEM image, currently as small as 0.3 to
0.5 nm. The electron diffraction pattern from this region will be
formed on the detector plane and may be observed and recorded if a
suitable two-dimensional detector system is provided. In principle
microanalysis of the selected region may be performed either by
detection and analysis of the characteristic X-ray emitted from
the atoms present or by using the electron energy loss spectrometer
to observe the characteristic absorption edges associated with the
excitation of the inner shell electrons of the constituent atoms.

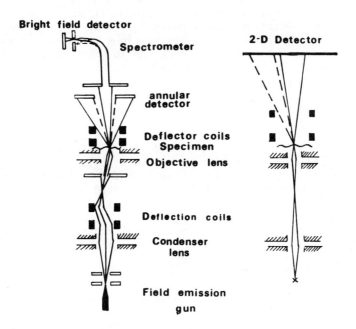

Figure 1.   Diagram of the essential components and electron paths of a STEM instrument. (Reproduced with permission from Ref. 25.)

Electron diffraction patterns have been observed and recorded at TV rates from regions 0.5 nm or less in diameter (8). Characteristic electron energy loss peaks have been observed from clusters of only a few atoms (9). However for quantitative micro-analysis, using either energy-dispersive X-ray spectrometry (EDS) or electron energy loss spectrometry (EELS), the requirements for adequate signal strength and, hence, high incident beam intensity, currently preclude the use of electron beams less than a few nm in diameter. The usual requirement that individual spots in the diffraction pattern should be clearly resolved implies that a small objective aperture should be employed (figure 1) with the result that microdiffraction patterns have more commonly been obtained with beam diameters at the specimen level of 1-2 nm. To make a distinction between this mode and the more conventional microdiffraction from regions 10-1000 nm in diameter we coin the term "nanodiffraction".

## STEM Instruments

To a limited extent, the type of results reported here could be obtained with a TEM instrument fitted with a STEM adapter, especially if a field-emission gun is used (10). We will refer here, however, only to the use of dedicated STEM instruments such as the HB5 or HB501 made by VG Microscopes Ltd. having a cold field emission gun. Most of the instruments of this type are specialized for microanalysis using EDS or ELS and some have been applied very effectively for compositional analysis and associated studies on catalyst particles, as reported elsewhere in this volume.

For the observation of nanodiffraction patterns it is inconvenient and ineffecient to use the Grigson technique of scanning the diffraction pattern over a small aperture. Instead we have employed a two-dimensional detector system of high efficiency. The diffraction pattern is formed on a fluorescent screen which is followed by an image intensifier and an optical analysis system. In the latter system, small mirrors are used to select portions of the diffraction pattern, reflecting the light to photomultipliers to give signals which may be used to produce any desired combination of simultaneous bright field and dark field images (11). The diffraction patterns are viewed with a low light level TV camera and may be recorded on video tape.

Many investigations of small particles or of other materials may involve the collection and analysis of diffraction patterns from very large numbers of individual specimen regions. For small metal particles, for example, it may not be sufficient to obtain diffraction patterns from just a few particles unless there is reason to believe that all particles are of the same composition, structure, orientation and size or unless these parameters are not of interest. More commonly, it is of interest to obtain statistics on the variability of these parameters. The collection of such

statistics may be assisted by the possibility of recording the
many patterns on videotape for subsequent frame-by-frame analysis
but, inevitably, the handling of sufficient data to provide good
statistics will be a lengthy and tedious task.  As a means for
alleviating the tedium of such investigations and also for
introducing further possibilities for data analysis, a pattern
recognition system has been devised and tested (12).

The incident beam in the STEM instrument is scanned over a
specimen area using a digital scan generator to give a regularly
spaced net of beam positions.  For each beam position the nanodiffraction pattern is produced and is compared with a calculated
pattern chosen to represent a configuration of atoms which is of
particular interest.  When there is a sufficient degree of
correlation between observed and calculated patterns, the observed
pattern is recorded in digital form in a computer memory by use of
a TV-rate digital image storage system.  The recorded patterns may
be subsequently compared, correlated or otherwise analysed to
provide the required information on questions such as:

1.  Where are regions giving a particular diffraction pattern
located?

2.  With what relative frequencies do particular crystal orientations occur?

3.  Are crystalline regions in one orientation commonly associated
with regions of some other orientation (e.g. because of the
occurrence of twinning or because of a crystal-substrate relationship)?

4.  Do crystallites of a particular structure often show ordered
superlattices and, if so, of what type? (12).

Nanodiffraction patterns from small metal particles

It is relatively straightforward to obtain diffraction patterns
from particles 2 nm or more in diameter.  If an incident beam of
1-2 nm diameter is used, bright field or dark field STEM images
can be obtained showing the particles in good contrast provided
that the substrate is not too thick or too highly structured.  For
example, Au and Pt particles of 2 nm diameter may be seen clearly
on amorphous silica, alumina or carbon films 5-20 nm thick or on
single crystal MgO or $Al_2O_3$ support of thickness up to 100 nm.  On
the STEM image display screen it is possible to place a cursor to
indicate where the beam will stop when the scan is switched off.
If this cursor is used to select a particle, the diffraction
pattern from that particle then appears and is recorded with the
beam stationary.

Particles of face-centered cubic metals of diameter 5 nm of
more have been studied extensively by high resolution electron
microscopy, diffraction and other methods.  It has been shown that
such particles are usually multiply twinned, often conforming
approximately the idealized models of decahedral and icosahedral
particles consisting of clusters of five or twenty tetrahedrally

shaped perfect crystal regions related by twinning on (111) planes (13,14). For such particles, it is possible to move a 1-2 nm diameter beam from one region to another within the particle, giving diffraction patterns which reveal the relative orientations.

For particles of diameter 2 to 5 nm, the high resolution TEM imaging techniques become more difficult to apply. Since for multiply twinned particles of this size, the individual single crystal regions are then 1-3 nm in diameter, nanodiffraction can show single crystal spot patterns or patterns indicating the existence of one, two or more twin planes within the illuminated area of the specimen. The pattern of figure 2(a) shows the presence of two (111) twin planes within the beam diameter. Such patterns were found to occur frequently for small gold particles, incorporated into a polyester film by co-sputtering (15). In order to establish the complete configuration of a particle in this size range it is possible to move the beam over the particle and examine the correlation of the nanodiffraction patterns from the different regions but this is a complicated procedure.

For particles in the 1-2 nm size range this situation is in some respects simpler but in other respects more difficult. Nanodiffraction patterns can be obtained from regions encompassing the whole particle so that evidence should be provided on the orientations and interrelations of all single crystal regions that are present.

However it is not so easy to locate particles of this size. If the same objective aperture size is used for the imaging as for the nanodiffraction the image resolution will be 1-2 nm so that the particles show very poor contrast. This difficulty has been overcome by using a large objective aperture size to obtain images of better than 1 nm resolution in order to locate the small particles relative to large ones. The larger particles are then used as reference points for locating the smaller particles in the images obtained with the smaller objective aperture so that a clear one-to-one correspondence of particle images and nanodiffraction patterns can be achieved.

A further point is that for a multiply-twinned particle of diameter 1 nm, for example, the constituent single crystal regions are half of this size or less and so contain only two or three planes of atoms. One can not expect, under these circumstances, that the diffraction pattern will be made up merely by addition of the intensities of the single crystal regions. Coherence interference effects from atoms in adjacent regions will become important. It is then necessary to compare the experimental patterns with patterns calculated for various model structures.

For particles of heavy atoms such as Au or Pt it is not sufficient to assume that the calculations of diffraction patterns can be made by use of the simple, single-scattering, kinematical approximation. This leads to results which are wrong to a qualitatively obvious extent (16). The calculations must be made using the full dynamical diffraction theory with the periodic continuation technique to allow the treatment of non-periodic objects and localized incident beams (17).

Figures 2 (b) and (c) show a diffraction pattern obtained from a particle of diameter 1.5 nm and a diffraction pattern calculated for a multiply twinned, decahedral particle. The conclusion drawn from the study of many such observed and calculated patterns obtained from gold particles in the size range of 1.5 to 2 nm contained in a plastic film is that very few particles are multiply twinned, many have one or two twin planes but more than half are untwinned (16). This suggests that, at least for this type of specimen, there is no confirmation of the theoretical prediction that the multiply twinned form is the equilibrium state for very small particles.

Particle-substrate orientational correlations

For small particles supported on thin films of amorphous or micro-crystalline materials it is not easy to determine whether there is any consistent correlation between the particle orientation and the orientation of the adjacent locally ordered region of the substrate. For some samples of Pt and Pd on gamma-alumina, for example, nanodiffraction shows that the support films have regions of local ordering of extent 2 to 5 nm. Patterns from the metal particles often contain spots from the alumina which appear to be consistently related to the metal diffraction spots.

In order to establish such a correlation, however, a statistical analysis of a very large number of patterns would be necessary. This is one possible area for application for the pattern recognition techniques mentioned above. For thin single crystal substrates, any epitaxial relationship of the metal particles to the support is clearly evidenced because the patterns are superimposed in nanodiffraction. A comparison can be made of the patterns obtained with the beam on and just off the particle.

In this way it has been shown that for Pd particles on single crystal alpha-alumina films and for Au particles on magnesium oxide smoke crystals there may be wide variations of behavior depending on factors which are not immediately apparent. In some cases, the particles are single crystals and show a strong correlation between their orientations and the orientation of the substrate. In other cases (sometimes for other areas of the same sample preparation) the particles are heavily twinned and show little correlation with the substrate orientation. It remains to be seen whether these differences may be associated with particular conditions, structures or chemical composition of the substrate surface.

For single crystal substrates which are not in the form of thin films, the techniques of transmission microscopy and nano-diffraction can not be used. For such cases, the techniques of reflection electron microscopy (REM) or its scanning variant (SREM) and reflection high energy electron diffraction (RHEED), in the selected area or convergent beam modes, may be applied (18).

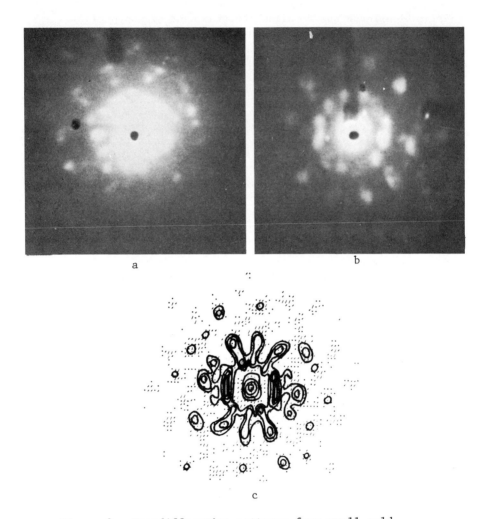

a                                                          b

c

Figure 2. Nanodiffraction patterns from small gold
particles for an incident beam diameter of 1-2 nm (a)
Observed for a particle of 2-3 nm diameter showing twinning
on two planes (b)  Observed for a multiply twinned particle
of 1.5 nm diameter. (c) Calculated for a model multiply
twinned particle.  The black spots in (a) and (b) are the
small mirrors in the optical analyser system used as
detectors for imaging.

For example, figure 3 (a) is an image of small gold crystals on the face of a large cubic crystal of MgO smoke, obtained in transmission through one 90° edge of the crystal. The region over which the gold particles are visible is severely limited because of the rapidly increasing crystal thickness. Figure 3 (b) is the image of the same specimen obtained by tilting the MgO crystal so that the electron beam is incident at a glancing angle on the crystal face, giving the reflection diffraction pattern of Figure 3 (c). Here the stronger spots are due to MgO, showing that the crystal is oriented to give the 008 reflection from the lattice planes parallel to the surface. The weaker, rectangular array of spots, comes from the small gold crystals almost perfectly aligned with the MgO lattice in [110] orientation. The image of figure 3 (b) is formed by placing the detector (the black spot) to collect the overlapping 008 reflections of MgO and Au and then scanning the incident beam over the surface.

The image of figure 3 (b) is formed mostly by diffraction from the small gold crystals since in this severely forshortened image the gold particles almost overlap in projection (compare figure 3(a)). Strong correlations in gold crystal positions are indicated by their alignment in rows, presumably corresponding to surface steps on the MgO.

The REM and SREM techniques have recently been shown to be very powerful for the study of flat surfaces of large crystals or bulk specimens (19,20). Single-atom surface steps may be seen clearly with a lateral resolution of 1 nm or better and the interactions of surface steps with bulk defects can be investigated. The study of surface reactions with the same resolution will follow the development of instrumentation for the adequate preparation and treatment of specimens under ultra-high vacuum and controlled atmosphere within the electron microscope.

Electron energy losses at surfaces

The use of EELS for microanalysis is becoming well established. This technique may be applied to the study of the composition of very small particles or of small surface features seen in STEM images although the difficulty of obtaining sufficiently high count rates from small specimen volumes, particularly in the presence of supporting film or bulk crystal, is always a limitation. In principle EELS may be combined with the imaging of surfaces of bulk samples in the reflection mode (SREM). The penetration of electrons into the surface of a perfect flat crystal face may be only a few nm so that the possibility should exist for chemical analysis of very thin surface layers. Experiments made to date (21) have not been encouraging. A high background in the EELS spectra may be produced by multiple inelastic scattering within the sample.

More success has been achieved in the observation of electron

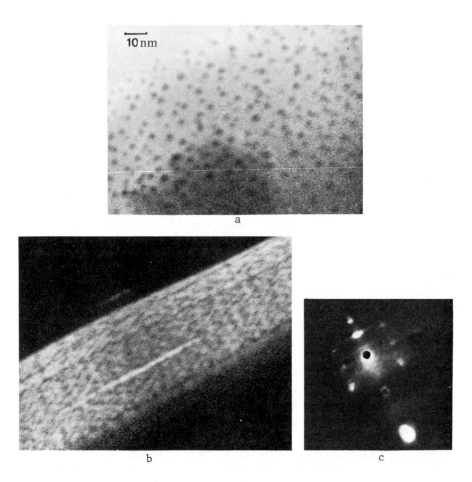

Figure 3. (a) Transmission STEM image of small gold
particles on a MgO smoke crystal. (b) Reflection image
(SREM) of same sample as for (a) with incident beam at
glancing incidence. (c) Diffraction pattern corresponding
to (b). The small black spot is the detector.

energy losses in the range of 0 - 50 eV, corresponding to the
excitation of outer shell electrons or the generation of
collective excitations of electrons.  For this range of energy
losses the inelastic scattering is relatively much stronger than
for the inner shell excitations so that it is possible to obtain
reasonable signal strengths even with very small beam diameters.

In transmission studies of small aluminum spheres, Batson
(22) has shown details of the generation of surface plasmons in
the aluminum and in the thin coating oxide layer.  Marks (23) and
Cowley (24) have examined the surface plasmons and surface state
excitations of small MgO smoke crystals.  It is clearly evident
that these excitations may be produced by electron beams passing
the crystal in the vacuum, 3 nm or more away from the surface.

The potential exists for the development of a powerful
technique for the study of variations of surface excitation
energies with a spatial resolution of 1 nm or better.

## Surface channelling and surface excitations

For crystals which have flat faces which extend for a fraction of
1 μm, a new type of phenomenon may be observed.  Electrons incident
at the edge of the crystal parallel to the surface may be
channelled along the surface.  The potential field of the crystal
extending into the vacuum deflects the electrons so that they tend
to enter the surface but they are scattered out of the crystal by
the surface atoms or by diffraction from the crystal lattice
planes parallel to the surface.  If the scattering angle is less
than the critical angle for total external reflection, the
scattered electrons can not surmount the external potential
barrier and are deflected back into the crystal (figure 4 (a)).

Evidence for this channelling process is provided by the
nanodiffraction pattern for an electron beam directed parallel to
the surface and within a distance of less than 1 nm from the
surface plane.  Strong additional spots from this source have been
observed in patterns from MgO, NiO and Au crystals (24,25).

Electrons which are channelled along a flat surface in this
way are particularly effective in exciting surface states of the
crystal.  Since they spend only a small fraction of their time
"inside" the crystal they are not greatly attenuated by bulk
scattering processes and their EELS spectra do not show bulk
excitation peaks.  They are therefore effective probes for studying
surface excitations.

The periodic channelling motion of the electrons does
influence the energy loss spectra, however.  If the periodicity of
this motion corresponds to the frequency for a strong surface
excitation, there will be a considerable enhancement of the
corresponding energy loss peak.  This is thought to account for
the strong peak at about 17 eV energy loss in figure 4 (b)
obtained with an incident beam paralleled to the flat face of a
MgO smoke crystal.  In this figure the multiple energy loss

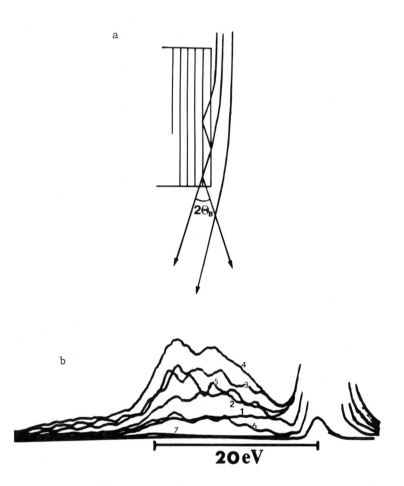

Figure 4.  (a)  Diagram suggesting the channelling of electrons along the surface of a crystal.  (b)  Electron energy loss spectra obtained from electron beams running parallel to the surface of a MgO crystal, as in (a).

spectra were recorded as the incident beam was moved, in steps of
a fraction of 1 nm, gradually closer to the surface.  The spectra
numbered 1-7 were obtained with the incident beam centered at
distances from the surface estimated to be 3, 1.5, 1.0, 0.6, 0.3,
0 and -1.0 nm, respectively.  When the beam is incident within 1
nm of the surface the surface channelling takes place,  the
intensity of the energy loss curve is enhanced and the 17eV peak
becomes prominent.  This energy loss corresponds to the frequency
of oscillation of electrons channelled along the surface.

Conclusion

The use of a STEM instrument allows the controlled movement of a
very fine electron beam in relation to the specimen and the
efficient detection of the scattering and energy losses of the
beam.  We have outlined here a few of the possible applications
arising from this capability.  Other applications, of increasing
sophistication and power will undoubtedly follow in time.   In
particular a range of phenomena, resulting from coherent interference effects in diffraction patterns produced by coherent
convergent beams, have been observed (26) but not yet exploited.
     For the types of specimen which are of particular interest
for catalysis studies, the advantages of being able to obtain
diffraction patterns and energy analysis detail from very small
particles are obvious.  These techniques can give information
concerning individual particles.  The interesting properties of
catalysts depend, of course, on the properties of large assemblies
of small individual particles, but techniques which give data
averaged over large assemblies of particles are often insufficient
because the averaging process involved in the observation often
obscures the information which is essential for consideration of
the catalytic action.
     Of course if the required information is gathered from one
particle at a time, it becomes necessary to make observations on
a very large number of particles in order to properly characterize
the assembly.  It is therefore fortunate that the scanning mode of
operation of the STEM instrument makes it ideally suited to the
automated collection of data.  Also the computer-based digital data
acquisition and analysis systems are now reaching the state of
sophistication which makes it possible to contemplate the
necessary handling of large numbers of one- or two-dimensional
arrays of data.

Acknowledgment

The original work reported in this article was supported by NSF
grant DMR-7926460 and Department of Energy Contract EY-76-S-02-2995
and made use of the resources of the Facility for High Resolution
Electron Microscopy within the Center for Solid State Science, ASU,
supported by NSF grant CHE-7916098.

Literature Cited

1.   Allpress, J.G.; Sanders, J.V. Surface Science 1967, 7, 1-25.
2.   Heinemann, K.; Yacaman, M.J.; Yang, C.Y.; Poppa, H., J.
     Crystal Growth 1979, 47, 177-186.
3.   Moodie, A.F.; Warble, C.E. Phil. Mag. 1967, 16, 891-904.
4.   Iijima, S. Optik 1977, 47, 437-452.
5.   Langmore, J.P.; Wall, J.; Isaacson, M. Optik 1973, 38
     335-350.
6.   Treacy, M.M.J.; Howie, A.; Wilson, C.J. Phil. Mag.1978, A38
     569-585.
7.   Treacy, M.M.J.; Howie, A.; Pennycook, S.J. in Electron
     Microscopy and Analysis 1979; Mulvey, T., Ed., Institute of
     Physics, London, 1980; pp. 261-264.
8.   Cowley, J.M. Ultramicroscopy 1981, 7, 19-26.
9.   Colliex, C., J. Microsc. Spectrosc. Electron. 1982, 7,
     525-542.
10.  Chan, I.Y.T.; Cowley, J.M.; Carpenter, R.W. in Analytical
     Electron Microscopy - 1981; Roy H. Geiss, Ed.; San
     Francisco Press, 1981; pp. 107-116.
11.  Cowley, J.M. in Scanning Electron Microscopy/1980, Vol 1;
     Johari, Om, Ed.; SEM Inc., Chicago, 1980; pp. 61-72.
12.  Monosmith, W.B.; Cowley, J.M. Ultramicroscopy 1983. In press.
13.  Ino, S.; Ogawa, S. J. Phys. Soc. Japan, 1967, 22, 1365-1374.
14.  Avalos-Borja, M.; Yacaman, M. J. Ultramicroscopy 1983, 10
     211-216.
15.  Cowley, J.M.; Roy, R.A. in Scanning Electron Microscopy/1981
     Johari, Om, Ed.; SEM Inc., Chicago, 1982; pp. 143-152.
16.  Monosmith, W.B.; Cowley, J.M. J. Catalysis 1983, In press.
17.  Cowley, J.M. "Diffraction Physics"; North Holland; Amsterdam;
     Second edit., 1981, Chapt. 13.
18.  Cowley, J.M. J. Microscopy, 1983, 129, 253-261.
19.  Hsu, Tung; Ultramicroscopy, 1983. In press.
20.  Cowley, J.M. in Microbeam Analysis - 1980; Wittry, D.B.,
     Ed.; San Francisco Press 1980; pp. 33-35.
21.  Krivanek, O.L.; Tanishiro, Y.; Takanayagi, K; Yagi, K.
     Ultramicroscopy, 1983, In press.
22.  Batson, P.E. Solid State Comm. 1980, 34, 477-480.
23.  Marks, L.D. Solid State Comm.1982, 43, 727-729.
24.  Cowley, J.M. Surface Science 1982, 114, 587-606.
25.  Cowley, J.M. Ultramicroscopy 1982, 9, 231-236.
26.  Cowley, J.M. Ultramicroscopy 1979, 4, 435-450.

RECEIVED December 2, 1983

# Atomic Number Imaging of Supported Catalyst Particles by Scanning Transmission Electron Microscope

M. M. J. TREACY

Exxon Chemical Company, Linden, NJ 07036

In this paper, the capabilities of the conventional and scanning transmission electron microscopes (CTEM and STEM respectively) as tools for studying supported catalysts are briefly compared. The advantages of $Z$ contrast, or the atomic number imaging technique in the STEM over conventional imaging techniques employed in the CTEM are emphasized. It is shown that $Z$ contrast is capable of detecting small clusters of heavy atoms such as Pt down to single atoms in size when supported on low atomic number supports such as charcoal or $\gamma$-$Al_2O_3$. The technique, however, is not reliable as a method for unambiguously identifying the chemical nature of such small clusters.

The transmission electron microscope is now well established as a useful tool for the characterization of supported hetero-geneous catalysts(1). Axial bright-field imaging in the conventional transmission electron microscope (CTEM) is routinely used to provide the catalyst chemist with details concerning particle size distributions(2,3), particle disposition over the support material(2-6) as well as particle morphology(7). Internal crystal structure(8-10), and elemental compositions(11) may be inferred by direct structure imaging.

However, because of the complexities of image formation, analyses based on bright field images are unreliable for particle sizes smaller than about 10Å(12,13). Below such sizes phase contrast effects dominate image contrasts, and images are sensitive to microscope aberrations such as objective lens defocus, astigmatism and spherical aberration. With care, and with favorable samples, the structure of very small clusters containing down to only a few atoms can be determined(6). However, background contrasts from the support material generally confuse particle images in this size range, leaving much ambiguity in image interpretation, and thus such analyses are generally too difficult or tedious to perform on a routine basis.

Dark field methods, where images are formed from scattered or diffracted electrons by placing an aperture over a portion of a diffraction ring, can offer some advantages over bright field imaging. Phase contrast, arising through interference between scattered and unscattered beams, is now eliminated. Of course, to maintain image resolution, tilted illumination must be used together with an axially positioned objective aperture (figure 1b). Tilted dark field imaging has been used with success for revealing the details of the inner structure of crystallites (14), detecting alloyed clusters in multi-metallic catalysts (11,15) and for determining the morphologies of larger clusters ($\geq$ 50 Å)(7).

However, the technique suffers the drawback that in real catalyst systems, particles are randomly distributed over the support and thus will not all be in the correct orientation for diffraction into the angular collection range subtended by the objective aperture, and that the smaller clusters diffract too weakly to be detected against the support. A further difficulty is that random superposition of atoms in amorphous support materials, such as charcoal or silica can give rise to "speckle" which may be easily confused with small catalyst clusters. (16,13).

The detection efficiency in dark field may be considerably improved by using "hollow cone" illumination. Hollow cone illumination is generated either by an annular condenser aperture(17), as in figure 1c or by electronically precessing the tilted illuminating beam (figure 1b) at a fixed angle to the optic axis(18). Use of such illumination increases the chances of finding a randomly oriented particle in a suitable diffraction condition. Furthermore, since hollow cone illumination is essentially incoherent, the confusing "speckle" from amorphous supports is suppressed(16,13). It has been demonstrated that in a catalyst comprising 30Å Pd clusters on $\gamma$-alumina, up to 90% of the Pd particles can be detected by allowing (111) and (220) diffracted beams to contribute to the image(2). However, it was found to be difficult to eliminate confusing diffraction from crystalline supports. Similar results can be obtained by use of an annular objective aperture which blocks the straight-through beam but selects complete diffraction rings(19). This "selected-zone" dark field method, however, is limited principally by the difficulties of constructing the small annular apertures necessary (typically ~ 50μm diameter), and does not eliminate "speckle".

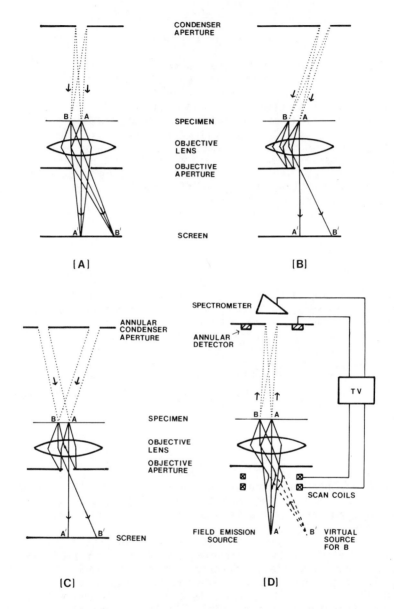

Figure 1: Diagrams showing the essential electron-optical configurations used for various imaging modes in CTEM and STEM as seen by two points A and B on the sample. (a) CTEM axial bright field, (b) CTEM tilted dark field, (c) CTEM hollow cone dark field, and (d) STEM with bright field and annular dark field detectors.

## The Scanning Transmission Electron Microscope (STEM)

Over the past decade increasing use has been made of the scanning transmission electron microscope (STEM) for examining catalysts(20). Owing to the increased efficiency with which scattered electrons and associated signals such as X-rays, secondary and Auger electrons may be collected, the STEM offers greater analytical flexibility compared with the CTEM (21).

In the STEM (see Fig. 1d) the electron beam is focused onto the sample by a condenser-objective lens, and the resulting electron probe can be moved across the sample by means of scan coils. Signals collected by various detectors positioned around the sample can be displayed on synchronously scanned video displays to build up images of the specimen, much in the same way as in the scanning electron microscope.

A remarkable aspect of the STEM is its equivalence to the CTEM because of the reciprocity theorem(22). The underlying principle is perhaps best seen by comparing figures 1a and 1d. Figure 1a shows schematically an electron ray diagram for bright field imaging in CTEM. In figure 1d, ray paths have been sketched for the case of electron detection in a small axial detector in STEM (in our case the detector is an electron energy loss spectrometer). Electron trajectories in the two cases are identical except that in STEM the ray directions are reversed. Provided the electron-optical configurations are identical and that the specimen is thin, the final STEM video image will be identical to the bright field image appearing on the CTEM phosphor screen. Similar arguments hold between CTEM hollow-cone illumination (figure 1c) and the STEM annular detector signal (see figure 1d, ray paths not indicated) again, provided the illumination/collection angles respectively are identical.

It should be noted that the various CTEM imaging modes are acquired simultaneously in the STEM. Furthermore, because samples are examined on a point-by-point basis, microanalytical information is potentially available with higher spatial resolution than is normally permissible in the CTEM. Thus the great attraction of the STEM over the CTEM is that it increases the information available, particularly from heterogeneous specimens.

Although the principle of reciprocity shows that images resulting from elastic scattering are similar in STEM and CTEM, the annular dark field detector can be far more effective in practice than the optically equivalent "hollow cone" technique in CTEM. The reason for this is that in STEM there need be no lenses after the specimen. This allows the use of a very large detector capable of efficient collection of electrons. It has been argued by Crewe et al.(23) that at very large angles electrons suffer mainly elastic scattering whereas small angle deflections result from both elastic and inelastic scattering. The large angle scattering is thus strongly dependent upon the atomic number of the specimen whereas the small angle scattering is not. Thus by electronically forming the ratio of the large angle elastic signal to the simultaneously collected small angle inelastic signal, images with contrast sensitive to atomic number but insensitive to specimen thickness can be formed. Such imaging has been termed Z contrast by Crewe and has been elegantly used to study the migration and interactions of single heavy atoms supported on thin carbon(24).

It is clear that since supported heterogeneous catalysts frequently comprise heavy atom clusters, such as Pt and Pd, distributed over light supports such as charcoal, silica and alumina, Z contrast imaging could be very useful in catalyst studies, particularly for detecting the smaller clusters (< 10Å) which are frequently missed by the conventional imaging methods.

## Atomic Number Imaging of Supported Catalysts

Figure 2 shows STEM images of a 8%wt Pd/2%wt Pt catalyst which is supported on charcoal. The sample was prepared for microscopy by embedding in epoxy resin and sectioning with a diamond knife in an ultramicrotome, and was examined in a Vacuum Generator's Ltd HB5 STEM, with a 5Å probe. The sample thickness is about 500Å.

Figure 2a is the STEM bright field image collected by the axial electron energy loss spectrometer and figure 2b is the annular detector image. Contrasts are reversed between the two images because electrons scattered away from the bright field detector are collected by the annular detector. Figure 2c is the electron energy loss (inelastic) signal collected by the spectrometer and differs from figure 2a, the bright field image in that the spectrometer has now been tuned to reject elastic scattering.

Figure 2d shows the "Z contrast" image formed from the ratio of the annular detector signal to inelastic signal. The ratio image has successfully enhanced the contrast of particles smaller than about 25Å, some of which are poorly visible in the bright field image because of the thickness of the support film. A 25Å Pd or Pt cluster contains approximately 400 atoms, thus it is clear that with the imaging conditions and sample thickness used for figure 2, single atom detection by Z contrast is out of the question. Clearly, sensitivity would increase if the support film were thinner.

It is worthwhile at this point, therefore, to examine in more detail some of the electron scattering mechanisms which govern the various signals involved and to show how detector geometries may be optimized for maximum sensitivity.

Scattering from Thin Amorphous Specimens. The Z contrast technique exploits the strong Z dependence of Rutherford scattering into the annular detector. Rutherford scattering is

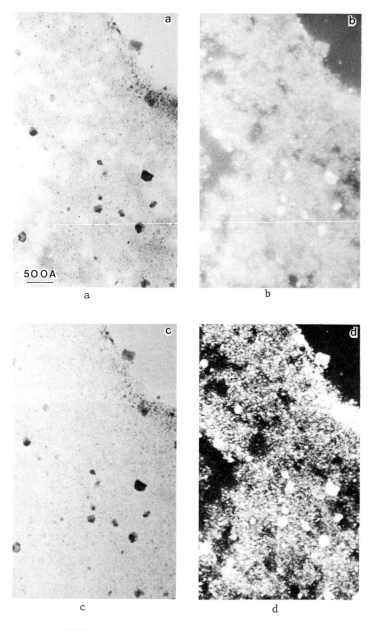

Figure 2: STEM images of a 8%wtPd/2%wtPt catalyst supported on charcoal. (a) Bright field, (b) annular dark field, (c) energy loss and (d) ratio of annular dark field/energy loss.

essentially elastic coulombic scattering from the screened charge of the atomic nucleus, and dominates at high scattering angles where it can be conveniently collected by the STEM annular detector. Some inelastic scattering, arising mainly through interactions with orbital electrons, will also be collected. The annular detector signal from an amorphous cluster of like atoms can be written as

$$I_{AD} = I_0 \rho t [\sigma_{el} + \sigma_{in}] \exp(-\mu_1 t) \qquad (1)$$

where $I_0$ is the incident beam current, $\rho$ the number density of atoms and t the specimen thickness. $\mu_1$ is an absorption factor allowing for multiple scattering of electrons out of the detector. For large solid angles of collection $\mu_1$ is small and may usually be ignored in thin film specimens. $\sigma_{el}$ is the cross section for elastic Rutherford scattering into the annular detector, which subtends an inner collection angle $\Theta_1$ and is assumed to have a large outer angle.

The important features of $\sigma_{el}$ are represented by the Thomas-Fermi model of the atom which assumes that orbital electrons screen exponentially the nuclear charge $Ze$. On this model we have,

$$\sigma_{el} \propto \frac{(Z/E_0)^2}{[(\theta/\lambda)^2 + 0.057 Z^{2/3})]} \qquad (2)$$

$E_0$ is the incident beam energy, and $\lambda$ is the de Broglie wavelength in Å of electrons at this energy. $\sigma_{el}$ is plotted in figure 3 (continuous lines) for 100kV electron as a function of $\Theta_1$ for Pt ($Z = 78$), Cu ($Z = 29$) and C ($Z = 6$). For small $\Theta_1$, $\sigma_{el}$ exhibits a $\sim Z^{4/3}$ dependence, whereas at high $\Theta_1$, $\sigma_{el}$ approaches the full $Z^2$ dependence of unscreened Rutherford scattering. This increase in $Z$ dependence is accompanied by a decrease in $\sigma_{el}$ which falls off rapidly ($\sim 1/\Theta_1^2$) at large angles.

The inelastic scattering cross section $\sigma_{in}$ is more complicated to calculate, and is best represented in terms of the complex dielectric response function $\varepsilon(\Theta, \Delta E)$ for an energy loss $\Delta E$. Inelastic scattering cross sections for Pt, Cu and C are also plotted in figure 3 (broken lines), and are based on the electron-gas statistical model of Ritchie and Howie(25). It should be noted that compared to the elastic component, inelastic scattering is strongly peaked in the forward direction (small $\Theta_1$) and is not so strongly Z-dependent. Furthermore, for low atomic number elements such as carbon, $\sigma_{in}$ can exceed $\sigma_{el}$ at small angles. Since in most Z contrast studies support

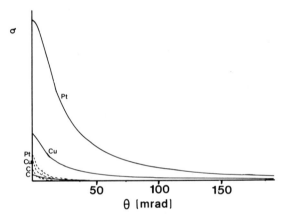

Figure 3: Elastic and inelastic scattering cross sections for scattering into an annular detector of inner collection angle $\Theta_1$, for 100kV electrons.

materials are usually of low Z such as charcoal, alumina or silica, it is desirable to avoid collecting the small angle scattering in the annular detector in order to minimize support signal.

The energy loss signal in the spectrometer from an amorphous material will be described approximately by

$$I_{in} = I_0 \rho t \sigma'_{in} \exp(-\mu_2 t) \tag{3}$$

where $\sigma'_{in}$ is the inelastic cross section for scattering into the spectrometer, between angles 0 to $\Theta_2$. Since the spectrometer collects electrons passing through the hole in the annular detector, then $\Theta_2 \leq \Theta_1$. $\mu_2$ is an absorption factor allowing for multiple scattering away from the spectrometer and into the annular detector. In non-crystalline specimens, it can be seen from equations (1) and (3) that the ratio signal $I_{AD}/I_{in}$ does not cancel the thickness dependence of support contrast as originally suggested by Crewe(23). Unlike $\mu_1$, $\mu_2$ can be large and cannot be ignored even in thin films. Indeed, it is clear from figure 2d that the thickness dependence of support contrast in the ratio image can be just as pronounced as in bright field (figure 2a)(26).

Scattering from Thin Crystals. The simple arguments given above hold only if samples are amorphous. Although some catalyst supports are non-crystalline, such as charcoal and silica, others such as alumina are not. Furthermore, the metal catalyst clusters themselves are generally crystalline and thus the above arguments must be modified to account for Bragg reflections from crystalline areas.

The behavior of diffracted electrons from crystals is best described by simple 2-beam dynamical theory of electron diffraction which gives the diffracted intensity $I_D$ from crystal planes of indices (hkl) in a crystal of thickness $t$ as

$$I_D = \frac{I_0}{\sqrt{(1 + w^2)}} \sin^2 \left[ \frac{\pi t}{\xi_{hkl}} \sqrt{(1 + w^2)} \right] \tag{4}$$

$\xi_{khl}$ is the extinction distance for hkl reflections and equals 206Å for Pd (111) reflections with 100KV electrons. $w = \xi_{hkl} \Delta\Theta/d_{hkl}$, where $d_{hkl}$ is the interplanar spacing and $\Delta\Theta$ the angular deviation from the exact Bragg condition. The diffracted beam is scattered through an angle $\lambda/d_{hkl}$ and may be collected by the annular detector if the inner collection angle $\Theta_1$ is small (if $\Theta_1 < \lambda/d_{hkl}$). The straight-through beam intensity, $1-I_D$ is collected by the spectrometer. In practice, the final detected intensities are averaged over the angles contained in the incident convergent probe.

Contrasts in the annular detector and spectrometer are complementary and sensitive to crystal orientation. Thus crystallites appear dark in both the bright field and inelastic signals (figures 2a and 2c) but particles of similar thickness do not necessarily show the same contrast within a given micrograph because of the orientation dependence. Consequently, the complementary nature of diffraction contrast in spectrometer and annular detector signals boosts contrast in the ratio image (figure 2d) but prohibits interpreting contrast variations between similarly sized particles in terms of atomic number variations.

If we refer back to figure 2, it is clear that although particle contrasts are usefully enhanced in the ratio image, it is impossible to deduce from this image alone whether the sample is a mixture of Pt and Pd particles or comprises alloyed Pt/Pd clusters.

Diffraction from Crystalline Supports. Although diffraction conveniently boosts particle contrast in ratio images, diffraction from crystalline supports such as $\gamma$-alumina, however, can seriously interfere with particle detection.

Figures 4a and 4b show STEM bright field and annular detector images respectively of a 5% wt Pt on $\gamma$-Al$_2$O$_3$ catalyst which was sectioned to ~ 200Å thickness after embedding in epoxy. The $\gamma$-Al$_2$O$_3$ is in the form of prismatic needles about 100Å wide and 500Å long. The mean Pt particle size is 25Å. The annular detector collection angles were 8 to 80 mrad. Some of the randomly oriented $\gamma$-Al$_2$O$_3$ crystals are diffracting into the annular detector and thus exhibit complementary contrasts in figures 4a and 4b. Indeed, some of the support is diffracting sufficiently strongly that some of the smaller Pt crystallites would not be detected if supported on these crystals such as in area A.

For such samples it is generally preferable to increase the inner collection angle $\Theta_1$, of the annular detector so that the strongest low angle diffracted beams are no longer collected (27). This angle is determined by the Debye-Waller attenuation factor for diffraction from the support and is approximately 80 mrad at room temperature for 100KV electrons for most thin crystals. Beyond this angle the scattering from crystals is mainly diffuse scattering which behaves like Rutherford-scattering at these angles, with an associated increase in effective Z dependence.

Figure 4c shows a high angle annular detector image of the same area, where the angular collection range is now 80 to 400 mrad. Support diffraction contrast has been attenuated and Pt particle contrast is significantly enhanced compared with that

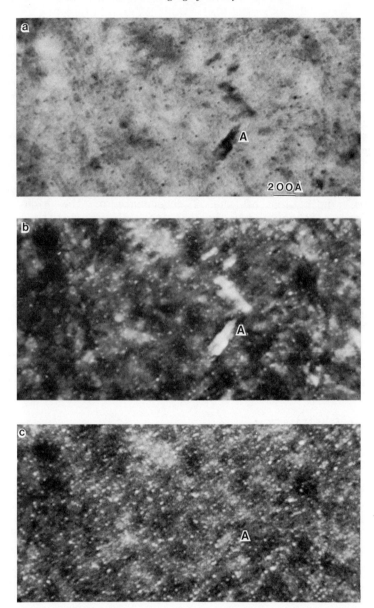

Figure 4: STEM images of a 5% wt Pt on $\gamma$-Al$_2$O$_3$ catalyst (a) Bright field, (b) low angle annular detector, (c) high angle annular detector.

in figure 4b, although contrast variations arising from support mass thickness fluctuations are not suppressed. Pt crystallites in area A are now clearly visible. Dividing by the inelastic signal in this case would reintroduce support diffraction contrast which would only interfere with Pt particle detection.

It follows, therefore, that the use of the unprocessed high angle detector signal can offer significant advantages for atomic number imaging, compared with the ratio image, when catalysts with crystalline supports are to be analyzed. However, the high angle detector suffers from the fact that signals are much weaker. The low angle detector image in figure 4b required a 20 second exposure, whereas the high angle signal, figure 4c, required 100 seconds and is still noticeably noisier.

Some indication of the Z dependence of the high angle Rutherford scattering may be obtained if we rewrite $\sigma_{el}$ of equation (2) in the empirical form $\sigma_{el} = A.Z^n$. A depends only on detector geometry $\theta_1$, whereas n can depend on both $\theta_1$ and Z. Figure 5 shows how n would be expected to vary as a function of $\theta_1$. Diffraction has been neglected. There is an initial rapid increase in n from n = 1.33 to n $\simeq$ 1.8 as $\theta_1$ increases from 0 mrad to 40 mrad. At $\theta_1 \simeq$ 80 mrad, n $\simeq$ 1.9 and beyond this angle, the increase in effective Z-dependence is small. Also, noteworthy is that n is not constant for all Z for a given detector geometry, tending to decrease as Z increases. A more sophisticated scattering model in place of the Thomas-Fermi model would also reveal local variations in n due to subtle changes in orbital configurations across the periodic table which in turn affect nuclear screening.

## Detection Sensitivity

The sensitivity of Z contrast for the detection of small clusters depends not only on the signal Z dependence, but also depends on the microscope resolution, which is governed by the probe size $\delta$. A small cluster containing N atoms of atomic number $Z_1$, supported on a film of effective atomic number $Z_2$ of thickness t, will have a contrast in the annular detector signal given by

$$ I_{Z_1}/I_{Z_2} = (4N/\pi\delta^2\rho t).(Z_1^{n_1}/Z_2^{n_2}) \quad (5) $$

assuming the cluster size is smaller than $\delta$. The factor $\pi\delta^2\rho t/4$ is equal to the number of support atoms contained within the probe. $n_1$ and $n_2$ are the effective Z dependences of the annular detector signal for atoms $Z_1$ and $Z_2$ respectively. It is clear that for maximum sensivity the support must be thin and the probe size $\delta$ must be small. In a STEM with a field emission source the probe size is limited by objective lens spherical aberration $C_s$ and is given by $\delta \simeq 0.43(C_s \lambda^3)^{1/4}$, when using a probe-forming objective aperture of optimum angular radius $(4\lambda/C_s)^{1/4}$.

Figure 6 shows a high magnification STEM annular detector image of Pt on thin $\gamma$-alumina. 20Å to 30Å diameter Pt clusters are visible as well as some bright areas of diffracting $\gamma$-alumina. In this image the annular detector inner collection angle $\Theta_1$ is about 30 mrad. Also visible in the image are small bright dots which measure down to 5Å in diameter, which is equal to the nominal probe size of the instrument. If we assume the mean thickness of the $\gamma$-$Al_2O_3$ in this micrograph to be 50Å, then putting $\delta = 5$Å, $\rho = 11$Å$^3$ and $n_1 = n_2 \simeq 1.8$, we can estimate that a single Pt atom should display an intensity approximately 40% that of the support. Likewise, 2-atom and 3-atom clusters should display about 80% and 120% contrasts respectively, thus providing a means of estimating number of atoms in a cluster.

Figure 7 shows a histogram of intensities for 23 of the $\sim$ 5Å dots, observed in figure 6 and nearby areas. Despite the poor statistics, there is some indication that dot intensities are quantized, peaks being clearest for arbitrary intensity units of 50 and 100, with perhaps a third peak near 170. It is concluded that these dots mostly represent 1-, 2- and 3- atom Pt clusters although substrate noise interferes with concluding with certainty whether or not some of the fainter dots are single Pt atoms. The reason for there being so few 3 atom clusters detected may be due to the fact that some of their image widths may exceed 5Å, depending on the cluster configuration, and so will have been excluded from the analysis.

Isaacson et al.(28) analyzed spot intensities of a specimen of co-evaporated Pt and Pd atoms, supported on thin carbon, using a STEM with probe size $\delta = 2.6$Å. With the smaller probe size, they were able to demonstrate unambiguously that, as well as single atoms being indentifiable, 2-atom clusters comprising Pd-Pd, Pt-Pt and Pd-Pt were also present.

However, it should be remembered that careful studies such as these are carried out on specimens of known composition. To estimate the atomic number of an unknown atom, intensities need to be compared under identical imaging conditions with a specimen of known composition acting as an internal standard. From the data of Isaacson et al. it appears that for elements with $Z \geqslant 46$, and $\delta = 2.6$Å, a certainty in Z of the order of $Z \pm 5$ can be achieved.

## Conclusions

The atomic number imaging technique in the STEM is a valuable method for locating very small imaging high Z catalyst clusters when supported on low Z supports. With electron probes of 5Å or less, single high Z atoms may be detected when supported on thin supports. Confusing support contrasts which plague most imaging

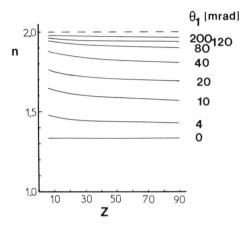

Figure 5:  Diagram showing how the effective Z-dependence
            of the annular detector changes with inner
            collection angle Θ1, as a function of Z.

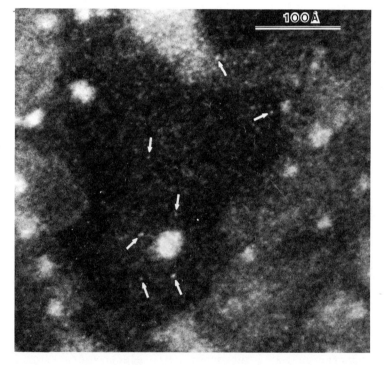

Figure 6:  STEM annular detector image of 5% wt Pt on γ-
            $Al_2O_3$.

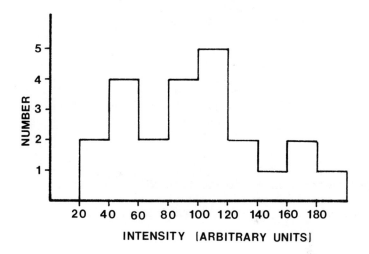

Figure 7: Histogram of intensities of some of the 5Å bright dots detected in figure 6.

studies in the CTEM may be suppressed either by forming the ratio of the annular detector signal to the inelastic signal in the case of an amorphous support, or by using a high angle detector signal alone when examining catalysts on crystalline supports. With care the technique can be used to estimate the number of atoms in small clusters by measuring intensities. However, the technique is not considered reliable as a means of unambiguously determining atomic numbers of small clusters. For this the technique should be used in conjunction with the other microanalytical outputs such as the electron energy loss and X-ray fluorescence spectra.

## Acknowledgments

Most of the results presented here were obtained whilst the author was a graduate student at the Cavendish Laboratory, Cambridge. The author is grateful to Prof. A. B. Pippard for provision of laboratory facilities and Dr. A. Howie for stimulating guidance. Catalyst samples were provided by Johnson Matthey Chemicals Ltd. Royston, England. The author is also grateful to P. Smith for help in the image intensity analysis for figure 7.

## Literature Cited

1.  Howie, A. in "Characterization of Catalysts," Thomas, J. M. and Lambert, R. M., eds., John Wiley and Sons, 1980, Chap. VI.
2.  Freeman, L. A., Howie, A. and Treacy, M. M. J., J. Microsc. 111, 165 (1977).
3.  Harris, P. J. F., Boyes, E. D. and Cairns J. A., J. Catal., 82, 127 (1983).
4.  Dautzenberg, F. M. and Wolters, H. B. M., J. Catal., 51, 26, (1978).
5.  Nakamura, M., Yamada, M., and Amano, A., J. Catal. 39, 125 (1975).
6.  Prestridge, E. B. and Yates, D. J. C., Nature (London) 234, 345, (1971).
7.  Yacaman, M. J. and Ocana Z, T., Phys. Stat. Sol. (a) 42, 571 (1977).
8.  Komoda, T., Japan J. Appl. Phys. 7, 27 (1968).
9.  Marks, L. D. and Howie, A., Nature (London) 282, 196 (1979).
10. Smith, D. J., White, D., Baird, T. and Fryer, J. R., J. Catal. 81, 107, (1983).
11. Yacaman, M. J., Zenith, J., and Contreras, J. L., Appl. Surf. Sci., 6, 1, (1980).
12. Flynn, P. C., Wanke, S. E. and Turner, P. S., J. Catal. 33, 233 (1974).
13. Treacy, M. M. J., and Howie, A., J. Catal. 63, 265 (1980).

14.   Yang, C. Y., Heinemann, K., Yacaman, M. J. and Poppa, H., Thin Solid Films, 58, 163 (1979).
15.   Diaz, G., Garin, F. and Maire, G., J. Catal. 82, 13, (1983).
16.   Gibson, J. M., Howie, A. and Stobbs, W. M., Inst. Phys. Conf. Ser. No. 36, p. 275 (1977).
17.   Heinemann, K. and Poppa, H., Appl. Phys. Lett., 16, 515 (1970).
18.   Krakow, W., and Howland, L. A., Ultramicroscopy, 2, 53 (1976).
19.   Heinemann, K. and Poppa, H., Appl. Phys. Lett, 20, 122 (1972).
20.   See for example the articles in this volume by J. M. Cowley, M. J. Yacaman and C. Lyman.
21.   Brown, L. M., J. Phys. F: Metal Phys., 11, 1 (1981).
22.   Cowley, J. M., Appl. Phys. Lett. 15 58 (1969).
23.   Crewe A. V., Langmore, J. P. and Isaacson, M. S., in "Physical Aspects of Electron Microscopy and Microbeam Analysis," (B. M. Siegel and D. R. Beaman, eds.) p. 47, Wiley, New York (1975).
24.   Isaacson, M. and Utlaut M., Optik, 50 219, (1979).
25.   Ritchie, R. H. and Howie, A., Phil. Mag., 36, 463, (1977).
26.   Treacy, M. M. J., Howie, A. and Wilson, C. J., Phil. Mag. A38, 569, (1978).
27.   Treacy, M. M. J., Howie, A and Pennycook, S. J., Inst. Phys. Conf. Ser. No. 52, 261, (1980).
28.   Isaacson, M. S., Kopf D. Ohtsuki, M. and Utlaut, M., Ultramicroscopy, 4, 101, (1979).

RECEIVED January 25, 1984

# NMR Techniques for Studying Platinum Catalysts

HAROLD T. STOKES

Department of Physics and Astronomy, Brigham Young University, Provo, UT 84602

We describe in some detail the techniques of nuclear magnetic resonance which are used for studying alumina-supported platinum catalysts. In particular, we describe the spin-echo technique from which the Pt lineshape can be obtained. We also discuss spin echo double resonance between surface Pt and chemisorbed molecules and show how the NMR resonance of the surface Pt can be separately studied. We present examples of experimental data and discuss their interpretation.

In recent years, the increased availability of superconducting magnets for nuclear magnetic resonance (NMR) has made possible many new types of studies requiring high sensitivity. Among these is the study of heterogeneous catalysis. An unusual example is the platinum-catalyst studies being carried out by the research group of Professor C. P. Slichter at the University of Illinois (1-6). In these studies, care has been taken to achieve the limits of sensitivity. In this paper, we will examine in some detail the NMR techniques required to carry out such studies. We will also show some examples of experimental data and discuss their interpretation.

## High Sensitivity in Solid-State NMR

During the last decade, several technological advances have enabled the experimentalist to greatly improve the sensitivity of NMR. Foremost among these are superconducting magnets which produce very high magnetic fields. These high fields affect the NMR sensitivity in two ways. First, the NMR signal is proportional to the nuclear magnetization $\vec{M}$, which, following Curie's law, is proportional to the magnetic field $\vec{H}_0$. This means that if we can double the magnetic field, then, from Curie's Law alone, we will double the NMR signal. Since the superconducting

0097–6156/84/0248–0385$06.00/0

magnets commonly available today produce fields 3 or 4 times as large as that of conventional iron-core magnets, we can see that a superconducting magnet would allow us to make substantial gains in signal sensitivity.

But high fields affect signal sensitivity even more substantially in another way. Since the NMR frequency $\omega_0$ is proportional to the field $H_0$, a higher field allows us to operate at a higher frequency. At higher frequencies, the electronic characteristics of a pulsed NMR spectrometer are vastly improved. For one thing, the spectrometer recovers from the high-power rf pulse in a much shorter time. This is an important factor in solid-state studies where the NMR signal often decays rapidly following the pulse.

Also, the problems of "acoustic ringing" of the sample coil virtually disappear at high frequencies. [Acoustic ringing is a poorly understood phenomenon associated with mechanical oscillations of the sample coil following an rf pulse (7).] In our spectrometer, the problem disappeared at frequencies greater than approximately 65 MHz. This was perhaps the greatest single advantage high fields afforded us in our studies of platinum catalysts. Our 85-kG superconducting magnet allowed us to operate at 74 MHz. Most of the data acquired at 74 MHz would be impossible to obtain at lower frequencies [even as high as 55 MHz, for example. We tried it (2) in some of our field-dependence studies!].

Besides the effect of high magnetic fields, the characteristics of the sample itself is an important consideration in improving sensitivity. A large number of nuclei are required to produce an observable NMR signal. Heterogeneous catalysis is a surface phenomenon. Thus, in NMR studies, we need to be able to observe the NMR signal from the nuclei at the surface. In macroscopic single crystals, the number of surface nuclei is very small, and NMR studies on this kind of sample are presently impossible. Thus, we are forced to use small-particle samples which have a very large surface area. Our catalyst samples consist of small platinum particles supported on eta alumina.

The particular NMR properties of Pt caused an additional problem. Due to the presence of surfaces near most of the nuclei, the NMR line is very broad (approximately 4 kG wide). This means that only a small fraction of the nuclear spins can be excited by an rf pulse and thus contribute to any given NMR signal. Given these various constraints, our NMR studies of platinum catalysts required 1-gram samples containing 5-10% Pt by weight.

## NMR Background

In order to better understand the NMR techniques described in this paper, let us first briefly review some fundamental concepts in NMR. (For more details, see Reference 8.) Throughout the discussion, we will use a classical treatment.

A nuclear spin has a magnetic moment $\vec{\mu}$ which, when placed in a magnetic field $\vec{H}$, obeys the equation of motion,

$$\frac{d\vec{\mu}}{dt} = \vec{\mu} \times \gamma \vec{H} \tag{1}$$

where $\gamma$ is the gyromagnetic ratio of that nucleus. If $\vec{H}$ is simply an external time-independent magnetic field $\vec{H}_0$, the solution to Equation 1 is simple. The moment $\vec{\mu}$ precesses about $\vec{H}_0$ with frequency $\omega_0 = \gamma H_0$ (see Figure 1a).

For more complicated cases, a very useful tool for solving Equation 1 is the rotating reference frame (RRF). Consider first of all the simple case already treated above: $\vec{H} = \vec{H}_0$. In a reference frame rotating about $\vec{H}_0$ with frequency $\omega_0$, the moment $\vec{\mu}$ appears to be motionless and behaves as though it were in zero field (see Figure 1b). If, however, we view the situation from a reference frame rotating more slowly at a frequency $\omega = \omega_0 - \Delta\omega$, the moment $\vec{\mu}$ appears to be precessing about $\vec{H}_0$ with a frequency $\Delta\omega$ and thus behaves as though it were in a field $h = \Delta\omega/\gamma$ in the direction of $\vec{H}_0$ (see Figure 1c). Similarly, in a reference frame rotating at a frequency $\omega = \omega_0 + \Delta\omega$, the moment $\vec{\mu}$ appears to be precessing <u>backwards</u> with a frequency $\Delta\omega$ and behaves as though it were in a field $h = \Delta\omega/\gamma$ which now points in the <u>opposite</u> direction of $\vec{H}_0$ (see Figure 1d).

Let us now apply this concept of the RRF to the case where an rf field $\vec{H}_1$ is present. We choose a Cartesian coordinate system with the z axis along the dc field $\vec{H}_0$ and the y axis along the rf field $\vec{H}_1$. The total field is given in the laboratory reference frame by

$$\vec{H} = H_0\hat{z} + 2H_1\hat{y} \cos \omega t \tag{2}$$

In a reference frame rotating about $\hat{z}$ with frequency $\omega$, the rf field $\vec{H}_1$ has two components: (1) a dc component $H_1\hat{y}$ and (2) a component oscillating with frequency $2\omega$. The $2\omega$-component has very little effect on $\vec{\mu}$ and can be discarded. If the rf field is "on resonance" ($\omega = \omega_0$), then, in the RRF, $\vec{H}_0$ disappears completely, and we are left with only the dc component of $\vec{H}_1$,

$$\vec{H} = H_1\hat{y} \tag{3}$$

The moment $\vec{\mu}$ appears to be in a <u>static</u> field $H_1$ which points along the y axis. The solution to Equation 1 in the RRF is simple. The moment $\vec{\mu}$ precesses about $\hat{y}$ with frequency $\gamma H_1$ (see Figure 2a).

## Free Induction Decay

Consider a large number of nuclear spins in some sample under study. Since Equation 1 is linear in $\vec{\mu}$, the nuclear magnetization $\vec{M} = \sum \vec{\mu}$ also obeys this equation of motion. If the sample is placed in a magnetic field $\vec{H}_0$, $\vec{M}$ follows Curie's Law and acquires a non-zero component along the z axis (the direction of $\vec{H}_0$). If we then apply an rf field and view $\vec{M}$ in the RRF, it will appear to precess about $\hat{y}$ with frequency $\gamma H_1$ (see Figure 2a). The angle through which it precesses is given by $\Delta\theta = \gamma H_1 \Delta t$. If we apply the rf only during a time interval $\Delta t = \pi/2\gamma H_1$, the moment $\vec{M}$ will have precessed through an angle $\Delta\theta = \pi/2$ and will be pointing along the x axis (in the RRF). Such a pulse of rf is commonly called a "90° pulse" (see Figure 2b). Similarly, if the length of the pulse is $\Delta t = \pi/\gamma H_1$, we have a "180° pulse," and $\vec{M}$ will be pointing along the negative z axis (see Figure 2c).

Now consider what happens following a 90° pulse. In the RRF, $\vec{M}$ is pointing along the x axis. In the laboratory reference frame, $\vec{M}$ is precessing about $\vec{H}_0$. This induces an rf voltage in a receiver coil around the sample and thus produces an "NMR signal" in the spectrometer. Such a signal is commonly called a "free-induction decay" (FID).

If all nuclear spins in the sample were in exactly the same field, this signal would not decay, as the name implies, but would persist "forever." In any real sample, though, local fields cause different nuclei to be in slightly different total fields, and their moments $\vec{\mu}$ precess at slightly different frequencies. Thus, following a 90° pulse, the moments $\vec{\mu}$ initially precess coherently but eventually get out of phase with each other, causing $\vec{M}$ to "decay" to zero.

In solids, local fields are often relatively large and thus cause very short decays. In order to observe the FID, the spectrometer must recover from the preceding rf pulse before $\vec{M}$ decays to zero. In most cases, this condition can be satisfied, and FID signals are normally observed. However, in the case of our Pt-catalyst samples, the distribution of local fields is very large. $\vec{M}$ decays to zero long before the spectrometer can recover from the rf pulse, and no FID signal can be observed.

## Spin Echo

If the local fields are static and depend only on position in the sample (this kind of situation is called "inhomogeneous broadening"), an observable NMR signal can be produced by a technique called spin echoes. Consider a moment $\vec{\mu}$ which is in a local field which causes it to precess faster than the RRF. In the RRF, the apparent field $\vec{h}$ in this case points along the positive z axis. Figure 3a shows the spin-echo sequence. First, at t = 0, we apply a 90° pulse which causes $\vec{\mu}$ to point along the x axis. Then $\vec{\mu}$ precesses about $\vec{h}$ through an angle $\Delta\theta = \gamma h\tau$ during a time interval $\tau$. At this point (t = $\tau$), we apply a 180° pulse which inverts the x-component of $\vec{\mu}$. The moment $\vec{\mu}$ is now at an angle $\pi - \Delta\theta$ with respect to its initial direction along the x axis. Following the 180° pulse, $\vec{\mu}$ precesses again through an angle $\Delta\theta$ during a second time interval $\tau$ so that at t = 2$\tau$, $\vec{\mu}$ is

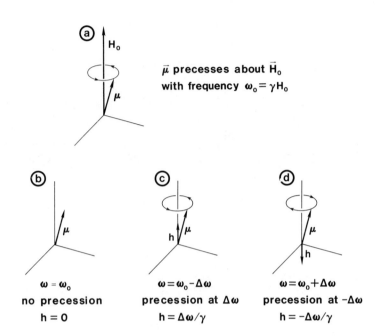

Figure 1.   Precessing magnetic moment in (a) the laboratory reference frame and in (b-d) the rotating reference frame.

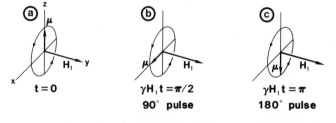

Figure 2.   Magnetic moment precessing about the rf field in the rotating reference frame.

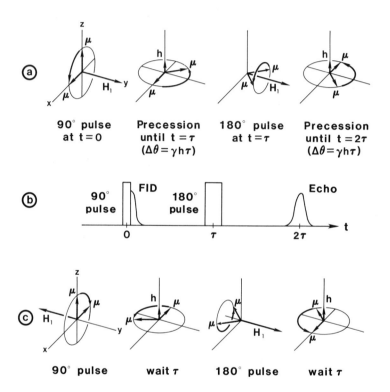

Figure 3.  Spin echo formation with (a-b) non-inverted and (c) inverted 90° pulse.

at an angle $\pi$, i.e., pointed along the negative x axis. Note that this result does not depend on the value of $\Delta\theta$. All moments $\vec{\mu}$, fast and slow, end up pointing along the negative x axis at the same time t = $2\tau$.

Thus the coherence among the moments which was lost in the FID following the initial 90° pulse is regained at t = $2\tau$. It does not last long, however, because the different moments $\vec{\mu}$ precessing at different frequencies soon get out of phase with each other again, and the signal decays to zero, just as in the FID. The signal has the appearance of an "echo" (see Figure 3b).

In Figure 4a, we show the envelope of such a spin echo obtained from one of our samples. The trace shown was obtained by averaging 20,000 signals. (The signals are digitized and then digitally added together.) The 90° and 180° pulses were separated by $\tau$ = 50 $\mu$s. The initial transients in the trace arise from recovery of the spectrometer from the 180° pulse. As can be seen, a major advantage of the spin-echo technique is that the signal can be moved away from those transients so that it is observable. By separating the two pulses by even a greater amount (increase $\tau$), the spin echo can be moved even further away from the 180° pulse. In our case, however, other effects cause the amplitude of the echo to strongly decrease with increasing $\tau$ so that we do not gain any advantage by increasing $\tau$ further than shown. (This effect is caused by the time-dependent part of the local field, i.e., "homogeneous broadening.")

The trace shown in Figure 4a is one of our strongest signals observed. In order to observe much weaker signals, we must average a larger number of signals. However, when we do that, the initial transients also grow larger and mask the smaller echo, no matter how many signals we average. We thus refine the spin-echo technique to remove the transients. Consider what happens if we invert the initial 90° pulse. (We do this by shifting the phase of the rf by 180°. In the RRF, $\vec{H}_1$ is then pointing along the negative y axis.) As can be seen in Figure 3c, the resulting $\vec{\mu}$ at t = $2\tau$ points along the positive x axis. This echo is thus 180° out of phase with the one produced by a non-inverted 90° pulse. With a spectrometer which uses phase detection to produce the envelope of the signal, the spin echo is "inverted," as seen in Figure 4b. The transients, though, are not inverted since they arise from the 180° pulse which is identical for the two cases. Thus, if we subtract the two traces from each other, the spin echoes add and the transients cancel, leaving us with a rather "clean-looking" spin echo shown in Figure 4c.

Up to this point, we have ignored the effect of $\vec{h}$ during the rf pulses. Of course, if h << $H_1$, then the effect is negligible. But in the case of very broad lines where h >> $H_1$ for many of the nuclear moments in the sample, the effect becomes very significant. The analytical solution to Equation 1 under these conditions is very complicated. In effect, the pulse sequence described above still produces a spin echo (as can be seen in

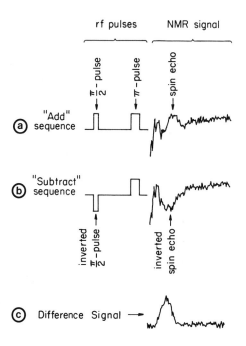

Figure 4. Spin-echo technique for removing initial transients from signal. Reproduced with permission from Ref. 1. Copyright 1982, The American Physical Society.

Figure 4), but only those nuclear moments $\vec{\mu}$ with small values of h $(\lesssim H_1)$ participate in the spin echo. (These are the moments $\vec{\mu}$ whose frequency of precession in the laboratory reference frame is near the rf frequency $\omega$.)

In a very broad line, then, the size of the spin echo is proportional to the number of nuclear moments near $\omega$. If we change $H_0$, we change h by the same amount and thus sample nuclear moments at a different local field. By measuring the size of the spin echo as a function of $H_0$, we obtain the NMR lineshape, i.e., in this case, the distribution of the local field among the nuclear spins in the sample.

## Experimental Data Using Spin Echoes

In Figure 5, we show some lineshapes obtained from some of our samples at 77 K. Each point represents the size of a spin echo obtained at that field. The dispersion of these samples (fraction of Pt atoms which are on the surface of the particles) were 4, 11, 15, 26, 46, and 58% for Figures 5a through 5f, respectively. These samples were untreated, i.e., exposed to air. The width of most of these lines is about 4 kG. Compared to $H_1 \cong 100$ G, these lines are indeed broad!

The source of such large local fields in our samples is the interaction between the Pt nuclei and the polarized conduction-electron spins. This interaction in Pt metal shifts the Pt NMR resonance to higher fields. This shift is called the Knight shift and is rather large compared to the shifts observed in non-metallic diamagnetic Pt compounds. In bulk Pt metal, the resonance is at $H_0/\nu_0 = 1.1380$ kG/MHz ($\nu_0 \equiv \omega_0/2\pi$, and therefore $H_0/\nu_0 = 2\pi/\gamma$, which is independent of field for a given substance), whereas the resonance in non-metallic Pt compounds ranges between approximately 1.085 and 1.100 kG/MHz (9-10), depending on the compound (see Figure 6). The shifts among the non-metallic compounds are called chemical shifts and arise from the interaction with polarized electron orbitals.

In a sample of bulk Pt metal, all of the nuclei have the same interaction with the conduction electrons and thus see the same local field. The resulting NMR line is quite narrow. However, in our samples of small Pt particles, many of the nuclei are near a surface where the state of the conduction electron is disturbed. This tends to reduce the Knight shift for these nuclei. Since the Pt particles in our samples are of many different sizes and shapes, this reduction in the Knight shift is not the same for every nuclear spin near a surface. Thus, we obtain a broad "smear" of Knight shifts resulting in the lineshapes of Figure 5.

These lineshapes exhibit some interesting features. Some of them have a peak at 1.138 kG/MHz, which is the position of the resonance in bulk Pt metal. This peak arises from Pt nuclei which are deep within the interior of the largest Pt particles. Their electronic environment looks very much like that of bulk Pt metal.

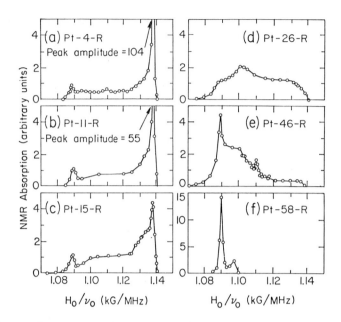

Figure 5.  Pt NMR lineshapes for six untreated samples.
Reproduced with permission from Ref. 1.  Copyright 1982,
The American Physical Society.

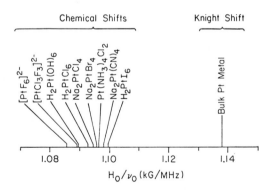

Figure 6.  Position of Pt NMR lines for various compounds.
Reproduced with permission from Ref. 1.  Copyright 1982,
The American Physical Society.

As the average particle size decreases (moving from Figure 5a through 5f), the intensity of this peak decreases since there are fewer large particles.

In Figure 5, we also see a peak at 1.089 kG/MHz which increases with decreasing particle size. From Figure 6, we find that this peak is at the position of the resonance in $H_2Pt(OH)_6$. Chemical shifts are very sensitive to the species of ligand forming the bonds as well as to the coordination of the ligand. This is especially true for the OH ligand which produces one of the largest paramagnetic shifts known for Pt. Thus, we feel it is reasonable to conclude that this peak in our lineshapes arises from surface Pt which are each bonded to six OH groups. Since there are more surface Pt in the samples of smaller particles, the intensity of this peak increases, as observed. The reactions with the atmosphere which cause this result are still under investigation. Note that our conclusion requires that the surface reconstruct and that all Pt-Pt bonds at the surface be broken. This result is consistent with recent EXAFS studies of Sinfelt, Via, and Lytle (11). In very high-dispersion Pt samples which had been exposed to air, they found that there are no Pt-Pt bonds.

## Spin Echo Double Resonance

We show in Figure 7 the Pt lineshapes (the solid line) obtained by Makowka (5-6) for two different samples (26% and 76% dispersion for Figures 7a and 7b, respectively) at 77 K which were cleaned by oxygen and hydrogen cycles at 300°C and then exposed to CO (90% enriched $^{13}C$) at room temperature. Approximately half a monolayer of CO was chemisorbed. As can be seen, this chemical treatment greatly affected the lineshape at the low-field end. (Compare the lineshape of the untreated 26%-dispersion sample in Figure 5d with that of the CO-treated sample in Figure 7a.) The electronic environment of the Pt nuclei near or at the surface of the particles is significantly altered by the presence of the CO molecules.

With C nuclei on the surface of the particles, we can use a technique called spin echo double resonance (SEDOR) to separate the resonance of surface Pt from the rest of the lineshape. The spin-spin interaction between Pt and C nuclear spins produces a local field $\Delta h$ at the Pt spin. During the first $\tau$-interval of the spin-echo pulse sequence (see Figure 3a), this additional field causes the Pt moment $\vec{\mu}$ to precess through an angle $\Delta\theta$ which is larger than before by $\gamma\Delta h\tau$. Now, consider the effect of applying a 180° pulse to the C nuclei. This inverts the C nuclear spins, causing $\Delta h$ to reverse direction, i.e., $\Delta h \to -\Delta h$. If we apply this 180° pulse to the C spins simultaneously with the 180° pulse applied to the Pt, then during the second $\tau$ interval, the Pt moment $\vec{\mu}$ will precess through an angle $\Delta\theta$ which is smaller than before by $\gamma\Delta h\tau$. At $t = 2\tau$, the moment $\vec{\mu}$ will not be at angle $\pi$ as before, but at angle $\pi - 2\gamma\Delta h\tau$. Since the C-Pt coupling randomly

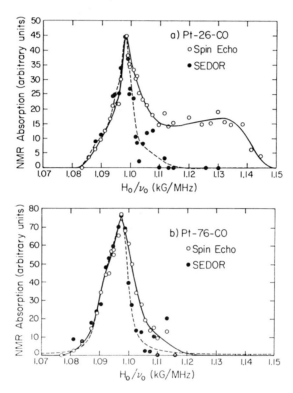

Figure 7.    Pt NMR lineshapes for two samples with
chemisorbed CO.    Reproduced with permission from Ref. 5.
Copyright 1982, The American Physical Society.

varies throughout the sample, so will the value of Δh.  The
direction of the Pt moments thus affected will be randomly
scattered about the angle π and will not refocus perfectly at
t = 2τ.  If Δh is large enough, we will observe a reduction in the
size of the spin echo.  Now, the size of Δh is large enough only
for nearest-neighbor Pt-C pairs.  Thus, any reduction in the spin
echo must come from surface Pt bonded to CO molecules.

The results are shown in Figure 7.  The lineshape indicated
by the dashed line is the difference between spin echoes obtained
with and without the 180° pulse applied to the C spins and
represents the lineshape due to the surface Pt alone.  Since this
difference was only 10-20% of the signal, a large number of
signals (more than a million!) needed to be averaged to obtain the
data shown.  Even so, a large amount of scatter in the data is
evident.  The "SEDOR" data was scaled to match the "spin echo"
data at the vicinity of the peak.

If we compare Figures 7a and 7b, we see that although the
lineshape due to all Pt nuclei in the sample are very different
(the solid lines), the lineshape due to surface Pt alone are
strikingly similar (the dashed lines).  The NMR characteristics of
the surface Pt are largely determined by the nature of the
chemical bonding to the nearby CO molecules and not by the size of
the particle itself.

We compared our results in Pt catalysts with that of
Pt-carbonyl molecules (12), which are often used as models of the
Pt catalytic surface.  The data for $[Pt_{38}(CO)_{44}]^{2-}$ was obtained
from a sample prepared by Dahl and Murphy at the University of
Wisconsin.  The data for the remaining three molecules was
obtained from Brown et al. (12).  As can be seen in Table I, we
found that the position of the surface Pt resonance in our
catalysts is very close to those in various Pt carbonyls.  (The
variations among the positions shown in Table I are small compared
to typical variations among Pt compounds.  See Figure 6.)  Thus,

Table I.  Position of NMR peaks in Pt catalysts
and various Pt-carbonyl molecules

| | $H_0/\nu_0$ (kG/MHz) | Reference |
|---|---|---|
| Pt catalysts | 1.096 | Our work |
| $[Pt_{38}(CO)_{44}]^{2-}$ | 1.096 | Our work (77 K) |
| $[Pt_3(CO)_6]^{2-}$ | 1.0975 | 12 |
| $[Pt_6(CO)_9]^{2-}$ | 1.0975 | 12 |
| $[Pt_9(CO)_{18}]^{2-}$ | 1.0975,1.0980 | 12 |

from the NMR data alone, we conclude that the Pt atoms in the Pt-carbonyl molecules behave very much like those on the surface of Pt-catalyst particles.

## Acknowledgments

I wish to acknowledge my co-workers at the Department of Physics, University of Illinois: C. P. Slichter, H. E. Rhodes, P.-K. Wang, C. D. Makowka, S. L. Rudaz, and J. P. Ansermet. I especially want to thank J. H. Sinfelt of Exxon Research Laboratories for providing the Pt catalyst samples used in these studies and also for the helpful advice he has given us. I also thank L. Dahl and M. Murphy of the Chemistry Department, University of Wisconsin, for providing us with samples of Pt carbonyls. This research was supported by the U.S. Department of Energy under Contract No. DE-AC02-76ER01198.

## Literature Cited

1.  Rhodes, H. E.; Wang, P.-K.; Stokes, H. T.; Slichter, C. P.; Sinfelt, J. H. Phys. Rev. B 1982, 26, 3559.
2.  Rhodes, H. E.; Wang, P.-K.; Makowka, C. D.; Rudaz, S. L.; Stokes, H. T.; Slichter, C. P.; Sinfelt, J. H. Phys. Rev. B 1982, 26, 3569.
3.  Stokes, H. T.; Rhodes, H. E.; Wang, P.-K.; Slichter, C. P.; Sinfelt, J. H. Phys. Rev. B 1982, 26, 3575.
4.  Rhodes, H. E. Ph.D. Thesis, University of Illinois, Urbana, 1981.
5.  Makowka, C. D.; Slichter, C. P.; Sinfelt, J. H. Phys. Rev. Lett. 1982, 49, 379.
6.  Makowka, C. D. Ph.D. Thesis, University of Illinois, Urbana, 1982.
7.  Fukushima, E.; Roeder, S. B. W. J. Mag. Reson. 1979, 33, 199.
8.  Slichter, C. P. "Principles of Magnetic Resonance"; Springer, New York, 1980.
9.  Carter, G. C.; Bennett, L. G.; Kahan, D. J. in "Progress in Material Science"; Chalmers, B.; Christian, J. W.; Massalski, T. B., Eds.; Pargamon, New York, 1977; Vol. 20, Part I, pp. 295-302.
10. Kidd, R. G.; Goodfellow, J. in "NMR and the Periodic Table"; Harris, R. K.; Mann, B. E., Eds.; Academic, New York, 1978; p. 251.
11. Sinfelt, J. H., personal communication.
12. Brown, C.; Heaton, B. T.; Chini, P.; Fumagalli, A.; Longini, G. J. Chem. Soc. Chem. Commun. 1977, 309.

RECEIVED September 26, 1983

# Photoacoustic Spectroscopy of Catalyst Surfaces

E. M. EYRING, S. M. RISEMAN, and F. E. MASSOTH

Departments of Chemistry and Fuels Engineering, University of Utah, Salt Lake City, UT 84112

Microphonic Fourier transform infrared photoacoustic spectroscopy (FT-IR/PAS) has emerged as a useful tool for characterizing fractions of a monolayer of organic species adsorbed on opaque, high surface area samples. Such a study of calcined and sulfided hydrodesulfurization catalysts will be discussed. Specifics such as indications that Bronsted acidity may be associated with polymolybdate structure and the observation of a low frequency feature at 1310 reciprocal centimeters will be described along with generalizations regarding the present limitations of this technique.

Although photoacoustic spectroscopy (PAS) was first conceived by Bell and his contemporaries over one hundred years ago (1-3), the applications of PAS to the study of surfaces have all emerged within the last ten years. The decisive factor in this belated renaissance of interest in PAS was the publication of the one-dimensional Rosencwaig-Gersho (R-G) model (4) of PAS with microphonic detection. In Figure 1 a cylindrical PAS sample cell is depicted by a rectangle one end of which is illuminated by a beam of light chopped at an audio frequency. The light beam traverses a tightly sealed, transparent window, passes through the transparent gas behind the window, and is incident upon a solid sample. Energy absorbed by the sample surface from the incident light beam may be converted by radiationless transitions to a thermal wave that returns (by thermal diffusion) to the sample surface and warms the thin layer of gas in contact with the surface. This layer of periodically heated gas acts as a "thermal piston" on the rest of the gas in the cell and causes a sound wave of the same frequency as that at which the light beam was chopped but of delayed phase. These acoustic waves in the gas impinge on a microphone located at the end of a duct (see Figure 2) that prevents scattered light from striking the microphone and producing spurious signals. Thermal properties of the material used as a backing to the sample can also influence the intensity and phase of the PA signal detected by the microphone.

0097–6156/84/0248–0399$06.00/0

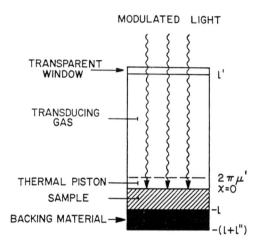

Figure 1. Schematic diagram of a photoacoustic cell used to develop the one-dimensional theory of microphonic PAS by Rosencwaig and Gersho.

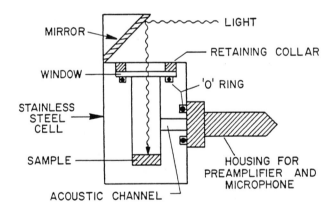

Figure 2. Schematic diagram of a photoacoustic cell for solid samples that depicts the acoustic channel (diameter exaggerated) to the microphone from the gas filled sample chamber.

The solution to this boundary value problem was approximated by Rosencwaig and Gersho for six different cases (4) one of which, a thermally thick but optically thin sample, often applies to layers adsorbed on heterogeneous catalysts. The photoacoustic signal arises from the chemisorbed species and the support. Optical properties of the chemisorbed monolayer are usually paramount, and this layer is much thinner than the substrate and support that experience heating. The photoacoustic signal intensity Q is given by the proportionality

$$Q \propto \beta\mu f^{-3/2} \tag{1}$$

where $\beta \equiv$ optical absorptivity ($cm^{-1}$), $\mu \equiv$ thermal diffusion length (cm), and $f \equiv$ beam chopping frequency ($s^{-1}$).

An important experimental insight follows from equation 1: The PA signal to noise ratio, S/N, rapidly diminishes with increasing chopping frequency. Thus microphonic PA measurements are often made at incident light chopping frequencies lower than 500Hz. High intensity of the incident light beam is also advantageous, and high wattage arc lamps are therefore frequently used for PAS at ultraviolet and visible wavelengths. Some advantage is found in using helium to carry the sound wave from the sample surface to the microphone (because of the high thermal conductivity of He), but a more important consideration in the choice of a gas is its transparency: The PA effect is much larger in gases absorbing electromagnetic radiation (where it is called the optacoustic effect) than in liquids or solids. Thus a trace of $CO_2$ (g), for example, can overwhelm PA signals from a surface in the 2310 to 2380 $cm^{-1}$ region of the infrared spectrum (see Figure 3).

The depth in the sample surface from which the PA signal comes depends on the beam chopping frequency. At low chopping frequencies spectral information comes from greater depths in the sample. In other words, if one speeds up the motor of the device, such as a fan blade, that is chopping the incident light beam, not only will S/N diminish, but the sample will also be probed at a shallower depth below its surface. This ability to yield subsurface spectral and thermal information is a peculiar advantage of PAS over reflectance and transmission spectroscopies that still remains to be widely exploited (5).

In detector noise limited spectroscopies such as PAS it is advantageous to enhance the throughput of energy (Jacquinot's advantage) by utilizing a Michelson interferometer. One then Fourier transforms (FTs) the resulting interferogram to yield a PA spectrum that qualitatively resembles an absorption spectrum. Thus while one never sees commercial FT spectrometers for ultraviolet-visible (UV-VIS) absorption measurements (because photomultiplier tubes are much quieter detectors than are microphones), FT-VIS/PA spectrometers have been built that permit speedier acquisition of high S/N photoacoustic spectra (6-7).

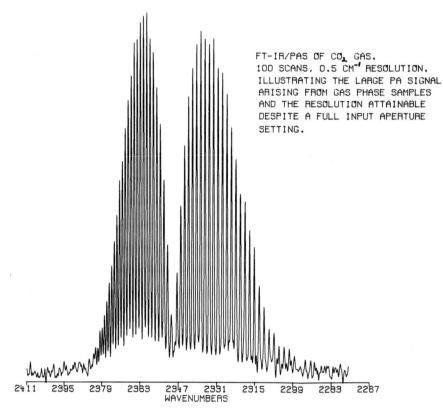

FT-IR/PAS OF CO₂ GAS.
100 SCANS, 0.5 CM⁻¹ RESOLUTION.
ILLUSTRATING THE LARGE PA SIGNAL
ARISING FROM GAS PHASE SAMPLES
AND THE RESOLUTION ATTAINABLE
DESPITE A FULL INPUT APERTURE
SETTING.

Figure 3.  Photoacoustic infrared spectrum of gaseous $CO_2$ obtained in a microphonic PAS cell for solid samples when the operator exhaled once into the cell before closing. 100 scans, 0.5 cm$^{-1}$ resolution.  This illustrates the large photoacoustic signal arising from gas phase samples and the high resolution attainable.

Microphonic detection can be used in such FT/PA experiments in which case the R-G theoretical arguments (4) still apply.

In situations where absorption of the incident radiation by the transducing gas is troublesome a piezoelectric transducer (made from barium titanate, for example) can be attached to the sample (or sample cuvette in the case of liquids) to detect the thermal wave generated in the sample by the modulated light (8,9). The low frequency, critically damped thermal wave bends the sample and transducer thus producing the piezoelectric response. The piezoelectric transducer will also respond to a sound wave in the solid or liquid but only efficiently at a resonant frequency of the transducer typically of the order of 10 to 100 KHz (see Figure 4). Thus neither in the case of microphonic nor piezoelectric detection is the PA effect strictly an acoustic phenomenon but rather a thermal diffusion phenomenon, and the term "photoacoustic" is a now well established misnomer.

The chemist generally finds infrared spectral data to be very much more informative than UV-VIS data for identifying species on surfaces. For this reason the discovery by Rockley (10) and Vidrine (11) that photoacoustic spectral measurements can be performed conveniently on commercial FT-IR spectrometers by substituting a microhponic (or piezolectric) PA detector for the usual deuterated triglycine sulfate (DTGS) infrared detector was of capital importance. A schematic representation of the adaptation at the University of Utah of a Nicolet 7199 FT-IR spectrometer for FT-IR/PAS is shown in Figure 5. Quality of the PA spectral data can be improved by setting the microphonic sample cell on vibration isolation mounts, foam rubber, or other damping materials to intercept otherwise troublesome low frequency vibrations arising from cryostats or other mechanically noisy equipment in the vicinity of the spectrometer.

No beam chopping device is shown in Figure 5. Motion of the moving mirror in the Michelson interferometer is equivalent to beam chopping and the frequency f is given by

$$f = 2 \nu v c^{-1} = 2 v \bar{\nu} \qquad (2)$$

where $v \equiv$ mirror speed (cm s$^{-1}$) and $\bar{\nu} \equiv$ infrared frequency (cm$^{-1}$). If a step-and-integrate mode is selected for the mirror motion, the photoacoustic measurements are all made at a single audio frequency. This has the advantage that the "absorbances" measured at all wavelengths of the IR spectrum are for the same depth below the sample surface. This also facilitates lock-in detection thus improving S/N. Unfortunately, the typical presently available commercial FT-IR spectrometer is "rapid scan" and the mirror sweeps with a continuous motion that produces a higher chopping frequency at shorter wavelengths. Thus, for example, when the interferometer mirror is moving at a speed of 0.112 cm s$^{-1}$ the chopping frequency is only 90 Hz at 400 cm$^{-1}$ but has increased to 900 Hz at 4000 cm$^{-1}$. Thus the photoacoustic signal is coming from distinctly different depths in the sample

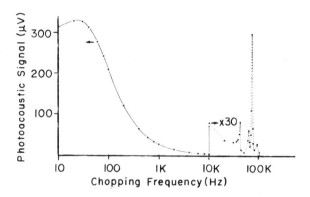

Figure 4. Photoacoustic signal measured in a sample liquid with an attached piezoelectric transducer having a resonant frequency of several tens of thousands of kilohertz. Note the change in scale of the amplitude and thus the much greater sensitivity of the detector at low light chopping frequencies. Argon ion laser light source, 400 mW, $\lambda$ = 488 nm; sample 25 $\mu$g/mL $BaSO_4$ powder suspended in aqueous glycerine. Reproduced with permission from Ref. 21 copyright 1980, American Chemical Society.

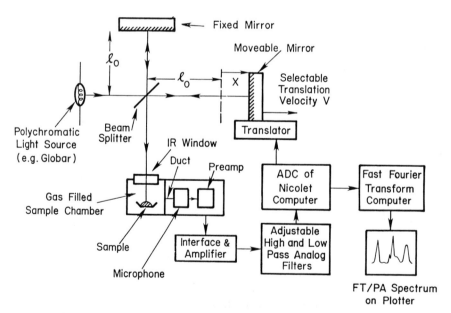

Figure 5. Schematic diagram of the adaptation of a Nicolet 7199 FT-IR spectrometer for photoacoustic measurements on solid samples.

depending upon incident wavelength and a complete mid-IR spectrum for a particular sample surface depth can only be made by changing the mirror motion to many different successive constant speeds and then combining information from different spectra to get a composite spectrum for a single sample depth. A serious limitation of the new, more affordable, lower resolution FT-IR spectrometers is that they often do not offer more than one rapid scan mirror speed. This poses no problem for ordinary FT-IR work but makes these units distinctly less attractive than the top of the line FT-IR spectrometers for PAS work.

One other operational detail merits brief mention before applications to surface spectroscopy are considered. Infrared sources decline markedly in intensity at longer wavelengths and therefore PA spectra must be source intensity normalized before peak heights can be ascribed any quantitative significance. It has sometimes been mistakenly supposed that the PA spectrum of graphite could be used to normalize infrared PA spectra. Depending on the source of the graphite, one obtains distinctly different IR/PA spectra (frequently caused by adsorbed species) and the response of the DTGS detector of an IR spectrometer turns out to be a more accurate measure of variable source intensity (12). A normalization technique (13) requiring measurement of the spectrum at two different mirror velocities and corrected by black body spectra taken at the same two velocities appears to be the best normalization method reported thus far.

Light scattering by the sample can cause correctable (14) errors in photoacoustic spectra, particularly at visible and shorter wavelengths. However, at mid-infrared wavelengths this is no longer an important consideration.

Methods of applying PAS to the study of liquids and highly transparent solids are now well established (9) but are inappropriate to the present discussion.

## APPLICATIONS

In seeking interesting applications of FT-IR/PAS one usually looks for samples of maximum suface area and high opacity. Not surprisingly many heterogenous catalytic systems qualify. In the first stage of such an investigation one prefers to examine a sample system that has been previously characterized successfully by conventional transmission-absorbance type spectral measurements.

Two such well studied systems are pyridine chemisorbed on alumina (15) and pyridine chemisorbed on silica-alumina (16). It had been previously shown that alumina contains only sites which adsorb pyridine in a Lewis acid-base fashion whereas silica-alumina has both Lewis and Bronsted acid sites. These two different kinds of sites are distinguishable by the characteristic vibrational bands of pyridine adducts at these sites (see Table I). Photoacoustic and transmission results are compared in Table II. Note that the PA signal strength depends on factors such as sample particle size and volumes of solid sample and transducing

Table I.  Assignments of Pyridine Chemisorbed on Silica-Alumina
          As Lewis Acid Sites (LPY) and Bronsted Acid Sites (BPY)

| vibrational assignment[a] | LPY,[b] $cm^{-1}$ | BPY,[b] $cm^{-1}$ | LPY,[c] $cm^{-1}$ | BPY,[c] $cm^{-1}$ |
|---|---|---|---|---|
| 8a  $\upsilon_{CC(N)}$ | 1620 | 1638 | 1621 | 1639 |
| 8b  $\upsilon_{CC(N)}$ | 1577 |      | 1578 |      |
| 19a $\upsilon_{CC(N)}$ | 1490 | 1490 | 1493 | 1493 |
| 19b $\upsilon_{CC(N)}$ | 1450 | 1545 | 1454 | 1547 |

[a]Kline, C.H.; Turkevich, J.J. Chem. Phys. 1944, 12, 300.
[b]Basila, M.R.; Kantner, T.R.; Rhee, K.H. J. Phys. Chem. 1964, 68, 3197.
[c]Riseman, S.M.; Massoth, F.E.; Dhar, G.M.; Eyring, E.M. J. Phys. Chem. 1982, 86, 1760.

Table II.  Vibrational Frequencies of Pyridine Chemisorbed on
           $\gamma$-Alumina, $cm^{-1}$

| transmission,[a] | 1453 | 1495 | 1578 | 1614 | 1622 |
|---|---|---|---|---|---|
| photoacoustic,[b] | 1447 | 1493 | 1578 | 1614 | 1621 |

[a]Moné, R. "Preparation of Catalysts", Delmon, B.; Jacobs, P.A.; Poncelet, G.; Eds.; Elsevier: Amsterdam, The Netherlands, 1976; pp. 381.
[b]Riseman, S.M.; Massoth, F.E.; Dhar, G.M.; Eyring, E.M. J. Phys. Chem. 1982, 86, 1760.

gas so that a simple correlation of absorptivity and PA signal magnitude is elusive.  However, relative ratios of absorptivities can be deduced for the PA data when the thermal properties of the samples are maintained invariant (17).

The principal advantage of PA over transmission spectroscopy lies in the determinaton of vibrational species chemisorbed on opaque, light scattering surfaces.  This we have demonstrated by obtaining PA spectra of pyridine chemisorbed on reduced and sulfided $Mo/Al_2O_3$ and $Co-Mo/Al_2O_3$ catalysts (18).  The black sulfided samples are opaque at both visible and infrared wavelengths, but good quality PA spectra of these surfaces are readily obtained.  Only Lewis acid sites are detected on these surfaces (See Figure 6).  In addition, the high surface sensitivity of this technique (a small fraction of a monolayer) permits PA detection of a surface cobalt-aluminate type of domain which is uninfluenced by the presence of molybdenum, is resistant to sulfiding, and is capable of adsorbing pyridine.  This PA band (at 1310 $cm^{-1}$) was not observed in transmission studies because such spectral measurements of attentuation of a beam passing through the sample lack the requisitive surface sensitivity.

There are situations in which the sensitivity to gases of a FT-IR/PAS sample cell intended for solids is advantageous.  By plotting PA intensity (ratioed to a silica PA internal standard in the region 866 to 767 $cm^{-1}$) versus the volume of CO(g) added to a special, microphonic PA cell one can develop a calibration curve.  This curve can then be used to deduce the residual gas phase CO when carbon monoxide is injected into a PA sample cell containing $Ni/SiO_2$ of predetermined surface area that, unlike pure $SiO_2$, tends to adsorb CO.  It was found (19) that 40% of the active sites on the $Ni/SiO_2$ catalyst had absorbed CO molecules (assuming a molecular cross section of 16 $Å^2$/CO molecule and single occupancy of surface sites.)

An inherent disadvantage of microphonic PA cells is their fragility for operation at the high temperatures and pressures typical of commercial catalytic processes.  While Helmholtz resonance sample cell configurations (20) can maintain a microphone at moderate temperatures while the PA sample is at very low or at elevated temperatures, the high gas pressure problem is not resolved in this fashion.  A most promising photothermal technique for infrared spectral measurements on high temperature-high pressure sample surfaces is photothermal deflection spectroscopy (PDS or sometimes also "mirage effect" spectroscopy) (21).  In a PDS experiment (see Figure 7) the illumination of a surface by the focused output from a Michelson interferometer gives rise to thermal gradients that in turn produce a time dependent thermal lens in the medium (gas or liquid) above the surface.  A small-diameter probe laser beam passing through this thermal lens and almost grazing the surface is then deflected through an angle whose magnitude and direction is measured with a position sensing detector.  In the special case of heterogeneous catalysts at high temperatures and pressures a high pressure,

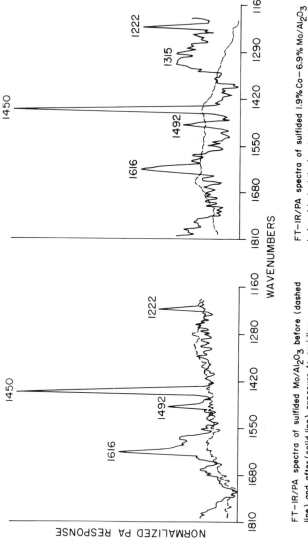

WAVENUMBERS

NORMALIZED PA RESPONSE

FT-IR/PA spectra of sulfided Mo/Al₂O₃ before (dashed line) and after (solid line) exposure to pyridine.

FT-IR/PA spectra of sulfided 1.9%Co–6.9% Mo/Al₂O₃ before (dashed line) and after exposure to pyridine.

Figure 6. Photoacoustic spectra of sulfided HDS catalysts. Frequencies (cm⁻¹) of the most prominent absorbance bands of pyridine on the sulfided Mo/Al₂O₃ and Co–Mo/Al₂O₃ are indicated. Only bands representative of Lewis acid sites are observed.

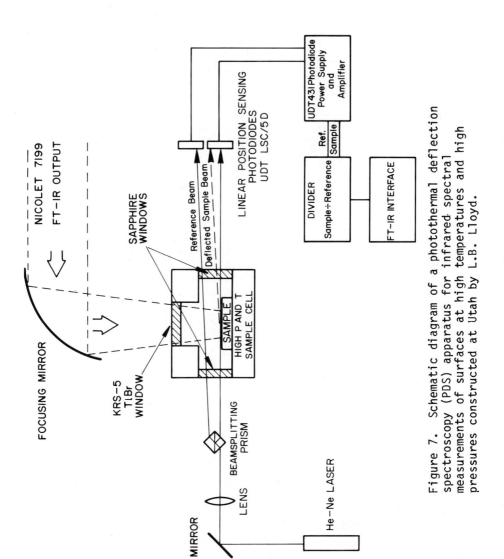

Figure 7. Schematic diagram of a photothermal deflection spectroscopy (PDS) apparatus for infrared spectral measurements of surfaces at high temperatures and high pressures constructed at Utah by L.B. Lloyd.

heated sample cell with three windows (and no microphone) as in Figure 7 permits infrared spectral measurements under conditions closely approximating "the real thing."

## Acknowledgment

Financial support of this work by a contract from the Department of Energy (Office of Basic Energy Sciences) is gratefully acknowledged.

## Literature Cited

1.  Bell, A.G. Am. J. Sci. 1880, 20, 305.
2.  Tyndall, J. Proc. Roy. Soc. London 1881, 31, 307.
3.  Rontgen, W.C. Philos. Mag. 1881, 11, 308.
4.  Rosencwaig, A.; Gersho, A. J. Appl. Phys. 1976, 47, 64.
5.  Helander, P; Lundstrom, I.; McQueen, D.J. Appl. Phys. 1981, 52, 1146.
6.  Farrow, M.M.; Burnham, R.K.; Eyring, E.M. Appl. Phys. Lett. 1978, 33, 735.
7.  Lloyd, L.B.; Burnham, R.K.; Chandler, W.L.; Eyring, E.M.; Farrow, M.M. Anal. Chem. 1980, 52, 1595.
8.  Farrow, M.M.; Burnham, R.K. Auzanneau, M.; Olsen, S.L.; Purdie, N.; Eyring, E.M. Appl. Optics 1978, 17, 1093.
9.  Patel, C.K.N.; Tam, A.C. Rev. Mod. Phys. 1981, 53, 517.
10. Rockley, M.G. Chem. Phys. Lett. 1979, 68, 455.
11. Vidrine, D.W. Appl. Spectrosc. 1980, 34, 314.
12. Riseman, S.M.; Eyring, E.M. Spect. Lett. 1981, 14, 163.
13. Teng, Y.C.; Royce, B.S.H. Appl. Optics 1982, 14, 163.
14. Burggraf, L.W.; Leyden, D.E. Anal. Chem. 1981, 53, 759.
15. Mone, R. in "Preparation of Catalysts," Delmon, B,; Jacobs, P.A.,; Poncelet, G., Eds.; Elsevier: Amsterdam, The Netherlands, 1976; p. 381.
16. Basila, M.R.; Kantner, T.R.; Rhee, K.H. J. Phys. Chem. 1964 68, 3197.
17. Riseman, D.M.; Massoth, F.E.; Dhar, G.M.; Eyring, E.M. J. Phys. Chem. 1982, 86, 1760.
18. Riseman, S.M.; Banyopadhyay,; Massoth, F.E.; E.M. Eyring, submitted for publication to J. Catalysis.
19. Gardella, J.A. Jr.; Jiang, D. -Z.; Eyring, E.M. Appl. Spectrosc. 1983, 37, 131.
20. Pelzl, J.; Klein, K; Nordhaus, O. Appl. Optics 1982, 21, 94.
21. For references see Aamodt, L.C.; Murphy, J.C. J. Appl. Phys. 1983, 54, 581.
22. Oda, S.; Sawada, T.; Moriguchi, T.; Kamada, H. Anal. Chem. 1980, 52, 650.

RECEIVED November 1, 1983

# IR Photothermal Beam Deflection Spectroscopy of Surfaces

M. J. D. LOW, C. MORTERRA, A. G. SEVERDIA, and J. M. D. TASCON

Department of Chemistry, New York University, New York, NY 10003

IR photothermal beam deflection spectroscopy (PBDS) and measurements of IR spectra of solids over the range 3950–450 cm$^{-1}$ made with an interferometer coupled with a detector which senses the photothermal effect by the deflection of a laser beam are described. PBDS is a general technique and requires no sample preparation; all that is needed is to hold the sample at the IR focus. The sample must have a flat spot about 2 mm in diameter accessible to the IR and laser beams. As no sample cells per se are needed, the solid to be examined can be very large, so that selected areas on an entire catalyst pellet can be examined. Examples are given, including an auto exhaust monolith. IR species can also be recorded of adsorbed species at submonolayer coverage and, significantly, this can be done under the rigorously controlled conditions normal to surface studies. Adsorbents and catalysts which scatter and/or absorb IR strongly can be studied, e.g., catalysts with high metal contents, carbons and chars, and carbon-supported metal catalysts. Such materials would be impossible to examine by the conventional IR transmission/absorption techniques. Examples are given such as surface species on carbon and spectra of CO chemisorbed on 50 weight % Ni, as well as SO$_2$ and pyridine on a sulfuric acid catalyst pellet.

In symposia such as the present one, and especially in topical meetings concerned with "surface analysis," it is becoming increasingly rare to find work stressing infrared transmission/-absorption spectroscopy (IR-T/A) per se. This is not because IR-T/A is no longer useful for surface studies; indeed, it is so useful and widely applied that it has lost its original status as

0097–6156/84/0248–0411$06.00/0

special investigative technique, warrants little special
attention, and is no longer fashionable. IR-T/A is applied
almost routinely to a wide variety of adsorbents and catalysts
and, routinely, furnishes valuable data. There remain, however,
certain problem areas where IR-T/A has not been useful, and
cannot be profitably applied. Special IR techniques are needed,
and it is the purpose of the present paper to outline such a
technique and to give some examples of its application.

## Experimental

The technique employed is IR-FT photothermal beam deflection
spectroscopy (PBDS). It is an off-shoot of photoacoustic
spectroscopy (PAS) [1] and is based on the "mirage" detection of
the photothermal effect invented by Boccara et al. [2] and shown
to result in a spectroscopic technique of remarkable versatility
and utility. Some applications of "mirage spectroscopy," mainly
in the visible, and theoretical treatments, have been described
[3-6]. The method has now been developed for use in the IR. The
spectrometer and techniques are described in detail elsewhere
[7], but it will be useful to give a brief outline of the
principles.

Radiation from a broad-band IR source (a Nernst glower) (Fig.
1) is modulated by passage through a scanning interferometer [8]
and is then thrown onto the sample. If radiation is absorbed it
will be degraded to heat (it is assumed that there is no
fluorescence, which is generally the case in the IR) and the
sample warms. Essentially, this is the photothermal effect; the
gas warms and expands. If the resulting pressure change (the
photoacoustic effect) is detected with a microphone or
transducer, PAS results. The warming and expansion also induces
changes in the refractive index of the gas, so that a light beam
passing over the sample's surface is deflected. The deflection d
(the "mirage" effect [2]) can be measured with a position sensing
detector, and PBDS results. In practice, a vibration-compensated
mirage system is used [9].

The signal which is produced, the photothermal interferogram,
is transformed using the data system (a computer with many
peripherals) much as with a T/A-FT spectrometer and is plotted to
yield a single-beam spectrum, S. The latter contains useful
information but becomes more useful when corrected for
"instrument function" and is compensated for the change in the
intensity of the emission of the IR source as a function of
wavelength. This can be done conveniently by recording the
spectrum So of a carbon reference, which is for the present
purpose assumed to be a flat black absorber [10]. The spectrum
of the sample S is then ratioed against that of the carbon, So,
to result in a pseudo-double-beam spectrum S/So, which is then
the compensated PBD spectrum. Note that in the latter an
absorption is a positive ordinate excursion, i.e., bands "point

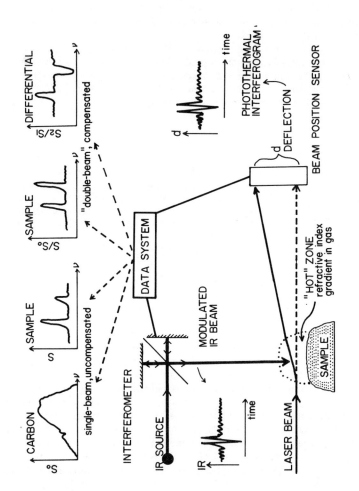

Figure 1. IR-FT-PBD Spectrometry

up," away from the abscissa. Another way of compensating, which
is better and particularly useful and applicable to surface
studies in which one particular sample is subjected to a series
of consecutive treatments A, B, C..., resulting in spectra $S_A$,
$S_B$, $S_C$..., is to compute spectra $S_B/S_B$, $S_C/S_B$, and so
on. Effectively, the sample acts as its own reference and
spectral differences are brought out by such differential
spectra. With the latter, negative bands (an absorption has
decreased) as well as positive bands (an absorption has
increased) may be obtained.

In order to increase the signal-to-noise ratio,
interferograms resulting from individual 1.5 sec. scans are added
coherently and then transformed. Typically, 800-1000 scans
suffice (measuring periods of the order of 20 min.). The spectra
shown were recorded at 8 $cm^{-1}$ resolution.

For surface studies the sample is confined in a cell of the
type in the largely self-explanatory Fig. 2. Samples can be
degassed and exposed to reactive gases, and so on, at
temperatures up to 1000°C under controlled conditions. After
the treatment, the cell is filled to 1 atm. with pure $N_2$ and
the spectrum is recorded. Coupling gases other than $N_2$ may be
used, and the technique becomes more sensitive at increased
pressure [11]. Note especially that the detector is outside the
sample cell, unlike PAS, where the detector is inside. As
detectors for PAS cannot be cleaned effectively, serious sample
contamination problems can arise with PAS.

Results

It has already been shown that PBDS can be usefully applied to a
variety of organic and inorganic solids, corrosion layers, and
surface species [12-15]. With some samples, IR-T/A techniques
would have yielded useful results, but only after extensive
sample preparation had been carried out. In contrast, PBDS
requires no sample preparation; no work is required, and the
danger of changing and/or contaminating a sample by the
preparation steps does not exist. With some other materials,
IR-T/A techniques would also have yielded useful results, but
only at the expense of carrying out a sampling procedure, i.e,
removing a small amount of material from an object, thus
destroying the integrity of the object and then subjecting the
sample to a preparation procedure. In contrast, PBDS is
nondestructive. The examples which are briefly given below
similarly point out the utility of PBDS, but in addition some of
the materials could not have been examined by IR-T/A at all.

Fig. 3 shows the spectra of a 50% Raney Ni catalyst (1000
scans, recorded with the sample in air). The uncompensated
spectrum S shows, as do other spectra S which follow, a prominent
negative band due to atmospheric $CO_2$; the "noise" near 3700
$cm^{-1}$ and in the 2000-1300 $cm^{-1}$ region is due to the negative

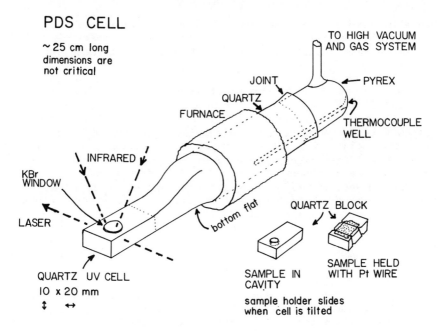

**PDS CELL**

~ 25 cm long
dimensions are
not critical

TO HIGH VACUUM
AND GAS SYSTEM

JOINT

QUARTZ

FURNACE

PYREX

THERMOCOUPLE
WELL

INFRARED

KBr
WINDOW

LASER

bottom flat

QUARTZ BLOCK

QUARTZ UV CELL
10 x 20 mm

SAMPLE IN
CAVITY

SAMPLE HELD
WITH Pt WIRE

sample holder slides
when cell is tilted

Figure 2. IR Cell

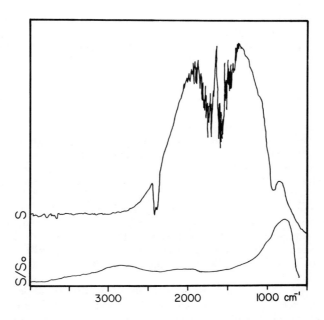

Figure 3. IR Spectra of Raney Nickel Catalyst

bands of atmospheric water vapor. When that spectrum is ratioed against the reference spectrum So of a highly degassed carbon, the compensated spectrum S/So is obtained; "instrument function" has been removed. The spectrum shows a broad absorption peaking near 2800 cm$^{-1}$, probably due to adsorbed hydrocarbons, but otherwise is much like the spectrum of alumina, i.e., what is observed is the oxidation layer on the Raney Ni particles; these were 400 mesh, i.e., relatively large metal particles. From this stage it is but a short step, although a relatively difficult one from the point of view of sample manipulation, to leach the catalyst and then expose the active catalyst to reactants. Raney catalysts have not been subjected to IR study before.

The subject of Figs. 4-6 involves another difficult topic: the examination of carbonaceous materials. In order to obtain carbons prepared under controlled conditions, the pyrolysis of cellulose was studied. Fig. 4 shows just three of many spectra of pyrolysis sequences. The numerous changes in the spectra, e.g., the decline of aliphatic C-H stretching absorptions just below 3000 cm$^{-1}$ accompanied by the build-up of aromatic C-H stretching absorptions just above 3000 cm$^{-1}$ as well as the growth of the trio of aromatic C-H out-of-plane absorptions in the 900-700 cm$^{-1}$ region, lead to the summary given in Fig. 5. The results are described and discussed in great detail elsewhere [16]. Two general observations or trends are to be noted.

The first trend involves the "blackening" of the char; while discrete absorptions decline in intensity as the temperature is increased, an IR continuum grows and begins to level off above about 700°C. The explanation, briefly, is that as the material is pyrolized, polyaromatic networks grow and tend in structure toward that of graphite. Following Kmeto [17], the band gap decreased drastically as the pyrolysis progresses and reaches about 0.1 eV near 700°C; see also the discussion of these effects by Delhaes and Carmona [18]. Essentially this means that carbon particles of this nature, even if they are very small, become totally absorbing so that IR-T/A studies are not possible. Some attempts have been made to grind high temperature carbons so finely that some IR transmission occurs, but IR observations made with such material are suspect because the grinding changes the sample; the topic is taken up in detail elsewhere [16].

The second observation is that discrete absorptions decline in intensity as the pyrolysis progresses and disappear near 700°C (the same trend is found with other carbons). It appears that no spectroscopically observable species remain after the high temperature pyrolysis or degassing. Some species, however, can be re-established [16].

Trace A of Fig. 6 is the spectrum of a carbon which had been degassed at 880°C, so that spectral features such as those of Fig. 3 had disappeared. After exposure to O$_2$ at 420°C (trace B) discrete absorptions of surface species re-appeared and became

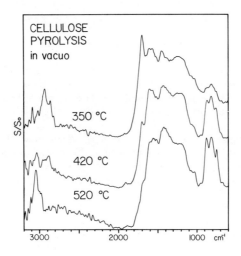

Figure 4.  IR Spectra of Cellulose Chars

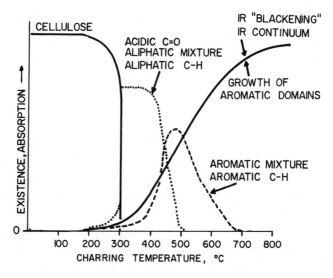

Figure 5.  Cellulose Pyrolysis Scheme

more prominent after further treatment with $O_2$ at 500°C (trace C); details are given elsewhere [16]. These and other results demonstrate that it is possible to examine carbons of various types and to do so under controlled conditions. The study of carbon adsorbents, catalysts, and catalyst supports thus becomes possible.

It is also possible to examine carbon-supported metals; an example is given in Fig. 7. The spectra are those of a 50 wt.% make Ni-on-carbon (Ni-C), prepared by the controlled charring of $Ni(NO_3)_2$-impregnated cellulose followed by reduction with $H_2$, before and after exposure to CO at room temperature. A scale-expanded segment is also shown. The new band at 2040 $cm^{-1}$ is unequivocally attributable to "linear" chemisorbed CO, i.e., to surface Ni-C-O species on the Ni portion of the catalyst. Analogous results have been obtained with other Ni-C samples, as well as with Fe-C, Pt-C, and Cu-C catalysts. It is pertinent to note that the carbon support itself was opaque over the entire spectral region so that the addition of the large amount of metal did not compound the transparency problem; it is intractible to T/A methods. The absorption of the chemisorbed CO is a very intense one, so that the observation of other chemisorbed species will present greater difficulties. Also, specular reflection affects PBDS in a totally negative way with respect to the intensity of the photothermal response so that some band distortions may occur with highly reflective samples [19]. However, it seems feasible to apply PBDS techniques to other systems so that an entire class of catalyst which IR studies have so far been forced to neglect can now be examined.

PBDS will also be useful in a related area for the examination of catalysts which are opaque not because of high unit absorption but because they are physically large, i.e., entire catalyst pellets. This is made possible by the favourable geometry of the apparatus and detection device. As indicated schematically in Fig. 1, the sample is merely placed at the focus of the IR beam (an off-axis elliptical mirror is used to focus the IR beam about 1 cm from the edge of the mirror) and a laser beam grazes the surface. The "sample space" of the spectrometer is thus of indefinite volume and can be made as large as needed to examine massive objects (in the present apparatus, a sphere of about 20 cm diameter could be accomodated). An example is shown in Fig. 8.

The object was an auto exhaust catalyst, a monolith cylinder 25 mm in length and 38 mm in diameter. The outside wall was broken away so that one of the 1 mm-wide channels became accessible to the IR and probe laser beams, and a portion of one channel was studied in the manner shown schematically in the insert of Fig. 8. The sample was examined in air, because a cell large enough to contain the monolith was not available. The spectrum shows the features of cordierite [20], the material from which honeycomb monoliths are usually made, a broad absorption in

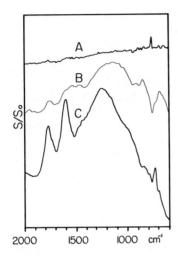

Figure 6. Spectra of Carbon

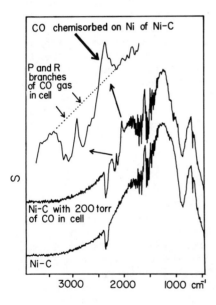

Figure 7. CO Chemisorption on Nickel-Carbon

the 3700-3200 cm$^{-1}$ region and a sharper feature near 1650 cm$^{-1}$ of adsorbed water, and a broad absorption near 3000 cm$^{-1}$ ascribable to adsorbed hydrocarbons.

Spectra of a spent bauxite-based desulfurization catalyst pellet ( 7 x 13 mm, examined in air) are shown in Fig. 9. The outside of the pellet was black and the single-beam spectrum S showed some of the continuum absorption found with chars. The compensated spectrum S/So, however, showed appreciable spectral structure. The broad band near 750 cm$^{-1}$ is probably due to the bauxite, and the absorptions near 3000, 1320 and 1000 cm$^{-1}$ to a mixture of hydrocarbons and thio species formed during the reaction. The feature near 1640 cm$^{-1}$ is probably caused by an olefinnic species.

Further examples are furnished by the spectra of Figs. 10 and 11. A single pellet of virgin catalyst ( 7 x 10mm) was placed in a cell (Fig. 2) and degassed at room temperature, and spectrum $S_2$ was recorded (the main spectral features are the strong absorptions of the kieselguhr support, but some sulfate absorptions can also be discerned). The catalyst was then exposed to 90 torr of $SO_2$ at room temperature and spectrum $S_1$ was recorded with $SO_2$ in the cell, when new features appeared. The ratioed, scale-expanded insert shows these more clearly. There is a negative band caused by the gaseous $SO_2$ upon which a positive doublet is superimposed. Pumping at room temperature caused the doublet to disappear. The 1375 and 1358 cm$^{-1}$ bands are attributed to $SO_2$ weakly coordinated to surface oxide ions [21]. In other experiments with the same pellet, pyridine (Py) was sorbed by the surface at room temperature and, after a few minutes, the residual gaseous Py was removed by pumping at room temperature. The compensated spectrum (Fig. 11) then showed the strong bands at 1542 and 1485 cm$^{-1}$ as well as minor bands, i.e., of a mixture of Py, H-Py, PyH$^+$, and L-Py [22] so that, with further work, it would be possible to obtain information about the nature of the surface.

## Discussion

In view of the examples shown, IR-PBDS would seem to be a versatile and useful technique for the study of adsorbents and catalysts. Indeed, entire areas of study which are inaccessible to IR-T/A techniques become possible by means of PBDS. It must be pointed out, however, that the results, which have been very good indeed, have so far been qualitative in nature.

There are at present some actual and potential problems of varying degrees of severity with PBDS. The simplest of these involves the obvious fact that the sample must be in contact with a gas which itself must be pure, unreactive, and have suitable refractive and thermal properties. There is thus a potential contamination problem which, however, can be controlled. A mechanical problem arises because of the need to position the

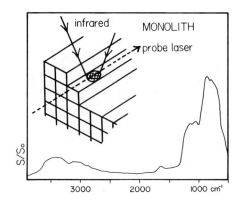

Figure 8.   Spectrum of Auto
Exhaust Monolith

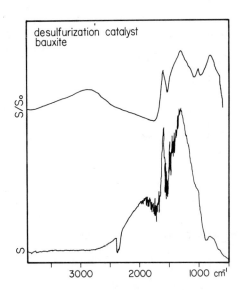

Figure 9.   Spectrum of Desulfurization Catalyst Pellet

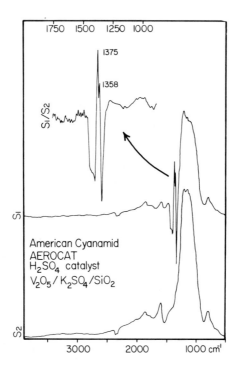

Figure 10. SO$_2$ Chemisorbed on Aerocat Catalyst Pellet

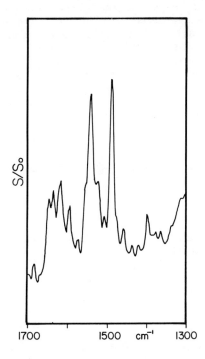

Figure 11. Pyridine Sorbed on Aerocat Catalyst Pellet

sample surface precisely in relation to the IR and laser probe beams in order to maximize the photothermal signal: if a sample is moved in order to carry out some treatment of the surface, re-positioning the sample precisely is very difficult, so that it is difficult to obtain quantitative data. Scattering can also influence the latter: PBDS is afflicted with some of the ills of diffuse reflection spectroscopy, in that particle size and shape, bed depth, packing and reflectivity influence the signal intensity [19,23], and the effects of Fresnel reflectivity may make themselves felt under certain conditions [19]. Sample heating may cause desorption of weakly-held species [14], and, in addition, potential problems may arise from inadequate source compensation [10], photothermal saturation [1], and with continuous scan interferometers, the continuous change in penetration depth [1] with wavenumber. There are thus some actual and potential problems concerned mainly with the quantitative aspects of the technique, but these are more than off-set by the ability of PBDS to permit the examination of materials it would be impossible to study by conventional means.

## Acknowledgment

Support by ARO contract DAAG29-83-K-0063 and NSF grant CPE-7922100 is gratefully acknowledged.

## Literature Cited

1. Rosencwaig, A., "Photoacoustics and Photoacoustic Spectroscopy", Wiley, New York, 1977.
2. Boccarra, A.C.; Fournier, D.; Badoz, J.; Appl. Phys. Lett. 1980, 36. 130.
3. Boccarra, A.C.; Fournier, D.; Jackson, W.; Amer, N.M.; Opt. Lett. 1980, 5, 377.
4. Fournier, D.; Boccarra, A.C.; Amer, N.M.; Gerlach, R., Appl. Phys. Lett. 1980, 37, 519.
5. Jackson, W.B.; Amer, N.M.; Boccarra, A.C.; Fournier, D., Appl. Opt. 1981, 20, 1333.
6. Debarre, D.; Boccarra, A.C.; Fournier, D., Appl. Opt. 1981, 20, 4281.
7. Low; M.J.D.; Lacroix, M., Infrared Phys. 1982, 22, 139.
8. Low M.J.D., "Encyclopedia of Industrial Chemical Analysis", Wiley: New York, 1971; Vol. 13, pp. 139 ff.
9. Low, M.J.D.; Parodi, G.A.; Lacroix, M., Chem. Biomed. Environ. Instrum. 1981, 11, 265.
10. Low, M.J.D.; Parodi, G.A., Spectrosc. Lett. 1980, 13, 633.
11. Low, M.J.D.; Arnold, T.H.; Severdia, A.G., Infrared Phys., in press.
12. Low, M.J.D.; Morterra, C.; Severdia, A.G.; Lacroix, M., Appl. Surf. Sci. 1982, 13, 429.

13. Low, M.J.D.; Lacroix, M.; Morterra, C., Spectrosc. Lett. 1982, 15, 57.
14. Low, M.J.D.; Morterra, C.; Lacroix, M., Spectrosc. Lett. 1982, 15, 159.
15. Low, M.J.D.; Lacroix M.; Morterra, C., Appl. Spectrosc. 1982, 36, 582.
16. Low, M.J.D.; Morterra, C., Carbon, in press.
17. Kmeto, E.A.; Phys. Rev. 1951, 82, 456.
18. Delhaes, P.; Carmona, F., Chemistry and Physics of Carbon 1981, 17, 89.
19. Morterra, C.; Low, M.J.D,; Severdia, A.G., Infrared Phys. 1982, 21, 221.
20. Keller, W.D.; Spotts, J.H.; Biggs, D.L., Am. J. Sci. 1952, 250, 453.
21. Goodsel, A.J.; Low, M.J.D.; Takezawa, N., Environ. Sci. Technol. 1972, 6, 268.
22. Knozinger, H., "Advances in Catalysis" Eley, D.D.; Pines, H.; Weisz, P.B., Eds.; Academic Press, New York, 1976; Vol. 25, p. 184.
23. Tilgner, R., Appl. Opt. 1981, 20, 378.

RECEIVED September 26, 1983

# Tunneling Spectroscopy of Organometallic Molecules

WILLIAM C. KASKA—Department of Chemistry, University of California, Santa Barbara, CA 93106

PAUL K. HANSMA—Department of Physics, University of California, Santa Barbara, CA 93106

ATIYE BAYMAN—Advanced Micro Devices, M-S# 111, Sunnyvale, CA 94086

RICHARD KROEKER—International Business Machines Corporation, San Jose, CA 95193

The presence of group frequencies or "finger print" regions in infrared spectra make vibrational spectroscopy a key analytical method in identifying classes of molecules.

Inelastic electron tunneling spectroscopy (IETS) takes advantage of the general applicability of vibrational spectroscopy by measuring the vibrational spectrum of molecules adsorbed on the insulation of a metal-insulator-metal junction (Figure 1).

The tunnel current which flows from one metal to the other when a potential difference is applied across the junction is mainly due to elastic tunneling. However, if the adsorbed molecules on the junction have a characteristic vibrational mode of energy $h\nu$, then an inelastic process can occur when $ev \geq h\nu$. Since the inelastic current is difficult to detect against the background of elastic current, the second derivative $d^2V/dI^2$ vs voltage is studied. Specifically, the tunneling spectrum $d^2I/dV^2$ vs V displays a peak which corresponds to the vibrational frequency of the molecule at $ev = h\nu$. The most important point about tunneling spectroscopy is that the inelastic conduction path only exists when the voltage across the junction is greater than $h\nu$ for the vibrating molecules. If this was not the case, then an electron from one metal could not lose energy $h\nu$ and still have enough energy to tunnel into an empty state in the second metal (1).

## Methods

Figure 2 shows the typical steps in junction preparation. The aluminum strip is evaporated onto a glass slide through a mask. After the surface is oxidized in air or $O_2$ glow discharge, it is

0097-6156/84/0248-0427$06.00/0
© 1984 American Chemical Society

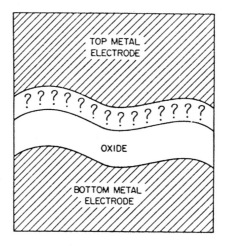

Figure 1.  Schematic representation of an inelastic electron tunnel junction.

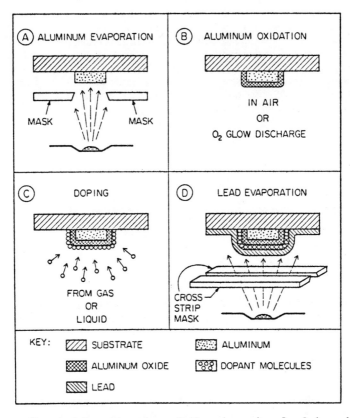

Figure 2.   A schematic view of the steps involved in making an inelastic tunnel junction.  In practice, the mask touches the substrate.  The lead mask is held 0.1 mm from the substrate by dimpling it with a center punch.  Reproduced with permission from "Tunneling Spectroscopy", Plenum Press, 1982.

treated with a second metal evaporation process where particles
of transition metals coat the alumina surface. The transition
metals generally appear as small hemispheres or islands with a
typical diameter of 25 - 40 Å. After or sometimes during deposition of the transition metal, a reactive gas like hydrogen, carbon monoxide, ethylene or acetylene is allowed to contact the
supported transition metal. The junction is completed with a
cross strip of lead (see Figures 3 and 4).

Figures 5 and 6 show one of the first examples of organometallic compounds adsorbed on the alumina of a metal-insulator-
metal junction. The carboxyl groups on the benzene ring and
cyclopentadienyl rings are most likely coordinated to the alumina.
The arrangement of the carbonyl groups with respect to the surface
would then be as shown below:

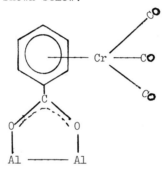

This arrangement may be the reason why the intensities of the CO
vibrations are much less than the other modes in the molecule because tunneling electrons generally interact more strongly with
dipoles perpendicular rather than parallel to the surface.

## Results and Discussion

Figure 7 is a differential tunneling spectrum of CO chemisorbed
on alumina supported rhodium particles. The identification of
the peaks is also shown below and consist of three separate
species. These are a gem dicarbonyl Rh $(CO)_2$, a linear carbonyl
RhCO and a bridging carbonyl $Rh_xCO$. The dicarbonyl is characterized by a peak at 413 $cm^{-1}$ and the linear species by a bending
mode at 465 $cm^{-1}$.

The modes at 413 and 465 $cm^{-1}$ are very sensitive to CO coverage. At very low coverage, $\theta < 0.1$ L,(L = Langmuir) the band
at 465 is the only peak present. At higher coverages, the band
at 465 remains about the same in intensity but the mode at
413 $cm^{-1}$ rapidly increases to about the same intensity as the
465 $cm^{-1}$ mode at 3 L coverage. A stretching mode at 1721 $cm^{-1}$
characterizes the bridging species and the peak at 1942 $cm^{-1}$ contains contributions (symmetric, asymmetric) from both the linear
and gem dicarbonyl species.

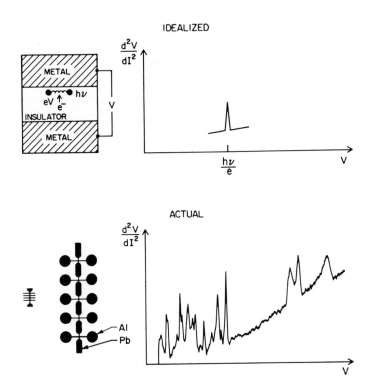

Figure 3. The harmonic oscillator in the idealized picture is one of the vibrational modes of a dopant molecule
in an actual junction. Each vibrational mode is revealed
as a peak in $d^2V/dI^2$ at a voltage of $V = h\nu/e$. The
tunneling spectrum can be compared to infrared and Raman
spectra: 0.1 V corresponds to 806.5 $cm^{-1}$. Reproduced
with permission from Catal Rev. <u>23</u> 553 (1981)(Marcel
Dekker, Inc.).

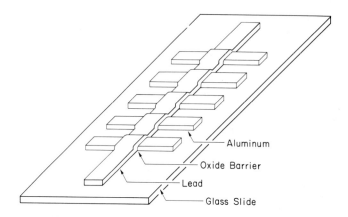

Figure 4.    Schematic view of a series of tunnel junctions.

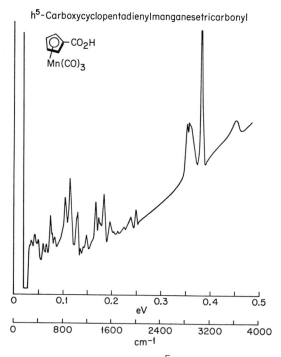

Figure 5.    Tunneling spectrum of $\eta^5$-cyclopentadienylcarbo-xymanganesetricarbonyl adsorbed on an Al-Oxide-Pb junction.

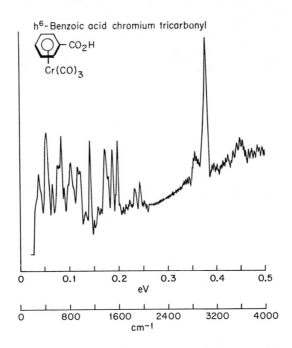

Figure 6. Tunneling spectrum of benzoic acid chromium tri-carbonyl adsorbed on an Al-Oxide-Pb junction.

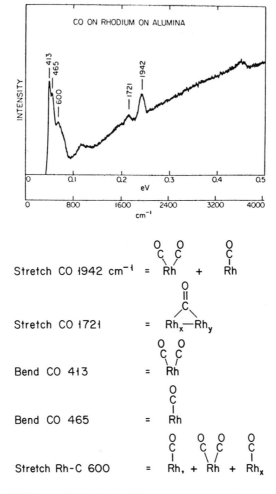

Figure 7. Differential tunneling spectrum of CO chemisorbed on alumina supported rhodium particles. Peak positions are not corrected for possible shifts due to the top
lead electrode. Peak positions vary with rhodium coverage
and CO exposure.

Figure 8 shows what happens when the completed junction (with lead top electrode) is heated in the presence of hydrogen. Hydrogen gas permeates the lead electrode and reacts with the CO to produce a dominant hydrocarbon on the rhodium. By using isotopic studies, the hydrocarbon is seen to form as a chemisorbed monolayer without the presence of gas phase CO. No surface species which contain oxygen can be observed. Since tunneling spectra are very sensitive, this is an interesting observation and suggests the resultant ethylidene species comes from polymerization of rhodium bridged methylene groups (2), rather than any OXO contained species.

Besides the interaction with CO, Figure 9 shows the most recent results of the simplest unsaturated hydrocarbon (acetylene) with alumina supported palladium. Here, however, the very small peak intensities indicate that tunneling spectroscopy shows little promise for studying complex hydrocarbons on supported metals.

The figure shows the complex number of peaks that can be formed when acetylene interacts with palladium.

Species A is the weakest held acetylene and probably involves π-coordination to the triple bond. Species B and C involve rehybridized acetylene and comparisons to known species in organometallic chemistry suggest that B corresponds to molecules like $C_2H_2Co_2(CO)_6$ (4) and (5)

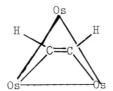

Species C could possibly be the metallocycle shown below because

the frequencies correspond to cis-1,2-dichloroethylene. Kroeker, et al. (6) have shown that halogen substituted hydrocarbons can be excellent models for molecules adsorbed on metal surfaces.

Species D is most likely an ethylidyne complex which forms from self-hydrogenation on the palladium surface. Such species along with species E have been suggested to be part of the compounds formed from platinum and acetylene.

When the junctions are heated in $H_2$, the most prominent compound formed is the formate ion instead of saturated hydrocarbons. Evidently the triple bond breaks at elevated

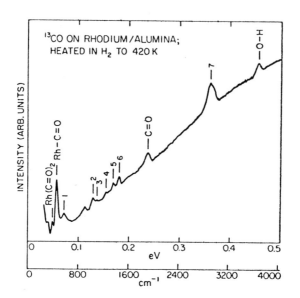

Hydrocarbon modes  1 – 7

Figure 8.  Differential tunneling spectrum of $^{13}CO$ on rhodium/alumina heated to 420° K in hydrogen.  Modes due to hydrocarbon are number 1 to 7.  The hydrocarbon species is identified as an ethylidene moiety.

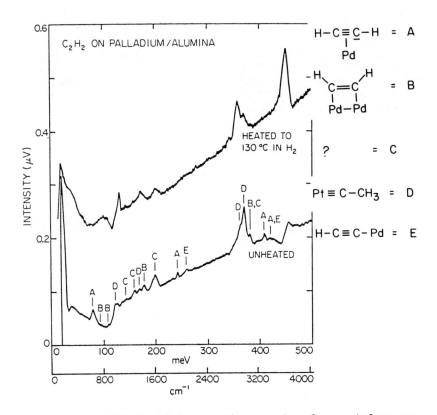

Figure 9. Differential tunneling spectra for acetylene on an Al-Oxide-Pb junction before (lower trace) and after (upper trace) heating in hydrogen. The possible identification of the peaks is shown by the lettered species.

temperatures and combines with surface OH groups to form formate ions which may be bonded to the palladium or may be migrating to the alumina substrate.

## Summary

The high sensitivity of tunneling spectroscopy and absence of strong selection rules allows infrared and Raman active modes to be observed for a monolayer or less of adsorbed molecules on metal supported alumina. Because tunneling spectroscopy includes problems with the top metal electrode, cryogenic temperatures and low intensity of some vibrations, model catalysts of evaporated metals have been studied with CO and acetylene as the reactive small molecules. Reactions of these molecules on rhodium and palladium have been studied and illustrate the potential of tunneling spectroscopy for modeling reactions on catalyst surfaces.

## Acknowledgments

We thank the Office of Naval Research for partial support of this research. Several items of equipment used in this research were obtained with partial support from National Science Foundation grant DMR79-25430 for which we are grateful.

## Literature Cited

1.  Hansma, P. K., ed. "Tunneling Spectroscopy", Plenum Press, 1982, for a recent update on various aspects of Tunneling Spectroscopy.
2.  Stone, F. G. A., West, R. A., "Adv. Organometallic Chem." Hemmann, W. A. 1982, 20, 159, Academic Press, 1982.
3.  Bayman, A; Hansma, P. K.; Kaska, W. C.; Dubois, L. H.; Appl. of Surf. Sci. 1983, 14, 194-208.
4.  Skinner, P.; Howard, M. W.; Oxton, I. A.; Kettle, S. F. A.; Powell, D. B.; Sheppard, N.; J. Chem. Soc. Faraday Trans. 1981, 2, 77, 1203.
5.  Anson, C. E.; Bandy, B. J.; Chesters, M. A.; Keiller, B.; Oxton, I. A.; Sheppard, N. A.; J. Electron Spectroscopy and Related Phenomena, 1983, 29, 315-316; Brundle, C. L.; Morawitz, H. eds.; "Vibrations at Surfaces", Proceedings of the Third International Conference, Asilomar, California, 1982, Elsevier Scientific Publishing Company, 1983, p. 315.
6.  Kroeker, R. M.; Kaska, W. C.; Hansma, P. K.; J. Catal. 1980, 63, 487-490.

RECEIVED October 31, 1983

# The Effect of Particle Size on the Reactivity of Supported Palladium

S. ICHIKAWA, H. POPPA, and M. BOUDART

Stanford-NASA/Ames Joint Institute for Surface and Microstructure Research, Department of Chemical Engineering, Stanford University, Stanford, CA 94305

Carbon monoxide adsorbed on sufficiently small palladium particles disproportionates to surface carbon and carbon dioxide. This does not occur on large particles. The $CO-O_2$ reaction is shown to be structure-insensitive provided the metal surface available for the reaction is estimated correctly. This varies with temperature for the small particles, as at low temperatures the deposited carbon eliminates sites for the reaction while the latter become available at higher temperatures at which surface carbon reacts away with $O_2$.

As a result of disproportionation of CO on small particles, the selectivity of the $CO-H_2$ reactions shifts from methanol on large particles to methane on small ones. The methanation activity increases as the metal particle size decreases, indicating that methanation is a structure-sensitive reaction on palladium.

The low pressure ($10^{-4}$ Pa) oxidation of carbon monoxide was studied recently on palladium particles that were vapor deposited on the $\{\bar{1}012\}$ face of a single crystal of $\alpha-Al_2O_3$ (1). In order to obtain turnover rates $v_t$ from the measured rate of production of $CO_2$, i.e., the total number of $CO_2$ molecules produced per second, $\dot{N}_{CO_2}$, use was made of temperature programmed desorption of CO from the Pd particles. As the size of Pd particles became smaller, a low temperature (LT) desorption peak started to appear next to the normal high temperature (HT) peak found exclusively on larger particles. In order to count the number of Pd sites,

$N_{Pd}$, for each sample, it was decided to use only the HT peak, count the number of molecules of CO under that peak $N_{CO,HT}$ and to multiply the latter quantity by the fraction of surface covered at saturation, $\Theta$, by CO molecules adsorbed under the HT peak. The value of $\Theta$ was determined to be approximately 0.45 for all samples. Thus the turnover rate $v_t$ was calculated to be:

$$v_t = \dot{N}_{CO_2}/N_{Pd} = \Theta \dot{N}_{CO_2}/N_{CO,HT} \tag{1}$$

With such a definition, it was found that $v_t$ at 445 K did not change within experimental error as the average Pd particle size, determined by transmission electron microscopy (TEM), was varied between 1.5 and 8.0 nm (Figure 1). Besides, this value of $v_t$ was also the same as that reported for the {111} face of a single crystal of Pd (2), the latter value being itself very much the same on other planes of Pd or on a polycrystalline wire (3).

These results provide a very strong evidence for the oxidation of CO to be a structure insensitive reaction, as it proceeds at almost the same rate, irrespectively of crystalline anisotropy or particle size. Yet there were some difficulties and others soon appeared. At a higher temperature, 518 K, the value of $v_t$ on the same samples of Pd clearly increased as particle size went down (Figure 1) to reach a value about 3 times larger on the smaller particles than on the larger ones. This was explained by an increase in accessibility of smaller particles to striking molecules at a temperature high enough, so that the reaction rate was essentially the sticking probability of CO. Details of the proposed explanation can be found in the original paper (1). The explanation was plausible but not convincing.

Soon after, evidence was presented, showing that dissociation of CO took place on small Pd particles though not on large ones and that the LT binding state of CO as studied by TPD was the one responsible for that unexpected phenomenon of CO dissociation on Pd (4,5).

The present work was undertaken to study further the dissociation of CO on small particles of Pd, not only on the type produced by vapor deposition, but also on typical catalytic material, Pd supported on silica gel (Pd/SiO$_2$). All experimental details will be published separately (6,7).

In this paper, the main results and conclusions of this work on CO dissociation on small Pd particles are presented and confronted with new data on CO oxidation at low pressure and CO hydrogenation. Again, all details of the catalytic experiments will be published separately (6,7).

## Results and Discussion

Low pressure studies: Adsorption of CO. The experiments were performed in an ultra-high vacuum system described previously (1). The data obtained on palladium particles with a size smaller than 2 nm or larger than 3 nm will be discussed in turn.

The results of temperature programmed desorption (TPD) of CO following its adsorption at room temperature (RT) on palladium particles vapor deposited on′ {0001} α-Al$_2$O$_3$ are shown in Figure 2. The average particle size of the palladium particles was 1.7 nm, as measured by electron microscopy. With a freshly prepared sample, the TPD spectrum following adsorption of CO (P$_{CO}$=10$^{-5}$ Pa) at RT shows a low temperature (LT) peak at 360 K and a high temperature (HT) peak at 420 K. When the adsorption-desorption cycle was repeated five times, the area under both peaks decayed progressively as shown by curves 2 to 5. After ten cycles, they converged to the dotted curve corresponding to a loss of adsorption capability of 80%. After ten cycles, temperature programmed reaction (TPR) with dihydrogen revealed surface carbon as an appreciable amount of methane was detected. The carbon deposited by CO also reacts with oxygen to form CO$_2$ but only at temperatures above 500 K. Thus, the broken curve in Figure 2 was obtained after treating the surface represented by the dotted curve with flowing dioxygen at 550 K. Carbon deposition seems to occur by

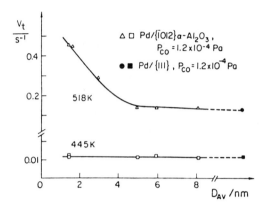

Figure 1.   Turnover rate vs. particle size at $(P_{O_2}/_{CO})$ = 1.1 and $P_{CO}$ = 1.2 x $10^{-4}$ Pa.   ( $\triangle \square$ ) Pd/{$\bar{1}$012} $\alpha$-$Al_2O_3$; ( $\bullet \blacksquare$ ) Pd {111}.

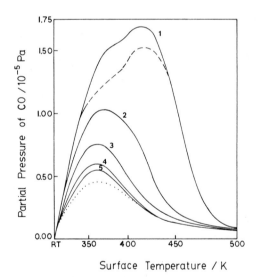

Figure 2.   TPD of CO on Pd/{0001} $\alpha$-$Al_2O_3$.   $D_{av}$ = 1.7 nm.
$\cdots$:  After 10 cycles of adsorption and desorption.
---:  After dioxygen treatment at 550 K, following the 10 cycles.

disproportionation ($2CO = C + CO_2$) of CO in the relatively weakly adsorbed state, since $CO_2$ was detected during CO desorption and the $CO_2$ peak temperature corresponded to the LT CO peak temperature in Figure 2.

Carbon deposition was also found on particles smaller than 1.0 nm supported on polycrystalline gamma alumina. In this case, the decay of TPD curves and the amount of carbon deposited was even more important than in the case discussed above of 1.7 nm particles on {0001} $\alpha$-$Al_2O_3$.

In contrast to the results on smaller particles, larger (3.6 nm) particles on {0001} $\alpha$-$Al_2O_3$ gave only one peak in the TPD curve which corresponded to the HT peak for 1.7 nm particles and the area under the curve decreased only slightly (less than 5%), even after five cycles. The amount of methane detected per CO adsorption site by TPR-$H_2$ was two orders of magnitude less than that observed on 1.7 nm particles.

It is noteworthy that surface carbon did not come from those CO molecules responsible for the HT peak but from sites that are able to disproportionate CO and correspond to the LT peak. Because the latter sites are important only on quite small particles, it is tempting to associate them with low coordination number surface metal atoms, the relative concentration of which increases rapidly as the particle size decreases below 2 nm (8). Thus, these atoms may be the sites responsible for the relatively weakly adsorbed state of CO. Results similar to our work were found on other Group VIII metals. In the case of a Ru/$SiO_2$ sample, Yamasaki et al. (9) have shown by infrared spectroscopy that the deposition of carbon occurs rapidly by CO disproportionation on the sites for weakly held CO. The disproportionation also occurred on a Rh/$Al_2O_3$ sample with 66% metal exposed so that appreciable concentrations of low coordination atoms are expected (10).

Reaction between carbon monoxide and dioxygen. The steady-state formation of $CO_2$ was measured on palladium particles vapor deposited on the {0001} face of $\alpha$-Al$_2$O$_3$. For 3.6 nm particles, surface carbon was not detected after the steady-state reaction at both low temperature (445 K) and high temperature (518 K). On 1.0 nm particles, however, an appreciable amount of carbon was found after the reaction at 445 K but not at 518 K.

The results of adsorption and desorption of CO mentioned above suggest that for the reaction at low temperature, the sites for relatively weakly chemisorbed CO are covered by the deposited carbon and the reaction occurs between molecularly adsorbed CO and oxygen on the carbon-free sites which are the sites for relatively strongly chemisorbed CO. Therefore, the definition of the turn-over rate at 445 K remains as given in Equation 1. For the reaction at 518 K, however, this definition becomes inappropriate for the smaller particles. Indeed, to obtain the total number of Pd sites available for reaction, we now need to take into consideration the number $N_{CO,LT}$ of CO molecules under the desorption peak. Furthermore, let us assume that disproportionation of CO takes place through reaction between two CO molecules adsorbed on two adjacent sites, and let us also assume that the coverage is unity for the CO molecules responsible for the LT desorption peak, since this was found to be approximately correct on 1.5 nm Pd on {$\bar{1}$012} $\alpha$-Al$_2$O$_3$ (1). Then, the number $N_{Pd}$ of palladium sites available for reaction at 518 K is given by $N_{CO,HT}/\Theta + 2N_{CO,LT}$ since the CO molecules under the LT desorption peak count only half of the available sites. Consequently, the turnover rate at 518 K should be defined as:

$$v_t = \dot{N}_{CO_2} / (N_{CO,HT}/\Theta + 2N_{CO,LT}) \qquad (2)$$

Now, if we use Equation 1 at 445 K and Equation 2 at 518 K, we obtain the results shown in Figure 3 which contains the data of this investigation and of the preceding one (1). At both reaction temperatures, the turnover rate remains almost constant when the particle size is varied between 1.0 nm and 8.0 nm.

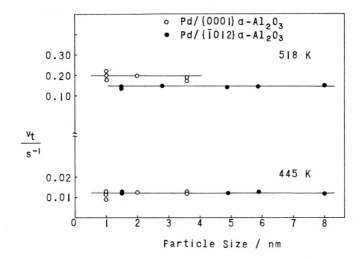

Figure 3. Reaction between CO and $O_2$ at steady-state. $P_{total}$ = 2.5 x $10^{-4}$ Pa, $P_{O_2}/P_{CO}$ = 1.1. (●) Pd/{$\bar{1}$012}α-$Al_2O_3$ (O) Pd/{0001} α-$Al_2O_3$.

Thus, although CO disproportionation is structure-sensitive, the $CO-O_2$ reaction appears to be structure-insensitive at both 445 K and 518 K, provided we define correctly the number of Pd sites available for reaction at both temperatures. It should also be noted that the turnover rate at 445 K is the same on metal particles supported on $\{\bar{1}012\}$ $\alpha-Al_2O_3$, $\{0001\}$ $\alpha-Al_2O_3$, and $\gamma-Al_2O_3$. At 518 K, the Pd particles supported on $\gamma-Al_2O_3$ and $\{0001\}$ $\alpha-Al_2O_3$ are slightly more active than those supported on $\{\bar{1}012\}$ $\alpha-Al_2O_3$. The difference is slight and its reason is not understood at the moment.

High pressure studies: Adsorption of CO.  Three samples (1.36% $Pd/SiO_2$, 1.88% $Pd/SiO_2$-A, 1.88% $Pd/SiO_2$-B) were prepared by cation exchange of palladium on acid cleaned silica gel. Dihydrogen chemisorption at RT gave the percentage metal exposed for these samples, viz., 100, 75.4 and 45.4 which correspond to particle sizes of 1.1, 1.5 and 2.5 nm, respectively. Bright-field electron microscopy gave corresponding particle size values of <1.0, 1.4 and 2.4 nm respectively in good agreement with the values obtained from chemisorption.  Carbon monoxide adsorption and desorption cycles were performed on these samples in a static closed system. Carbon monoxide was adsorbed at RT ($P_{CO}$ = 50 kPa) followed by evacuation to $10^{-4}$ Pa leaving only chemisorbed CO.  Then, the system was closed and cycles of heating to 673 K and cooling back to RT were performed.  The pressures at 673 K decreased while the pressure at 298 K increased as the heating cycle was repeated, which indicated disappearance of CO and blocking of CO adsorption sites.  Characterization before and after seven heating cycles by CO chemisorption is summarized in Table I.  The loss of CO chemisorption sites were observed to increase as the particle size decreased.  However, electron microscopy showed that the particles maintained their original sizes.  It should be noted that electron microscopy can be reliable for size determination when chemisorption is ambiguous and "percentage metal exposed" is more

appropriate than the more popular "dispersion". With the
palladium particles of 1.1 nm and 1.5 nm, gas phase analysis after
three cycles showed appreciable amounts of $CO_2$ which corresponded
to 24% and 13% of total surface palladium in each case. Also, for
the latter sample, the amount of carbon detected as methane
following treatment in $H_2$ was almost equal to the $CO_2$ amount.
These data indicate that carbon deposition occurs by disproportionation of CO on small particles of palladium.

Table I. Chemisorption of CO before and after cycles
of CO adsorption and desorption.

| Sample of Pd/SiO₂ Pd particle size/nm | *Percentage metal exposed before the cycles/% | *Percentage metal exposed after seven cycles/% | Particle size determined by BFEM+ after seven cycles/nm |
|---|---|---|---|
| 1.1 | 88.4 | 21.4 | <1.0 |
| 1.5 | 79.3 | 43.7 | 1.3 |
| 2.5 | 37.9 | 38.0 | 2.4 |

* Adsorption at room temperature. Stoichiometry CO/Pd = 1.
+ Bright-field electron microscopy.

Chemisorbed CO was examined by Fourier transform IR spectroscopy
(Figure 4). All spectra were obtained between 400 and 4000 $cm^{-1}$
by taking 250 scans at 1 $cm^{-1}$ resolution. Spectra on 1.1 nm
particles were obtained after CO adsorption at 6.67 kPa and
evacuation ($10^{-5}$ Pa) for 1/2 h, all at RT. Spectrum (a) was
obtained after reduction and evacuation at 673 K. After heating
to 673 K under 7.76 kPa of CO and evacuation at this temperature,
spectrum (b) was obtained. Then, the sample was heated to 673 K
in vacuum, dihydrogen was flowed for 1/4 h at atmospheric pressure
and the sample was evacuated for 2 h at 673 K. This resulted in
spectrum (c). In comparison with spectrum (a), the intensities
of all five absorbance peaks (2084, 2069, 2055, 1920, 1859 $cm^{-1}$)
decreased together in spectrum (b) and increased back together in
spectrum (c) for which the peak at 2084 $cm^{-1}$ was fully recovered.
These results indicate the loss of CO chemisorption sites after
heating under CO and regeneration of the sites by dihydrogen

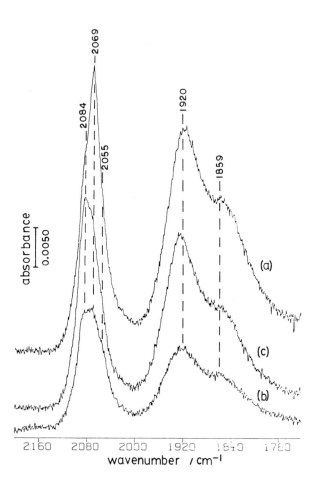

Figure 4.  IR spectra for CO chemisorbed at room temperature
on 1.1 nm Pd/SiO$_2$.  (a) After reduction and evacuation at
683 K.  T = 299 K.  (b) After heating to 673 K under
7.76 kPa of CO and evacuating at 673 K.  T = 298 K.
(c) After dihydrogen treatment and evacuation at 673 K.
T = 297 K.

treatment. These observations are consistent with the results of CO TPD (Figure 2). Infrared spectra were also obtained as above between 200 K and RT, but there was no decrease in any of the peaks (2090, 2082, 2051, 1936, 1853 $cm^{-1}$) for CO chemisorbed at 200 K after warming to RT in 7.8 kPa of CO. This suggests that carbon deposition does not occur below RT.

Reaction between carbon monoxide and dihydrogen. The catalysts used were the $Pd/SiO_2$ samples described earlier in this paper. The steady-state reaction was first studied at atmospheric pressure in a flow system (Table II). Under the conditions of this work, selectivity was 100% to methane with all catalysts. The site time yield for methanation, $STY_{CH_4}$, is defined as the number of $CH_4$ molecules produced per second per site where the total number of sites is measured by dihydrogen chemisorption at RT before use, assuming H/Pd = 1. The values of $STY_{CH_4}$ increased almost three times as the particle size decreased. The data obtained by Vannice et al. (11,12) are included in Table II and we can see that the methanation reaction on palladium is structure-sensitive. It must also be noted that no increase of $STY_{CH_4}$ occurred by adding methanol to the feed stream which indicates that methane did not come from methanol.

Table II. Reaction between CO and $H_2$ at 548 K, $H_2/CO = 3$ and atmospheric pressure.

| $Pd/SiO_2$ Catalyst wt/% Pd | Percentage metal exposed/% | $STY_{CH_4}/10^{-4} s^{-1}$ | Reference |
|---|---|---|---|
| 1.36 | 100 | 14.7 | This work |
| 1.88-A | 75.4 | 14.5 | This work |
| 1.88-B | 45.4 | 5.0 | This work |
| 4.75 | 46 | 3.2 | (11) |
| 1.93 | 20 | 1.2 | (12) |

Since the equilibrium conversion of CO to methanol under the conditions of Table II is low, the reaction was further studied under higher pressures. The results at 535 kPa are shown in Table III. The catalysts having smaller palladium particle sizes

still showed 100% selectivity to methane but the catalyst with
larger particles gave 17% selectivity to methanol.

Table III.  Reaction between CO and $H_2$
at 673 K, $H_2/CO$ = 3.6 and 535 kPa.

| Pd/SiO$_2$ catalyst Pd particle size/nm | $STY_{CH_4}/s^{-1}$ | $STY_{CH_3OH}/s^{-1}$ | Selectivity |
|---|---|---|---|
| 1.1 | * | methanol not detected | 100% to methane |
| 1.5 | * | methanol not detected | 100% to methane |
| 2.5 | 6.36 x 10$^{-3}$ | 1.03 x 10$^{-3}$ | ** 0.17 |

*Methane formed but not measured quantitatively.
**Equal to $STY_{CH_3OH}/STY_{CH_4}$.

The above results, together with the results of adsorption-
desorption cycles discussed earlier, suggest that on palladium,
the methanation reaction occurs via hydrogenation of the deposited
carbon on small particles. However, on larger particles, methanol
can be formed by hydrogenation of undissociated adsorbed CO, at
pressures high enough to shift the equilibrium towards methanol
formation.  If we review the published work on palladium from the
standpoint of particle size effect, we find data consistent with
the trends reported in Table II and Table III.  On Pd/$\eta$-Al$_2$O$_3$
having 100 percentage metal exposed, methanol was not detected
even at 2 MPa and only methane was observed (13). However, almost
100% selectivity to methanol was observed under similar pressure
condition on Pd/SiO$_2$ and Pd/$\gamma$-Al$_2$O$_3$ having 26 and 27 percentage
metal exposed respectively, and the methanol formation rates were
almost the same with both catalysts (14).  These results can be
explained simply by the shift in selectivity due to particle size.
Of course, we do not imply ruling out the effect of other factors
on selectivity to methanol.

## Conclusion

Small particles (1 to 2 nm) of palladium behave differently than larger ones or large single crystals with respect to the chemisorption of carbon monoxide. Small particles are capable of disproportionating CO to surface carbon and $CO_2$ at temperatures at which such reaction does not occur on large particles. As a result, the selectivity of CO-$H_2$ reactions shifts from methanol on large particles to methane on small ones. When disproportionation is taken into account, together with the fact that deposited carbon reacts away in $O_2$ at high temperatures but not at low temperatures, the turnover rate for CO oxidation does not change with particle size at either high or low temperature. Thus, the CO oxidation on palladium is indeed structure-insensitive as claimed previously ([1]). Besides, there is no need to postulate a facilitated accessibility of smaller particles, as was done previously ([1]) to explain a seemingly higher turnover rate at high temperatures on small particles, an effect which disappears when disproportionation is taken into account.

## Literature Cited

1. Ladas, S.; Poppa, H.; Boudart, M. Surf. Sci., 1981, 102, 151.
2. Engel, T.; Ertl, G. J. Chem. Phys., 1968, 69, 1267.
3. Ertl, G.; Koch, J. Proc. 5th Intl. Congr. on Catalysis, Hightower, J., Ed., North Holland, 1973, p. 969.
4. Doering, D.L.; Poppa, H.; Dickinson, J.T. J. Vac. Sci. Technol., 1980, 17, 198.
5. Doering, D.L.; Poppa, H.: Dickinson, J.T. J. Catal., 1982, 73, 91.
6. Ichikawa, S.; Poppa, H.; Boudart, M. Surf. Sci., to be published.
7. Ichikawa, S.; Poppa, H.; Boudart, M. J. Catal., to be published.
8. van Hardeveld, R.; Hartog, F. Adv. Catal. 1972, 22, 75.
9. Yamasaki, H.; Kobori, Y.; Naito, S.; Onishi, T.; Tamaru, K. J. Chem. Soc., Faraday Trans. I, 1981, 77, 2913.
10. Solymosi, F.; Erdohelyi, A. Surf. Sci., 1981, 110, L630.
11. Vannice, M.A. J. Catal., 1975, 40, 129.
12. Vannice, M.A.; Wang, S-Y.; Moon, S.H. J. Catal., 1981, 71, 152.
13. Vannice, M.A.; Garten, R.L. Ing. Eng. Chem. Prod. Res. Dev., 1979, 18, 186.
14. Poutsma, M.L.; Elek, L.F. Ibarbia, P.A.; Risch, A.P.; Rabo, J.A. J. Catal., 1978, 52, 157.

RECEIVED November 17, 1983

# INDEXES

# Author Index

# Subject Index

455

*Production and indexing by Deborah Corson*
*Jacket design by Anne G. Bigler*

*Elements typeset by Hot Type Ltd., Washington, D.C.*
*Printed and bound by Maple Press Co., York, Pa.*

RECENT ACS BOOKS

"NMR and Macromolecules:
Sequence, Dynamic, and Domain Structure"
Edited by James C. Randall
ACS SYMPOSIUM SERIES 247; 282 pp.; ISBN 0-8412-0829-8

"Geochemical Behavior of Disposed Radioactive Waste"
Edited by G. Scott Barney, James D. Navratil, and W. W. Schulz
ACS SYMPOSIUM SERIES 246; 413 pp.; ISBN 0-8412-0827-1

"Size Exclusion Chromatography: Methodology and
Characterization of Polymers and Related Materials"
Edited by Theodore Provder
ACS SYMPOSIUM SERIES 245; 392 pp.; ISBN 0-8412-0826-3

"Industrial-Academic Interfacing"
Edited by Dennis J. Runser
ACS SYMPOSIUM SERIES 244; 176 pp.; ISBN 0-8412-0825-5

"Characterization of Highly Cross-linked Polymers"
Edited by S. S. Labana and Ray A. Dickie
ACS SYMPOSIUM SERIES 243; 324 pp.; ISBN 0-8412-0824-9

"Polymers in Electronics"
Edited by Theodore Davidson
ACS SYMPOSIUM SERIES 242; 584 pp.; ISBN 0-8412-0823-9

"Radionuclide Generators: New Systems
for Nuclear Medicine Applications"
Edited by F. F. Knapp, Jr., and Thomas A. Butler
ACS SYMPOSIUM SERIES 241; 240 pp.; ISBN 0-8412-0822-0

"Polymer Adsorption and Dispersion Stability"
Edited by E. D. Goddard and B. Vincent
ACS SYMPOSIUM SERIES 240; 477 pp.; ISBN 0-8412-0820-4

"Assessment and Management of Chemical Risks"
Edited by Joseph V. Rodricks and Robert C. Tardiff
ACS SYMPOSIUM SERIES 239; 192 pp.; ISBN 0-8412-0821-2

"Chemical and Biological Controls in Forestry"
Edited by Willa Y. Garner and John Harvey, Jr.
ACS SYMPOSIUM SERIES 238; 406 pp.; ISBN 0-8412-0818-2

"Archaeological Chemistry--III"
Edited by Joseph B. Lambert
ADVANCES IN CHEMISTRY SERIES 205; 324 pp.; ISBN 0-8412-0767-4

"Molecular-Based Study of Fluids"
Edited by J. M. Haile and G. A. Mansoori
ADVANCES IN CHEMISTRY SERIES 204; 524 pp.; ISBN 0-8412-0720-8